THE STRUCTURE OF VOLATILE SULPHUR COMPOUNDS

THE STRUCTURE OF VOLATILE SULPHUR COMPOUNDS

ISTVÁN HARGITTAI

Professor of Chemistry
Department of Structural Studies
Research Laboratory for Inorganic Chemistry
Hungarian Academy of Sciences
Budapest, Hungary

D. Reidel Publishing Company

A MEMBER OF THE KLUWER ACADEMIC PUBLISHERS GROUP

Dordrecht / Boston / Lancaster

Library of Congress Cataloging in Publication Data

CIP

Hargittai, István.
The structure of volatile sulphur compounds.
"Revised English edition of 'Illékony kénvegyületek szerkezete' in the series of 'A kémia újabb eredményei' published by Akadémiai Kiadó, Budapest"—T. p. verso.
Bibliography: p.
Includes index.
1. Organosulphur compounds. I. Title.
QD412.S1H3713 1985 547'.06 84-6896
ISBN 90-277-1395-2 (Reidel)

Distributors for the U.S.A. and Canada
Kluwer Academic Publishers,
190 Old Derby Street, Hingham, MA 02043, U.S.A.

Distributors for Albania, Bulgaria, the People's Republic of China, the Republic of Cuba, Czechoslovakia, the German Democratic Republic, Hungary, the Democratic People's Republic of Korea, the People's Republic of Mongolia, Poland, Rumania, the U.S.S.R., Vietnam, and Yugoslavia
Kultura Hungarian Foreign Trading Company,
P.O.B. 149, H-1389 Budapest 62, Hungary

Distributors for all remaining countries
Kluwer Academic Publishers Group,
P.O. Box 322, 3300 AH Dordrecht, Holland

Joint edition published by
D. Reidel Publishing Company, Dordrecht, Holland
and
Akadémiai Kiadó, The Publishing House of the Hungarian Academy of Sciences, Budapest, Hungary

Printed in Hungary

ACKNOWLEDGEMENTS AND DEDICATION

Acknowledgements. The material of this book accumulated from about 15 years' research on the structure of volatile sulphur compounds. In these studies I have benefited from the cooperation with numerous people. I was pleasantly amazed to find that my co-authors on the relevant papers came from no less than 12 countries. I feel privileged that I could share work and interest in the field of sulphur stereochemistry with E. Augdahl, A. Baranyi, C. Bliefert, J. E. Boggs, Ch. Britt, J. Brunvoll, P. Bulcke, D. W. J. Cruickshank, B. N. Cyvin, S. J. Cyvin, S. Dobos, O. Exner, D. C. Gregory, G. Hagen, M. Hargittai, M. Kolonits, J.-F. Labarre, I. Mayer, F. C. Mijlhoff, J. H. Moore, B. Nagel, K. P. R. Nair, G. C. Pappalardo, P. Pulay, T. Radnai, G. Robinet, B. Rozsondai, Gy. Schultz, R. Seip, S. Skaarup, T. Székely, A. Szőke, E. Vajda, B. P. van Eijck, and L. V. Vilkov.

The Hungarian version of this book was written for the series "A kémia újabb eredményei" (Recent Results in Chemistry) at the suggestion of the series editor Professor B. Csákvári. It appeared as Vol. 50 in 1981.

The present English version or parts of it were read by Professors H. Jones, V. S. Mastryukov, I. C. Paul, and Gy. Varsányi, and all of them made valuable comments.

Professor H. Jones kindly corrected the English of the entire manuscript.

Drs. B. Mez-Starck and R. Mutter greatly assisted in the literature search.

Mrs. K. Kállay performed careful editorial work in preparing the manuscript for the printers.

Dr. B. Rozsondai corrected the proofs, compiled the indexes and ably and tirelessly assisted in the production of the book. He took on this burden as this process coincided with my visiting professorship at The University of Connecticut during the academic years 1983/85.

The drawing of the cover depicting part of an imagined spherical Periodic System was prepared by Eng. T. Mihalek.

I express my thanks and appreciation to all mentioned above.

Dedication. This book is dedicated to the memory of my father, Dr. Jenő Wilhelm (1901–1942). He came from a large family and worked his way to become a lawyer. He was killed in a forced labour camp. He is author of a book on unfair competition*.

* Pajor, R. and Wilhelm, J.: *A tisztességtelen verseny anyagi és alaki jogszabályai.* (Material and formal rules of unfair competition). Fővárosi Könyvkiadó Rt. Budapest, 1934.

CONTENTS

INTRODUCTION

The structural chemistry of sulphur compounds has great importance both for our knowledge on the chemical bond and for practical chemistry. The present book deals with the geometrical aspects of molecules containing a sulphur atom. The restriction to the *volatile* sulphur compounds is motivated by the advantage of dealing with relatively simple molecules in which mostly (though not always) sulphur is the central atom, and by the availability of a relatively large amount of accurate experimental data with a well-defined physical meaning. The special importance of vapour-phase data originates, of course, from the fact that they refer to the structure determined solely by *intra*molecular forces, i.e. to free molecules.

The intention of this work is the compilation, critical survey and systematization of available structural results and the search for characteristic variations and trends among them. With all this it is hoped that this book will first of all contribute to sulphur stereochemistry, will demonstrate an approach to dealing with the increasing amount of structural experimental data and finally will indicate the present possibilities and limitations of the physical techniques in molecular structure research.

The molecular geometry is defined by the relative spatial arrangement of the atoms or more specifically atomic nuclei in the molecule. The molecular geometry is usually described by bond lengths, bond angles and angles of internal rotation. The two main physical tools for the determination of the geometry of free molecules are electron diffraction and microwave spectroscopy. For sources of references see, for example, Hargittai (1974a; 1981) and Hargittai and Orville-Thomas (1981) on electron diffraction and Gordy and Cook (1970), Kroto (1975) and Wollrab (1967) on microwave spectroscopy.

The structure-dependent part of the scattered electron intensity of an N-atomic molecule is given by the expression

$$sM(s) = \text{const.} \sum_{i=1}^{N} \sum_{\substack{j=1 \\ i \neq j}}^{N} g_{ij}(s) \exp\left(-\frac{1}{2} l_{ij}^2 s^2\right) \sin\left[s(r_{ij} - \kappa_{ij} s^2)\right]/r_{ij}$$

The internuclear distances are denoted by r_{ij}, their mean amplitudes of vibration by l_{ij}

and the asymmetry parameters related to anharmonicity by κ_{ij}. The function $g_{ij}(s)$ contains the electron scattering amplitudes and are assumed to be known in the usual structure analyses. The angular variable s is a function of the electron wavelength and the scattering angle. An especially descriptive function is a modified sine Fourier transform of the above molecular intensity function,

$$f(r) = \text{const.} \int_0^{s_{max}} sM(s) \exp(-as^2) \sin(sr) ds$$

related to the probability density function, $P(r)$, $f(r) = \Sigma P_{ij}(r)/r$, where $P_{ij}(r)dr$ is the probability that the distance between the ith and jth atoms has a value in the interval of r and $r + dr$. The value of s_{max} is the upper limit of the observed intensities and the value of a in $\exp(-as^2)$ is such as to make the integral converge at s_{max}. Radial distributions of several molecules will be shown in the subsequent sections.

The determination of the structural parameters is based on comparison of the experimental molecular intensities and radial distributions with the corresponding theoretical functions calculated for various models. In early days the intensity pattern was estimated visually hence the term "visual technique". In modern studies there is a quantitative treatment of the experimental intensities and usually a least-squares procedure is employed to determine the structural parameters.

The parameter corresponding to the position of the centre of gravity of the $P(r)$ function is called r_g and is the average value of the internuclear distance. The r_a distance is obtained directly from the electron diffraction data and it is related to r_g to a good approximation by the following expression: $r_g \approx r_a + l^2/r_a$. The electron diffraction intensities are obtained as averaged from the molecules distributed among the vibrational states, and so are the corresponding internuclear distances. The average internuclear distance is related to the equilibrium internuclear distance (r_e) in the following way:

$$r_g = r_e + \delta r + \langle \Delta z \rangle + K + \ldots$$

This relationship refers to a Cartesian system whose z axis coincides with the equilibrium internuclear axis and whose origin is the equilibrium position of one of the two atoms. The equilibrium distance corresponds to the position of the minimum of the potential energy function. The term δr arises from the centrifugal distortions, $\langle \Delta z \rangle$ is a consequence of anharmonicity of vibrations and K expresses the effect of perpendicular vibrations, $K = (\langle \Delta x^2 \rangle + \langle \Delta y^2 \rangle)/2r_e$. Two further important parameters are the following: r_α^0 is the distance between average nuclear positions in the ground vibrational state and r_α (or r_α^T) is the corresponding one in thermal equilibrium at a given temperature T. They differ from the r_e parameter only because of anharmonicity, $r_\alpha = r_e + \langle \Delta z \rangle$.

The r_e parameter is the most unambiguous representation of molecular geometry, however, it is usually not obtainable experimentally. The r_g parameter is the most convenient characterization of the length of a chemical bond, but has no such descriptive meaning for the distances between non-bonded atoms because of the perpendicular vibrations causing the so-called shrinkage effect. An important merit of the r_α parameter is that it is free from the shrinkage effect. Thus the bond angles are expressed preferably in terms of r_α parameters. The shrinkage effect is increasingly important when the molecule is performing large-amplitude motion, whose magnitude may then be several times larger than the experimental uncertainties.

In microwave spectroscopy the rotational constants are obtained from the frequencies of the rotational transitions occurring in the microwave region. Since the rotational constants are related to the principal molecular moments of inertia, the molecular geometry may be deduced through them. The rotational transitions corresponding to different vibrational states are usually resolved in the rotational spectrum, and consequently the internuclear distances corresponding to different vibrational states can be determined. Up to three rotational constants may be obtained from the rotational spectrum of an isotopic species. The molecular geometry of molecules larger than the simplest is determined, however, by a larger number of independent data. To overcome this problem further data from the rotational spectra of isotopically substituted species are utilized, but since the molecular vibrations are mass-dependent, their influence will be somewhat different for the various species measured. The internuclear distance obtained directly from the rotational constants of the isotopic species is called r_0 and it is not invariant to the particular choice of isotopic species utilized. If the isotopic substitution is consequently and consistently performed for each atom of the molecule, and the so-called substitution coordinates of each atom are obtained, the r_s substitution structure can be calculated. The r_s parameters are very close to the equilibrium parameters. Neither the r_0 nor the r_s distances have, however, well-defined physical meaning. It is possible to produce the distances between the average nuclear positions by applying vibrational corrections to the rotational constants. The r_z internuclear distance obtained in this way from the microwave spectroscopic data has the same physical meaning as the r_α^0 parameter from electron diffraction. Finally the parameter type r_{av} is mentioned, which is another average internuclear distance corrected for the influence of intramolecular motion and usually obtained from a combined analysis of electron diffraction and microwave spectroscopic data.

The quantum chemical calculations refer to the minimum of the potential energy, thus their results on molecular geometry are r_e equilibrium parameters.

The uncertainties cited in the following discussion, either parenthesized referring to the last digit of the parameter or written as an error limit, will be those communicated in the original sources. Their meaning, however, may vary widely. The major error sources in an electron diffraction determination are the experimental systematic error, the standard deviation from the least-squares refinement, and the influence of the various assumptions used in the analysis. Many workers quote only the standard deviation or a multiple of it in their results, others also include experimental error, but it appears that only a minority take all three into account in the error estimates. The main error sources in a microwave spectroscopic analysis are the experimental error of the rotational constants, the standard deviation of the fitting procedures, the influence of intramolecular vibrations and the possible consequences of various assumptions. Again, very often only the first two are considered in the error estimates of the original papers.

Maximum effort has been exercised to consider all available electron diffraction and microwave spectroscopic data on the geometry of sulphur-containing molecules. It is hoped that a reasonably full coverage of the literature has been achieved. There has also been a reasonable effort to collect and consider experimental data from other techniques as well as the results of quantum chemical calculations for the same or related molecules representing complementary information.

The classes of sulphur compounds that have representatives whose molecular geometries in the vapour phase have been elucidated to date are listed in Table 1.

In systematizing the available data and searching for characteristic variations and trends among them, two qualitative models have proved to be very useful, *viz.* the valence shell electron pair repulsion or VSEPR model (Gillespie, 1972) and considerations of the atom–atom non-bonded interactions (Bartell, 1960; 1962; 1968;

TABLE 1
Typical classes of sulphur compounds from the following discussion

One-coordinated sulphur
$X_2C{=}S$ $XNC{=}S$ $X_3P{=}S$
Two-coordinated sulphur
XSX, XSY sulphides
XSSX disulphanes
cyclic sulphides
$X{=}S{=}Y$
Three-coordinated sulphur
$X_2S{=}O$ sulphoxides
cyclic sulphites
$RN{=}SX_2$ thionylimides
Four-coordinated sulphur
sulphuranes SF_4
XSO_2Y sulphones
iminosulphur derivatives $X_2S(NH)_2$
M_2SO_4 sulphates
Five-coordinated sulphur
$F_4S{=}O$
Six-coordinated sulphur
SF_6, XSF_5

Jacob, Thompson and Bartell 1967; Glidewell, 1973; 1975; 1976; 1977; 1979). These are, of course, only models and rather qualitative ones at that, which put emphasis on one or a few of the many complex interactions governing molecular structure. Their application appears to be very useful, however, for two reasons.

Their descriptiveness provides a basis for an understanding between the structural parameters determined from the experimental data and the language of the chemist. Also, the real predictive ability of these qualitative models has been proven already in numerous cases. At the same time it is important to realize the limitations of these qualitative models since it is only with a knowledge of these limitations that the application of the models will be made more reliable.

According to the VSEPR model the bond configuration of the central atom in the molecule is determined by the number of the electron pairs, both bonding and lone pairs, in its valence shell and their relative size and shape. The arrangement of the valence shell electron pairs is such that it maximizes their distances apart. The effects influencing the bond configuration of the central atom are summarized in the following four basic postulates:

1. A lone pair exercises greater repulsions towards the neighbouring electron pairs than a bonding pair.
2. The repulsion from a bonding pair diminishes with increasing ligand electronegativity; this is the "electronegativity effect" or "electronegativity rule".
3. Multiple bonds exercise greater repulsions towards the neighbouring electron pairs than single bonds; this is the "multiple bond effect" or "multiple bond rule".
4. Repulsions in completely filled valence shells are greater than in incompletely filled valence shells.

The above "rules" are very simple and it is just this simplicity combined with applicability to a wide variety of classes of compounds that provides the beauty and effectiveness of the VSEPR model. Although the Pauli exclusion principle has been suggested to provide the foundation for the VSEPR model, there is apparently no rigorous quantum mechanical basis for it. Therefore, it is especially interesting to compare the predictions of this model with some plots of localized molecular orbitals as shown in Fig. 1 after Schmiedekamp (1976) and Schmiedekamp, Cruickshank, Skaarup, Pulay, Hargittai and Boggs (1979). It is indeed seen that the lone pair of electrons occupies more space than the bonding pairs in the vicinity of the central atom. Also, a bond to a more electronegative ligand such as fluorine occupies less space in the vicinity of the central atom than a bond to a less electronegative ligand such as hydrogen, or a double bond occupies more space than a single bond. The angular ranges of the corresponding contours on the electron density plots are all in good agreement with the postulates of the VSEPR model, the difference between the lone pairs and bonding pairs being especially large.

Experience shows that the importance of valence shell electron pair repulsions diminishes with the increasing size of the ligands relative to the size of the central atom.

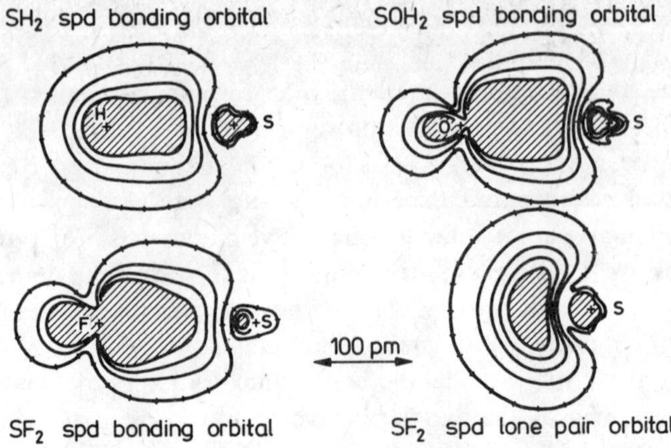

Fig. 1. Pictorial representation of localized orbitals after Schmiedekamp, Cruickshank, Skaarup, Pulay, Hargittai and Boggs (1979). The contour lines denote the electron densities 0.02, 0.04, 0.06, etc., electron $Bohr^{-3}$

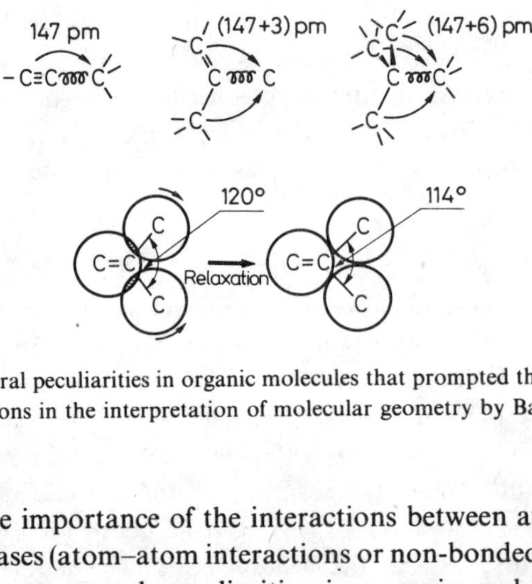

Fig. 2. Two of the structural peculiarities in organic molecules that prompted the examination of the non-bonded interactions in the interpretation of molecular geometry by Bartell (1960; 1968)

At the same time the importance of the interactions between atoms not linked by a chemical bond increases (atom–atom interactions or non-bonded interactions). Figure 2 shows two of the structural peculiarities in organic molecules that originally prompted the examination of non-bonded interactions in the interpretation of molecular geometry by Bartell (1968). On the basis of experimental data on molecular geometries, so-called intramolecular non-bonded radii, or intramolecular van der Waals radii, or 1,3 radii have been postulated (Bartell, 1960; Glidewell, 1976). The name "1,3 radii" indicates that the interaction considered here refers to two atoms separated by one between them. A series of non-bonded radii together with covalent radii and van der Waals radii are given in Table 2.

TABLE 2
Covalent, 1,3 non-bonded and van der Waals radii (pm)

Element	Covalent radii Pauling (1960)	Covalent radii Glidewell (1976)	1,3 Non-bonded radii Bartell (1960) Glidewell (1976)	Van der Waals radii Pauling (1960)
B		81.7	133	
C	77.2	76.7	125	
N	70	72.6	114	150
O	66	74.5	113	140
F	64	70.9	108	135
Al		120.2	166	
Si	117	116.9	155	
P	110	110.3	145	190
S	104	102.0	145	185
Cl	99	99.4	144	180
Ga		126	172	
Ge	122	120.2	158	
As	121	121.8	161	200
Se	117	116.3	158	200
Br	114	114.2	159	195
In		147	195	
Sn	140	138.6	188	
Sb	141	133.6	188	220
Te	137	135.6	187	220
I	133	134.0	186	215

The valence shell electron pair repulsions and the atom–atom non-bonded interactions should, of course, be considered as acting together in building the actual molecular structure. Their relative magnitude determines, which of them prevails in each given case. It should be kept in mind that in reality the intramolecular forces governing the structure are certainly much more complex than the two rather simplified interactions that were selected for convenience in the discussion.

The overall molecular shape, or rather the shape of the sulphur bond configuration is easily established from the general starting principle of the VSEPR model. Using the notation X and E for ligands and lone electron pairs, respectively, we shall consider some basic types of sulphur-containing molecules from Table 1. There are two lone pairs of sulphur in sulphides; thus, they may be written as SX_2E_2. The arrangement is tetrahedral and the X—S—X bond configuration is accordingly highly bent with C_{2v} symmetry.

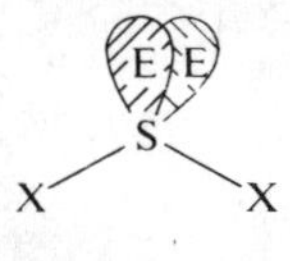

The sulphur dioxide type molecules may be written as SX_2E and the planar arrangement

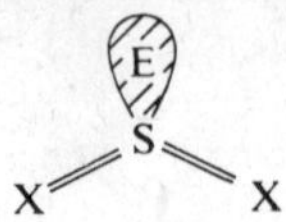

corresponds to a bent X=S=X bond configuration with C_{2v} symmetry. The sulphoxides may be expressed by the formula SX_3E, and the sulphones by SX_4, thus the arrangement is tetrahedral in both cases

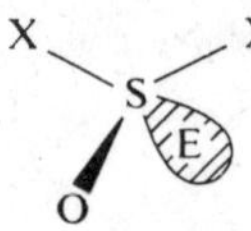

Sulphoxide

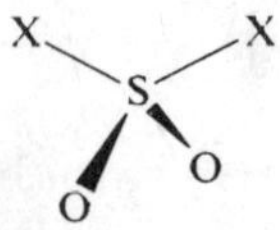

Sulphone

with pyramidal $X{=}SX_2$ (C_s symmetry) and tetrahedral $(X{=})_2SX_2$ (C_{2v} symmetry) bond configurations, respectively. The bond configuration of SX_3 sulphur trioxide is trigonal planar with D_{3h} symmetry

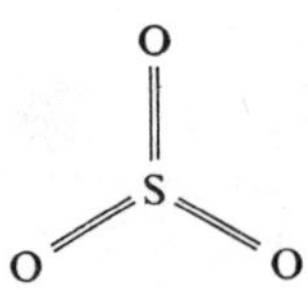

The sulphuranes may be written as SX_4E and accordingly the arrangement is trigonal bipyramidal

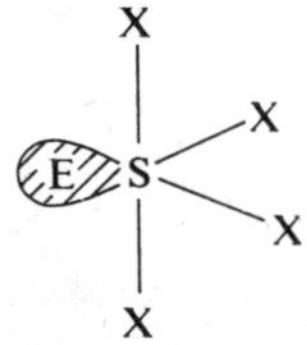

and the lone pair is in the equatorial plane because of its large space requirement. The SX_4 bond configuration of sulphuranes has, accordingly, C_{2v} symmetry. There is a similar bond configuration in thionyl tetrafluoride corresponding to the formula SX_5 with the double bond replacing the lone pair. The molecules SX_6 will have octahedral configuration

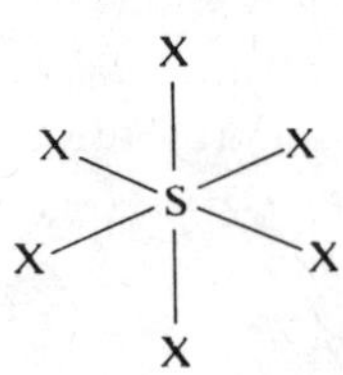

with O_h symmetry.

In the foregoing section, the molecular shapes and sulphur bond configurations have been established from the general principle of the VSEPR model without applying its various postulates.* More subtle structural details will be considered in the following discussion.

*

Note on the units: All parameters are given in SI units in this book. Data from the original sources have been converted accordingly.

* Except for the case of the sulphuranes where it is not possible to distinguish between the axial and equatorial positions for the lone pair without involving the postulate about the space requirement difference between lone pairs and bonding pairs.

Chapter 1

MOLECULES WITH ONE-COORDINATED SULPHUR

A one-coordinated sulphur cannot be the central atom in a molecule. Therefore, the relevant structural studies are reviewed in somewhat less detail. Our discussion is primarily concerned with the length of the bond involving the one-coordinated sulphur atom, the reliability of its determination and the conformational properties of the molecules involved.

The molecules with one-coordinated sulphur are grouped in the following way: molecules containing a C═S bond involving an sp^2 hybridized carbon atom; molecules containing a C═S bond involving an sp hybridized carbon atom, including isothiocyanates; molecules of type XYZM═S where X, Y, and Z are identical or different ligands and M is phosphorus or arsenic; finally the molecules F_2S═S, HB═S, S═BSB═S, and S_2, and the radicals NS and SO.

1.1. C═S BONDS WITH AN sp^2 CARBON ATOM

Recently the molecular geometry for a relatively large series of simple molecules in which the sulphur atom is linked to an sp^2 hybridized carbon atom has been determined with up-to-date techniques. The geometrical parameters of interest for our discussion are collected in Tables 3–5. All the parameters presented originate from experimental studies except those for O═CFCl. Extensive *ab initio* molecular orbital calculations have been carried out on simple carbonyl and thiocarbonyl molecules (Oberhammer and Boggs, 1979a; Kapur, Steer and Mezey, 1978; 1979a; 1979b). The calculated equilibrium geometries of the ground state are in good agreement with the experimental data where available. The only exception was O═CFCl, where on the basis of the calculated structure and comparison with related geometries, it was strongly suggested (Oberhammer and Boggs, 1979a) that the assumptions utilized in the microwave spectroscopic analysis (Mirri, Guarnieri, Favero and Zuliani, 1962) resulted in erroneous bond angles.

The assumptions concerning the planarity of the molecules with four atoms have been found to be in complete agreement with the experimental data. The C—O—C═S

TABLE 3
Bond lengths and bond angles of thioformaldehyde, thioformamide, and methyl thioformate and their oxygen analogues

Parameter	Molecule					
	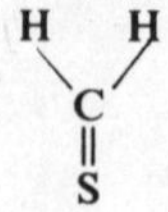	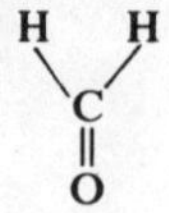	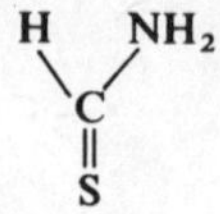	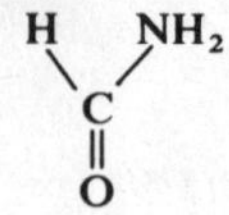	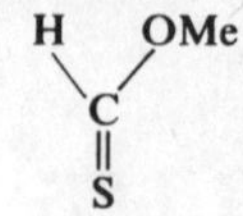	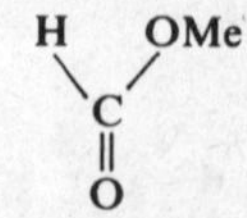
r(C=S), pm	161.1(2)		162.6(2)		161.2(3)	
r(C=O), pm		120.8(3) 120.9(3)		121.9(2)		120.0(10)
r(C—H), pm	109.2(2)		110(8)	109.8(10)	111.4(9)	110.1(10)
r(C—N), pm			135.8(3)	135.2(12)		
r(C—O), pm					136.9(3)	133.4(10)
∢H—C—H (°)	116.9(1)	116.5(7) —				
∢H—C=S (°)	(121.55)		127(5)			
∢H—C=O (°)		(121.75) —		122.5(20)		
∢H—C—N (°)			108(5)	112.7(20)		
∢N—C=S (°)			125°16′(15′)			
∢N—C=O (°)				124.7(3)		
∢H—C—O (°)					106.7(28)	109.3(10)
∢O—C=S (°)					126.6(5)	
∢O—C=O (°)						125.9(10)
Technique *(a)*	MW	MW ED	MW	MW	ED	MW
Parameter type	r_s	r_s r_g	r_s	r_s	r_a	r_s
Reference	*(b)*	*(c)* *(d)*	*(e)*	*(f)*	*(g)*	*(h)*

(a) MW = microwave spectroscopy; ED = electron diffraction
(b) Johnson, Powell and Kirchhoff (1971)
(c) Takagi and Oka (1963)
(d) Kato, Konaka, Iijima and Kimura (1969)
(e) Sugisaki, Tanaka and Hirota (1974)
(f) Hirota, Sugisaki, Nielsen and Sørensen (1974)
(g) De Rooij, Mijlhoff and Renes (1975)
(h) Curl (1959)

chain was found to have a nearly planar *syn* orientation in *methyl thioformate*

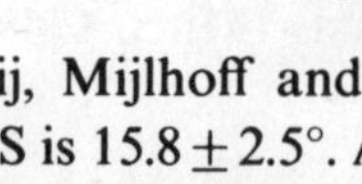

by electron diffraction (De Rooij, Mijlhoff and Renes, 1975). The dihedral angle between the planes COC and OCS is 15.8 ± 2.5°. A microwave spectroscopic analysis showed the COCO chain of methyl formate to be planar (Curl, 1959).

TABLE 4
Bond lengths and bond angles of thiocarbonyl halides and carbonyl halides

Parameter	Molecule				
$X(Y)C{=}A$	$F_2C{=}A$	$F(Cl)C{=}A$		$Cl_2C{=}A$	$Br_2C{=}A$
1. Thiocarbonyl halides, A=S					
r(C=S), pm	158.9(2)	159.3(1)	159.5	160.1(3)	159.7(5)
r(C—F), pm	131.6(2)	133.9(1)	132.6		
r(C—Cl), pm		171.8(1)	171.5	172.9(3)	
r(C—Br), pm					189.4(9)
∢X—C—Y (°)	107.1(2)	(108.6)	(109.1)	111.2(2)	111.6
∢X—C=S (°)	}(126.45)	123.9(1)	123.8	}(124.4)	}(124.2)
∢Y—C=S (°)		127.5(1)	127.1		
Technique *(a)*	MW+ED	ED+MW	MW	MW	ED
Parameter type	r_{av}	r_g, ∢$^0_\alpha$	r_0	r_z	r^0_α
Reference	*(b)*	*(c)*	*(d)*	*(e)*	*(f)*
2. Carbonyl halides, A=O					
r(C=O), pm	117.2(1)	117.5	117.3(2)	117.6(3)	117.8(9)
r(C—F), pm	131.6(1)	134.5	133.4(2)		
r(C—Cl), pm		173.1	172.5(2)	173.8(2)	
r(C—Br), pm					192.3(5)
∢X—C—Y (°)	107.7(1)	(109.9)	(108.8)	111.8(2)	112.3(4)
∢X—C=O (°)	}(126.2)	123.5	123.7(2)	}(124.1)	}(123.8)
∢Y—C=O (°)		126.6	127.5(3)		
Technique *(a)*	MW+ED	Calc.	ED+MW	MW+ED	ED
Parameter type	r_{av}	r_e	r_{av}	r_e	r_g, ∢$^0_\alpha$
Reference	*(g)*	*(h)*	*(i)*	*(j)*	*(k)*

(a) MW = microwave spectroscopy; ED = electron diffraction; Calc. = *ab initio* calculations

(b) Careless, Croto and Landsberg (1973); Christen, Oberhammer, Zeil, Haas and Darmadi (1980)

(c) Gleisberg, Haberl and Zeil (1975)

(d) Kohrmann and Zeil (1975)

(e) Carpenter, Rimmer, Smith and Whiffen (1975)

(f) Christen (1980)

(g) Carpenter (1974); Nakata, Kohata, Fukuyama, Kuchitsu and Wilkins (1980)

(h) Oberhammer and Boggs (1979a); Oberhammer (1980)

(i) Oberhammer (1980)

(j) Nakata, Fukuyama, Kuchitsu, Takeo and Matsumura (1980); Nakata, Kohata, Fukuyama and Kuchitsu (1980)

(k) Nakata, Fukuyama, Wilkins and Kuchitsu (1981)

The most important forms of internal rotation to be considered for *dimethyl trithiocarbonate*, $S{=}C(SCH_3)_2$, are shown in Fig. 3. On the basis of the electron diffraction data (Almenningen, Fernholt, Seip and Henriksen, 1974) it was possible to exclude the *syn–anti* form as well as the *syn–gauche* and *gauche–gauche* forms. The experimental data could be well approximated with the *syn–syn* conformer. Vibrational spectroscopic studies on liquid-phase samples indicated the presence of two forms with the *syn–syn* conformer being the major component (Herzog, Steger,

TABLE 5
Geometrical parameters (r_a) of dimethyl trithiocarbonate, dimethyl dithiocarbonate and 1,1-bis(methylthio)ethylene from electron diffraction

Parameter	$S{=}C(SCH_3)_2$ *(a)*	$O{=}C(SCH_3)_2$ *(b)*	$H_2C{=}C(SCH_3)_2$ *(c1)*	*(c2)*
r(S=C), pm	163.4(5)			
r(O=C), pm		120.6(4)		
r(C=C), pm			134.8(5)	134.8(5)
r(=C—S), pm	175.2(4)	177.7(3)	176.2(5)	177.2(5)
r(C—S), pm	180.0(6)	180.2(4)	181.7(5)	181.2(5)
∢S=C—S (°)	126.2(2)			
∢O=C—S (°)		124.9(2)		
∢C=C—S (°)			122.4(10)	122.1(11)
∢S—C—S (°)	103.3(4)	99.3(4)	102.1(1)	103.1(20)
Angle of rotation *(d)* (°)			122(3)	124(5)

(a) Almenningen, Fernholt, Seip and Henriksen (1974)
(b) Auberg, Samdal and Seip (1979)
(c) Jandal, Seip and Torgrimsen (1976); Nozzle temperatures: *(c1)* 60 °C; *(c2)* 200 °C
(d) Angle of rotation around the =C—S bond

Rosmus, Scheithauer and Mayer, 1969; Dräger and Gattow, 1971). The *S,S'-dimethyl dithiocarbonate*, $O{=}C(SCH_3)_2$, molecules were found to have conformational properties similar to those of the trithio derivative (Auberg, Samdal and Seip, 1979). In fact only the *syn–syn* form could be detected but the presence of a small amount (less than 15%) of a *syn–gauche* form could not be ruled out. For *O,O'*-dimethyl carbonate, $O{=}C(OCH_3)_2$, Mijlhoff (1977) established the *syn–syn* conformation by electron diffraction. The geometrical parameters of dimethyl trithiocarbonate are collected in Table 5.

The unstable *thioacetaldehyde*

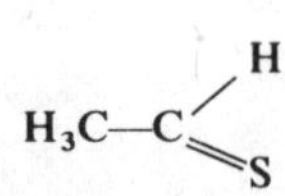

was characterized by the following geometrical parameters (r_s) from a microwave spectroscopic study (Kroto and Landsberg, 1976):

C=S 161.0(9) pm C—C 150.6(7) pm
∢C—C=S 125.3(11)°

A relatively long C=S bond, 167.1 pm, was determined in

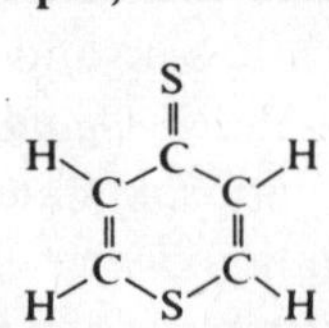

by Corkill, Cox and Ewart (1976).

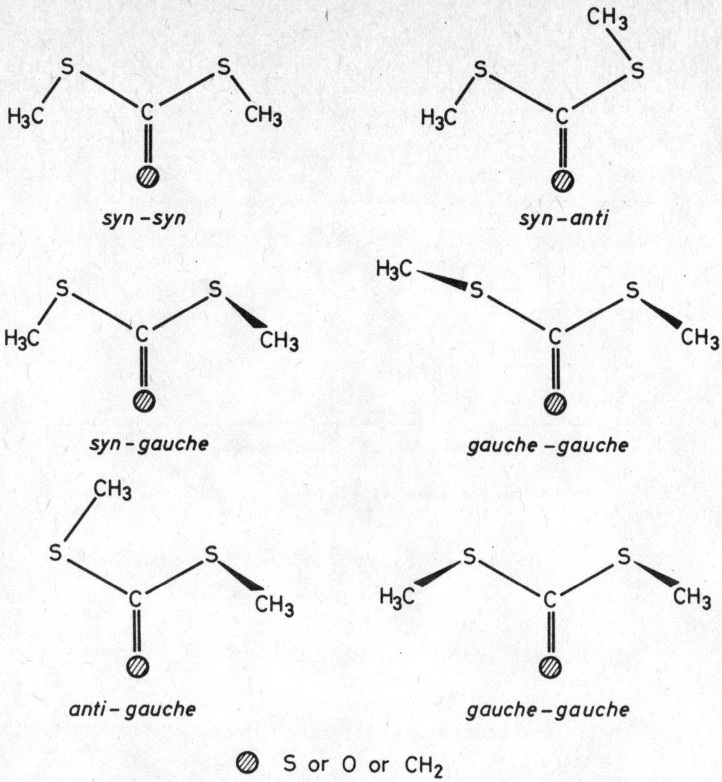

Fig. 3. The most important forms of internal rotation to be considered for dimethyl trithiocarbonate, $S{=}C(SCH_3)_2$, *S,S'*-dimethyl dithiocarbonate, $O{=}C(SCH_3)_2$, and 1,1-bis(methylthio)ethylene, $H_2C{=}C(SCH_3)_2$

Table 6 contains mean amplitudes of vibration (l) for $C(sp^2){=}S$ bonds. The relatively small l values indicate the rigidity of these bonds. There are three different C—S bonds in dimethyl trithiocarbonate (*cf.* Table 5). With increasing bond length a gradual increase can be observed in the corresponding l values, *viz.* 4.8, 5.2, and 5.3 pm. The respective bond stretching force constants decrease in the same order, *viz.* 360, 300, and 220 N m^{-1} (Almenningen, Fernholt, Seip and Henriksen, 1974). It should be noted that the relatively large changes in the force constants are accompanied by rather small changes in the vibrational amplitudes. Other molecules that contain $C(sp^2){=}S$ bonds and for which the force field has been investigated include

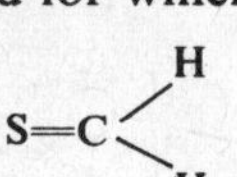 (Randhawa, 1974)

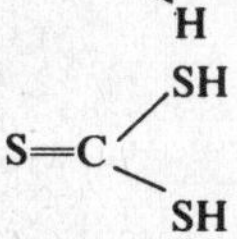 (Müller, Krebs and Gattow, 1967)

TABLE 6
Mean amplitudes of vibration (l, pm)
of the C=S bond determined from electron diffraction (ED)
and spectroscopic (SP) data

Compound	l (ED)	l (SP)	
		0 K	298 K
$S{=}CF_2$		3.83 *(c)*	3.85 *(c)*
			3.92 *(d)*
S=CFCl	4.8(2) *(a)*		4.1 *(a)* [303 K]
$S{=}CCl_2$		4.16 *(c)*	4.20 *(c)*
			4.09 *(d)*
$S{=}CH(OCH_3)$	5.2(5) *(b)*		
$S{=}C(SCH_3)_2$			4.9 *(e)*

(a) Experimental temperature 303 K; Gleisberg, Haberl and Zeil (1975)
(b) Experimental temperature 293 K; De Rooij, Mijlhoff and Renes (1975)
(c) Müller, Krebs, Fadini, Glemser, Cyvin, Brunvoll, Cyvin, Elvebredd, Hagen and Vizi (1968)
(d) Müller and Nagarajan (1967b)
(e) Probably room temperature; Almenningen, Fernholt, Seip and Henriksen (1974)

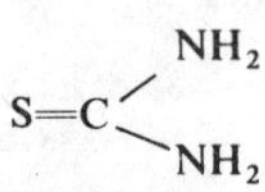

(Hadži, Kidrič, Keževic and Barlič, 1976)

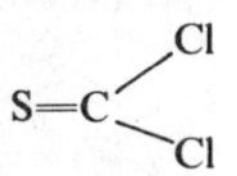

(Brand, Callomon, Moule, Tyrre and Goodwin, 1965; Frenzel, Blick, Bennett and Niedenzu, 1970)

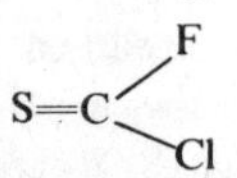

(Hamm, 1979)

In connection with the geometrical characterization of the $S{=}C\genfrac{}{}{0pt}{}{X}{Y}$ bond configuration, it is of interest to examine the variations with changing ligands X and whether any conclusion can be drawn from the comparison of the analogous bond configurations $S{=}C\genfrac{}{}{0pt}{}{X}{Y}$ and $O{=}C\genfrac{}{}{0pt}{}{X}{Y}$. The most striking geometrical feature of these bond configurations is that the X—C—Y bond angle is invariably smaller than 120°. This is, of course, in complete agreement with the VSEPR model or rather with its postulate on the multiple bonds exercising greater repulsion towards the neighbouring electron pairs than do single bonds. The C=S (as well as the C=O bonds) of these structures are considered formally as double bonds. In fact the C=S bonds of the $S{=}C\genfrac{}{}{0pt}{}{X}{Y}$ are among the shortest carbon–sulphur bonds ever observed. The carbon–halogen bonds are also very short in these molecules with X or Y or both being halogen. For comparison, a series of carbon–halogen bond lengths for molecules involving sp^3, sp^2, or sp hybridized carbon are shown in Table 7.

TABLE 7
Carbon–halogen bond lengths (pm)

Bond type	X=F *(a)* Compound	r(C—F)	X=Cl Compound	r(C—Cl)	X=Br Compound	r(C—Br)
C(sp³)—X	$FC(CH_3)_3$	142.5(8) r_a	$ClCH_3$	178.5(1) r_0 *(c)*	Br_2CH_2	192.7(10) r_s *(q)*
	FCH_3	138.5 r_s	Cl_3CH	175.8(2) r_s *(d)*		192.1(2) r_a *(r)*
	F_2CH_2	136.0(5) r_a	Cl_3CSO_2Cl	175.8(10) r_g *(e)*	Br_3CH	193.0(3) r_0 *(s)*
	F_3CH	133.6(5) r_g	Cl_3CGeCl_3	177.1(5) r_g *(f)*		
	F_3CSO_2Cl	132.6(4) r_g *(b)*	CCl_4	176.7(3) r_g *(g)*		
	CF_4	131.9(5) r_g	Cl_3CNO_2	172.6(10) r_a *(h)*		
C(sp²)—X	C_6H_5F	135.4 r_s	$CH_2{=}CH{-}CClO$	181.6(30) r_0 *(i)*	$Br_2C{=}CBr_2$	188.1(3) r_a *(t)*
	$CH_2{=}CHF$	134.8(4) r_g	CH_3CClO	178.9(5) r_s *(j)*	$BrOC{-}COBr$	192.5(4) r_a *(u)*
	C_6F_6	132.4(6) r_g		179.8(2) r_g *(k)*		
	$CH_2{=}CF_2$	132.4(3) r_g	$ClOC{-}COCl$	174.4(2) r_a *(l)*		
			$HClC{=}CClH$	171.8(10) r_g *(m)*		
			$Cl_2C{=}CCl_2$	171.8(3) r_a *(n)*		
C(sp)—X	$FC{\equiv}CH$	127.9 r_s	$ClC{\equiv}CBr$	162.4(1) r_0 *(o)*	$BrC{\equiv}CCl$	178.9(5) r_0 *(o)*
	FCN	126.2 r_s	$ClCN$	162.9(6) r_e *(p)*	$BrC{\equiv}C{-}C{\equiv}CH$	179.0(5) r_a *(v)*

(a) From the compilation of Yokozeki and Bauer (1975)
(b) Brunvoll, Hargittai and Kolonits (1978)
(c) Duncan, McKean, Mallinson and McCulloch (1973)
(d) Jen and Lide (1962)
(e) Brunvoll, Hargittai and Seip (1978a)
(f) Vajda, Hargittai, Maltsev and Nefedov (1974)
(g) Morino, Nakamura and Iijima (1960)
(h) Knudsen, George and Karle (1966)
(i) Kewley, Hemphill and Curl (1972)
(j) Sinnott (1961)
(k) Tsuchiya and Iijima (1972)
(l) Hagen and Hedberg (1973a)
(m) Davis and Hanson (1965)
(n) Strand (1967b)
(o) Bjørseth, Kloster-Jensen, Marstokk and Møllendal (1970)
(p) Lafferty, Lide and Toth (1965)
(q) Chadwick and Millen (1971)
(r) Beagley, Brown and Freeman (1974)
(s) Williams, Cox and Gordy (1952)
(t) Strand (1967a)
(u) Hagen and Hedberg (1973b)
(v) Almenningen, Hargittai, Kloster-Jensen and Stølevik (1970)

It has already been noted that the shape of the S=CXY molecules is in complete agreement with the "multiple bond rule" of the VSEPR model. As the electronegativity of the X and Y ligands decreases, the X—C—Y bond angle opens somewhat in accordance with the "electronegativity rule" of the VSEPR model. The atom–atom non-bonded interactions have apparently little importance in these structures since the bond angles in thiocarbonyl chloride and thiocarbonyl bromide are approximately the same (*cf.* Table 4) and since the halogen–carbon–halogen bond angles are considerably smaller than the H—C—H bond angle in thioformaldehyde (Table 3). A similar pattern emerges from the bond angle data on the O=CXY carbonyl derivatives.

Finally, the structural variations accompanying the substitution of the carbonyl group by the thiocarbonyl one will be examined. As the electronegativity of sulphur is smaller than that of oxygen one might expect a lengthening of the other bonds and an opening of the bond angles involving the carbon–sulphur bond. However, no such trend, or any other trend for that matter, can be observed from the actual data. Oberhammer and Boggs (1979a) advanced the following explanation for this phenomenon. They determined the localized orbitals from *ab initio* calculations and found the distance between the centre of the localized double bond orbitals and the carbon atom to increase a mere 8 pm upon the oxygen to sulphur substitution, while the bond length increased by 40 pm. Accordingly, the electron distribution of the bonds in the vicinity of the carbon atom remains essentially unchanged.

The discussion on thiocarbonyl derivatives is concluded by giving the results of a structure determination of *selenocarbonyl fluoride*, $Se{=}C{<}^{F}_{F}$ (Christen, Oberhammer, Zeil, Haas and Darmadi, 1980):

$$r_\alpha^0(\text{Se=C})\ \ 174.3(3)\ \text{pm} \qquad r_\alpha^0(\text{C—F})\ \ 131.4(3)\ \text{pm}$$
$$\angle_\alpha^0\,\text{F—C—F}\ \ 107.5(4)^\circ$$

Thus the C—F bond is even somewhat shorter in this molecule than those in the thio derivatives.

1.2. C=S BONDS WITH AN sp CARBON ATOM

Carbonyl sulphide has been repeatedly investigated by microwave spectroscopy as it is an excellent model. The relevant literature is only referred to here (Landolt–Börnstein, 1976). The bond lengths of the thio, seleno and telluro derivatives of carbon dioxide are compiled in Table 8. All these molecules have a linear configuration and are rather rigid in accordance with being formed from two double bonds. The relative rigidity of these molecules is also demonstrated by the small mean amplitudes of vibration and the small shrinkage effects listed in Table 9. Two more papers are mentioned here dealing with the force field (Jones, 1967 and Williams and Ladd, 1968).

TABLE 8
Bond lengths (pm) in carbon dioxide and its sulphur, selenium and tellurium analogues*

Compound		O=C	S=C	Se=C	Te=C
O=C=O	(r_e)	115.979			
O=C=S	(r_e)	115.7(2)	156.1(2)		
O=C=Se	(r_s)	115.7(2)		170.8(2)	
S=C=S	(r_e)		155.256		
Se=C=S	(r_0)		155.3(2)	169.3(2)	
Te=C=S	(r_0)		155.7(5)		190.4(5)

* The original sources can be found in Landolt–Börnstein (1976). The Se=C bond length of OCSe is cited directly from Morino and Matsumura (1967)

TABLE 9
Mean amplitudes of vibration (l, pm) and shrinkage effects (δ, pm) of carbon disulphide and analogous molecules from electron diffraction (ED) and spectroscopic calculations (SP)

Compound	Atomic pair	l(ED)	l(SP) *(b)*		δ(ED)	δ(SP)
			0 K	298 K		
CS_2	C=S	3.6(4) *(a)*		3.9 *(a)*		
	S...S	4.5(4) *(a)*		4.1 *(a)*	0.6(4) *(a)*	0.7 *(a)*
OCS	C=O					
	C=S					
	O...S					0.6 *(c)*
OCSe	C=O		3.5	3.5		
	C=Se		3.8	3.9		
	O...Se		4.0	4.2		0.7 *(c)*
SCSe	C=S		3.8	3.9		
	C=Se		3.8	4.0		
	S...Se		3.8	4.1		0.8 *(c)*
SCTe	C=S		3.9	3.9		
	C=Te		4.0	4.3		
	S...Te		3.8	4.3		0.8 *(c)*

(a) Morino and Iijima (1962)
(b) Cyvin (1969a)
(c) Cyvin (1961; 1962)

The unstable *thioketene*, $H_2C{=}C{=}S$, is another molecule with a C (sp)=S bond and its geometry was determined in a microwave spectroscopic investigation by Georgiou, Kroto and Landsberg (1979). The molecular dimensions of thioketene and ketene itself are very similar, except the lengths of the thiocarbonyl and carbonyl bonds:

	$H_2C{=}C{=}S$ (r_s)	$H_2C{=}C{=}O$	
C=S	155.4(3) pm	116.14 pm (r_s)	C=O
C=C	131.4(3) pm	131.42 pm (r_s)	
C—H	109.0(6) pm	107.68 pm (r_0)	
H—C—H	120.3(5)°	122.24° (r_0)	

For sources of the data on ketene, see Landolt–Börnstein (1976).

1.3. ISOTHIOCYANATES

The isothiocyanates are treated as a separate class of compounds here, as there is a relatively large number of molecules with the —N=C=S bonding system whose vapour-phase structures have been determined. In addition to the isothiocyanates listed in Table 10, a limited amount of structural information has been determined for ethyl isothiocyanate, CH_3CH_2NCS (Sakaizumi, Ohashi and Yamaguchi, 1976) and for phenyl isothiocyanate, C_6H_5NCS (Onda, Kambayashi, Sakaizumi and Yamaguchi, 1976). The structural chemistry with special focus on the molecular geometry of the

TABLE 10
The N=C and C=S bond lengths (pm) in vapour-phase isothiocyanate molecules

Compound	r(N=C)	r(C=S)	Parameter	Reference
HNCS	121.6(2)	156.1(2)	r_0(N=C) r_s(C=S)	*(a1)*
CH_3NCS	119.2(6)	159.7(5)	r_a	*(b1)*
SiH_3NCS	119.7(7)	156.3(6)	r_a	*(b2)*
	122.08(5)	157.45(5)	r_s	*(a2)*
$(CH_3)_3SiNCS$	118(1)	156(1)	r_g	*(b3)*
GeH_3NCS	114.4(15)	154.2(20)	r_s(N=C)	*(a3)*
F_2PNCS	122.1(6)	155.3	r_a	*(b4)*

(a) Microwave spectroscopy
(b) Electron diffraction
(a1) Dousmanis, Sanders, Townes and Zeiger (1953)
(a2) Dössel and Sutter (1977)
(a3) Durig, Li and Sullivan (1979)
(b1) Anderson, Rankin and Robertson (1972)
(b2) Glidewell, Robiette and Sheldrick (1972)
(b3) Kimura, Katada and Bauer (1966)
(b4) Rankin and Cyvin (1972)

isothiocyanates has been summarized by Hargittai and Paul (1977) together with that of the isocyanates and related compounds. For a discussion of some structural features and detailed structural information on the analogous isocyanates the reader is referred to this review. The microwave spectroscopic studies on CH_3CH_2NCS and C_6H_5NCS (see references above) as well as that on GeH_3NCS (Durig, Li and Sullivan, 1979) and an extensive microwave spectroscopic reinvestigation of SiH_3NCS (Dössel and Robiette, 1977; Dössel and Sutter, 1977; 1979) are new additions to this field.

The X—N=C=S molecules or skeletons have been shown to be planar; for some simple molecules coplanarity is an automatic consequence of the linearity of the NCS chain

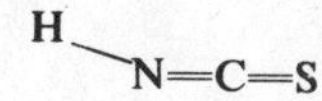

A staggered conformation has been determined for *methyl isothiocyanate*

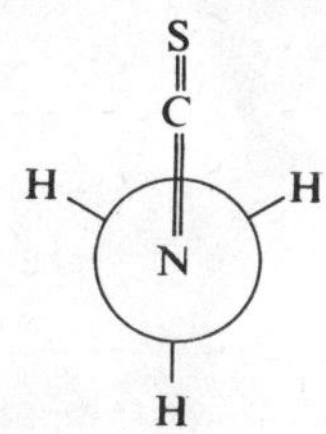

by electron diffraction (Anderson, Rankin and Robertson, 1972). The barrier to internal rotation was reported to be 1.3 ± 0.2 kJ mol^{-1} (Lett and Flygare, 1967). In the electron diffraction study the symmetry axis of the methyl group and the H_3C—N bond were found to coincide within experimental error. With a linear Si—N=C=S chain, the *silyl isothiocyanate* molecule has C_{3v} symmetry which makes all conformations equivalent. If the Si—N=C bond angle differs from 180°, the most likely conformation is the staggered form. For the *trimethylsilyl isothiocyanate* molecule C_s symmetry and a staggered conformation were

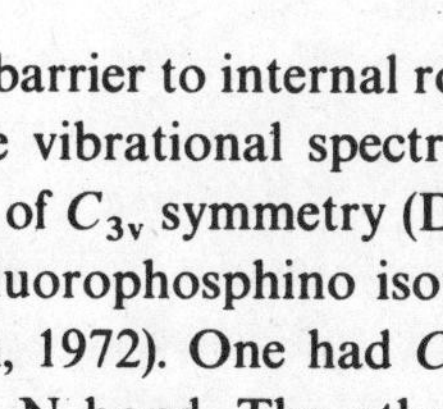

established with a 5.4 kJ mol^{-1} barrier to internal rotation (Durig, Craven, Mulligan, Hawley and Bragin, 1973). The vibrational spectra of germyl isothiocyanate were consistent only with a molecule of C_{3v} symmetry (Durig, Li and Sullivan, 1979). The electron diffraction data on difluorophosphino isothiocyanate were consistent with two models (Rankin and Cyvin, 1972). One had C_s symmetry and large-amplitude torsional motion about the P—N bond. The other model had somewhat smaller-

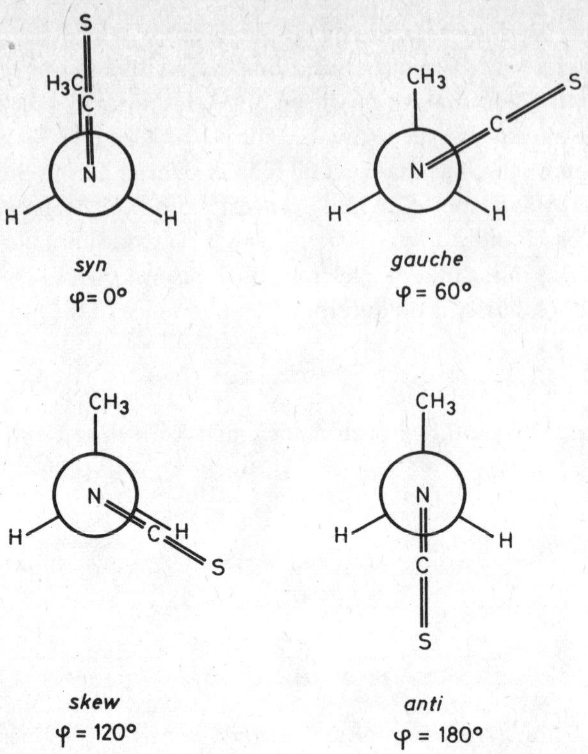

Fig. 4. Four rotational isomers of ethyl isothiocyanate shown by Newman projections representing views along the N—CH_2 bond

amplitude torsional motion but there was some deviation (ca. 14°) from the C_s symmetry.

Four rotational isomers of *ethyl isothiocyanate* with respect to rotation about the H_2C—N bond are shown in Fig. 4. From the microwave spectra Sakaizumi, Ohashi and Yamaguchi (1976) have established that the molecule has a symmetry plane thus the H—C—C—N=C=S chain is planar. In this case only four hydrogen atoms are situated out of this plane at a distance c_H from the plane and the following relationship holds for a rigid molecule:

$$I_a + I_b - I_c = 8\, m_H c_H^2$$

where m_H is the mass of hydrogen. From the experimental data $8m_Hc_H^2 = 6.159 \times \times 10^4$ amu pm^2, while 6.345×10^4 amu pm^2 was obtained assuming r(C—H) $= 109.1$ pm and ∢H—C—H $= 108.8°$. Of course, both the *syn* and the *anti* forms have symmetry planes. Of these two, the *syn* form was shown to be present on the basis of the comparisons of the experimental and calculated rotational constants. Assuming reasonable values for the other parameters, the following bond length and bond angles were determined:

r(C—N=) 145.4 pm
∢C—C—N 109.6° ∢C—N=C 151.0°

Although other conformers could not be identified, the rich rotational spectra indicated the presence of further forms. Earlier Kniseley, Hirshmann and Fassel (1967) presumed the presence of two conformers on the basis of the infrared spectra.

Phenyl isothiocyanate has been known to have a bent C(1)—N=CS chain from vibrational spectroscopy (*cf.* Varsányi, 1973) although some controversy had persisted for some time (*cf.* Stephenson, Coburn and Wilcox, 1961; Ham and Wills, 1960). The bent configuration has also been confirmed by low-resolution microwave spectroscopy (Higgins, Combs, Malloy and Cook, 1975). Further unambiguous evidence for the bent planar structure of phenyl isothiocyanate came from the high-resolution microwave spectroscopic investigation of Onda, Kambayashi, Sakaizumi and Yamaguchi (1976). The planarity was demonstrated by the very small inertial defect (0.055×10^4 amu pm^2). The bent geometry follows from the considerably large I_a value (98.862×10^4 amu pm^2) of C_6H_5NCS as compared with the usual I_a ca. 89×10^4 amu pm^2 for monosubstituted benzene derivatives with C_{2v} symmetry. In the latter the moment of inertia originates solely from the *ortho* and *meta* carbons and hydrogens. Assuming standard values for the other parameters, Higgins, Combs, Malloy and Cook (1975) determined the C(1)—N=C bond angle to be about 145–147°. Similarly, Onda, Kambayashi, Sakaizumi and Yamaguchi (1976) gave r(C(1)—N) 137.0(5) pm and ∢(C(1)—N=C) 148.0(5)°. All these numerical results, however, should only be considered to be tentative since a large number of uncertain assumptions are used to provide them. The overall structure of phenyl isothiocyanate is similar to that of phenyl isocyanate (Bouchy and Roussy, 1973).

Only a partial geometry could be determined from the available experimental data for *germyl isothiocyanate* (Durig, Li and Sullivan, 1979). For the Ge—N and N=C bonds r_s distances were obtained. However, their uncertainty is increased by the nitrogen atom being very close to the centre of mass of the molecule. Assumptions had to be used in estimating the other geometrical parameters. Hence the relatively large error limits given by Durig, Li and Sullivan (1979):

r_s(Ge—N)	181.7(15) pm	r(Ge—H)	152.0 pm (assumed)
r_s(N=C)	114.4(15) pm	r(C=S)	154.2(20) pm

∢H—Ge—N 106.9(5)°

According to the data of Table 10, the C=S bond seems to be rather insensitive to changes of the X ligand linked to the N=C=S chain. A much wider variation is observed in the N=C bond lengths with a strikingly short bond in the germyl derivative. There is a very large difference of 8 pm in the r_s(C=N) bond distances of silyl isothiocyanate and germyl isothiocyanate. It would be of considerable interest to compare the C═N bond lengths of analogous isocyanates and isothiocyanates. However, the electron diffraction results for this bond length in isocyanates should be treated cautiously as they may be strongly affected by the strong correlation originating from the similarity of the N═C and C═O bond lengths. The estimated uncertainties do not always reflect the possible effect of this correlation (see more on this in the discussion on the sulphonyl chloride isocyanate structure).

The geometrical variations in the rest of the molecule will now be examined in analogous isothiocyanates and isocyanates. The relevant data are collected in Table 11. The X—N bond of the X—N═C═E (E = S or O) chain is unchanged or is slightly shorter in isocyanates than in isothiocyanates. The X—N═C bond angle seems to be invariably smaller, though to varying extent, in the isocyanates. When the HN═C═O, HN═C═S, HN═C═Se series is considered, there is indeed a gradual increase in the H—N═C bond angles (see Table 11 for the two former while it is 143° for the selenium derivative). The structure of *isoselenocyanic acid*, HNCSe, was determined by

TABLE 11
Comparison of geometrical parameters of isothiocyanates and isocyanates

Compound	r(X—N), pm	r(N═C), pm	∢X—N═C (°)	Source
HNCO	99(1)	121(1)	128.1(5)	*(d)*
HNCS	98.9(3)	121.6(2)	135.0(2)	*(e)*
Δ	0	−1	−6.9	
CH_3NCO	145.0(4)	116.8(5)		*(f)*
CH_3NCS	147.9(8)	119.2(6)		*(f)*
Δ	−3	−2		
C_6H_5NCO	137.0	[121]	140.6	*(g)*
C_6H_5NCS	137.0(5)	[121]	148.0(5)	*(h)*
Δ	0	—	−7.4	
SiH_3NCO	169.9	115.0	180	*(i)*
	170.3(4)	121.6(9); 120.0(5)	152.2(12)	*(j)*
SiH_3NCS	171.4(10)	121.1(10)	180	*(k)*
	170.4(6)	119.7(7)	163.8(26)	*(j)*
Δ	−1 *(a)*	−6 *(b)*	0 *(b)*; −12 *(c)*	
$SiMe_3NCO$	176(2)	120(1)	150(3)	*(l)*
$SiMe_3NCS$	178(2)	118(1)	154(2)	*(l)*
Δ	−2	2	−4	
GeH_3NCO	183.1(4)	119.0(7)	141.3(3)	*(m)*
GeH_3NCS	181.7(15)	114.4(15)	180	*(n)*
Δ	1.4	4.6	−39	

(a) Calculated from the mean values
(b) Calculated from the microwave spectroscopic data
(c) Calculated from the electron diffraction data
(d) Jones, Shoolery, Shulman and Yost (1950)
(e) Dousmanis, Sanders, Townes and Zeiger (1953)
(f) Anderson, Rankin and Robertson (1972)
(g) Bouchy and Roussy (1973)
(h) Onda, Kambayashi, Sakaizumi and Yamaguchi (1976)
(i) Gerry, Thompson and Sugden (1966)
(j) Glidewell, Robiette and Sheldrick (1972)
(k) Jenkins, Kewley and Sugden (1962)
(l) Kimura, Katada and Bauer (1966)
(m) Murdoch, Rankin and Beagley (1976)
(n) Durig, Li and Sullivan (1979)

TABLE 12
Mulliken gross atomic charges *(I)* and overlap populations *(II)* for isocyanic and isothiocyanic acid after Howell, Absar and Van Wazer (1973)

Molecule	*(I)*			
	H	N	C	O(S)
HNCO	0.309	−0.624	0.688	−0.373
HNCS without d orbitals	0.321	−0.561	0.085	0.155
with d orbitals	0.315	−0.653	0.559	−0.220

	(II)		
	H—N	N—C	C—O(S)
HNCO	0.726	1.067	1.211
HNCS without d orbitals	0.715	1.289	0.762
with d orbitals	0.717	1.161	1.117

Landsberg (1979) from microwave spectroscopic data. The geometry of this quasilinear (see later) molecule was characterized by the following parameters: H—N 99, N=C 119.5, C=Se 171.7 pm, and H—N=C 143°.

Howell, Absar and Van Wazer (1973) compared the electronic structures of isothiocyanic acid and isocyanic acid primarily in order to examine the role of the sulphur d orbitals. The microwave spectroscopic geometries were assumed in the *ab initio* molecular orbital calculations. Only a few of the interesting results are cited here. Table 12 shows the Mulliken gross atomic charges and the overlap populations. The atomic charges differ considerably depending on whether d orbitals were included or not in the calculations. Even the overall sign of the sulphur charge changes. Another feature is that the overlap population of the C=O and C=S bonds are similar only when the sulphur d orbitals are included. One of the illustrations of electron density plots is reproduced in Fig. 5 showing the total electron densities for both HNCO and HNCS and a plot for HNCS showing the difference between including and omitting the d orbitals in the basis set. The latter demonstrates very well the increase of the C=S bond electron density when d orbitals are included. The two most important conclusions of Howell, Absar and Van Wazer (1973) may be summarized as follows: (1) The participation of the nitrogen lone electron pair in the π-electron system of HNCS is stabilized by the relatively high-energy CS π-bond.* On the other hand, the relatively high-energy nitrogen lone electron pair increases the stability of the CO π-bond which,

* By π-bonding, Howell, Absar and Van Wazer (1973) denote the electron distribution containing a nodal plane passing or nearly passing the two nuclei participating in the bond in question. This nodal plane is not necessarily a symmetry plane of the molecule.

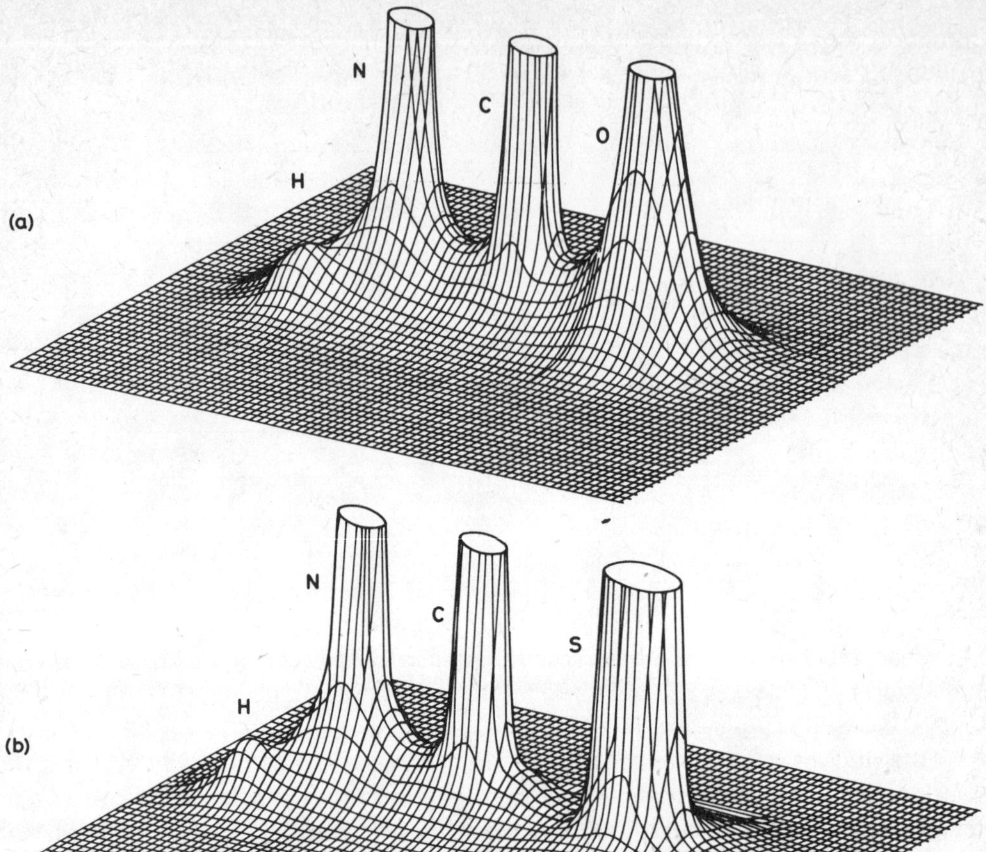

Fig. 5. Representation of the electron density distribution of the molecules HNCS and HNCO after Howell, Absar and Van Wazer (1973). The electron density plots are given for the plane of the molecule

(a) HNCO, total electron density
(b) HNCS, total electron density calculated with d orbitals
(c) HNCS, difference plot of total density with d orbitals included minus total density with d orbitals omitted from the basis set (at a fivefold magnification)

in turn, stabilizes the π-electron system of HNCO. (2) The sulphur bond orbitals accept electron density in a σ-bond manner rather than as π-bonding.

The discussion of the isothiocyanate structures is concluded by comments on the apparent disagreement between the microwave spectroscopic and electron diffraction results on the configuration of the Si—N═C═S chain (*cf.* Table 11). The difference is related to the large-amplitude intermolecular motion, *viz.* bending vibrations of these chains and the different averaging procedure over this motion by the various physical techniques. The isothiocyanates and isocyanates are, of course, very similar in this respect. The microwave spectroscopic results establishing a linear Si—N═C═S(O) skeleton are derived from rotational spectra characteristic of symmetric top molecules (Gerry, Thompson and Sugden, 1966; Dössel and Sutter, 1977; Duckett, Robiette and Mills, 1976). On the other hand, the low-frequency large-amplitude bending motion in these molecules produces large shrinkage effects that result in a definitely non-linear average configuration as determined by electron diffraction (Glidewell, Robiette and Sheldrick, 1972; Cyvin, Brunvoll and Robiette, 1971). Further vibrational spectroscopic evidence for the linearity and low-frequency bending (quasi-linear character) of the Si—N═C═S(O) chains has been communicated by Durig, Kalasinsky and Kalasinsky (1978a; 1978b) and by Durig, Sullivan, Cox and Streusand (1978), and for a similar behaviour of germyl isothiocyanate, by Durig, Li and Sullivan (1979).

The molecular geometries of $PF_2(NCSe)$ and $SiH_3(NCSe)$ were investigated by gas electron diffraction by Cradock, Laurenson and Rankin (1981) and Fraser, Laurenson and Rankin (1981), respectively. Some of the geometrical parameters of interest are:

$PF_2(NCSe)$		$SiH_3(NCSe)$	
r_a (C═Se)	170.0 (10) pm	r_a (C═Se)	175.9 (7) pm
$\measuredangle_a$PNC	143.9 (13)°	$\measuredangle_a$SiNC	158.9 (6)°
$\measuredangle_\alpha$PNC	149.0 (15)°	$\measuredangle_\alpha$SiNC	180.0°

1.4. MOLECULES WITH P═S OR As═S BOND

First the conformational properties of some molecules containing P═S bond are discussed.

Khaikin, Vilkov, Vasil'ev, Mel'nikov, Tulyakova and Anashkin (1972) determined the geometry of vapour-phase $ClCH_2P(S)Cl_2$ by electron diffraction. The experimental data were consistent with the presence of two conformers by rotation about the C—P bond

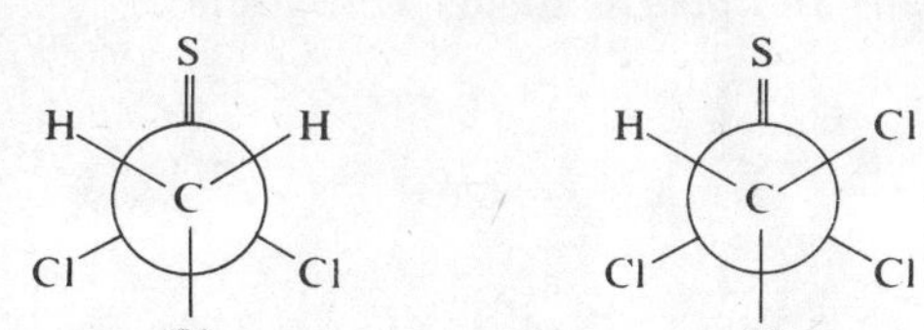

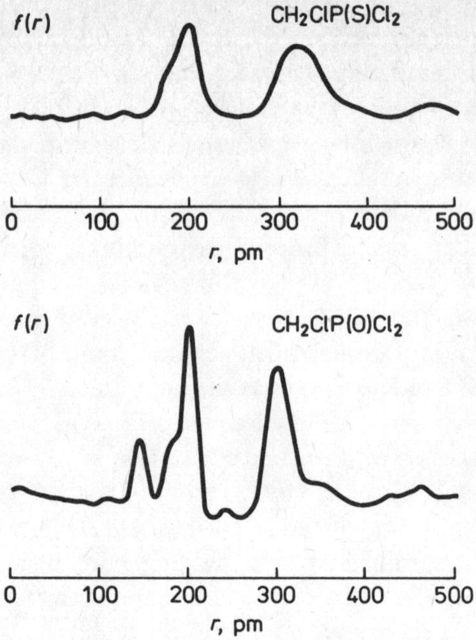

Fig. 6. Electron diffraction radial distributions of $CH_2ClP(S)Cl_2$ after Khaikin, Vilkov, Vasil'ev, Mel'nikov, Tulyakova and Anashkin (1972) and $CH_2ClP(O)Cl_2$ after Vajda, Kolonits, Hargittai and Szőke (1976)

A detailed conformational analysis was hindered by the strong correlation of the rotation-dependent parameters. An additional difficulty was the approximately equal electron scattering ability of sulphur and chlorine since that made the distinguishing of the rotation-dependent Cl . . . Cl and Cl . . . S interactions practically impossible. The structure analysis was performed under the assumption of equal amounts of the two rotational isomers. An electron diffraction study (ca. 85°C) of the analogous $ClCH_2P(O)Cl_2$ confirmed the above assumption (Vajda, Kolonits, Hargittai and Szőke, 1976). A better resolution of the internuclear distances and the differing contribution to the total scattering from the rotation-dependent Cl . . . Cl and Cl . . . O interactions make this molecule much better suited for electron diffraction structure analysis than its thio analog. This is well illustrated by the two radial distribution curves shown in Fig. 6. The presence of the two rotational forms in about equal amounts was also established for both compounds by a vibrational spectroscopic investigation on liquid-phase and solution samples (Steger, Řehák and Faltus, 1965).

For $CH_3OP(S)Cl_2$ *anti* and *gauche* forms were found

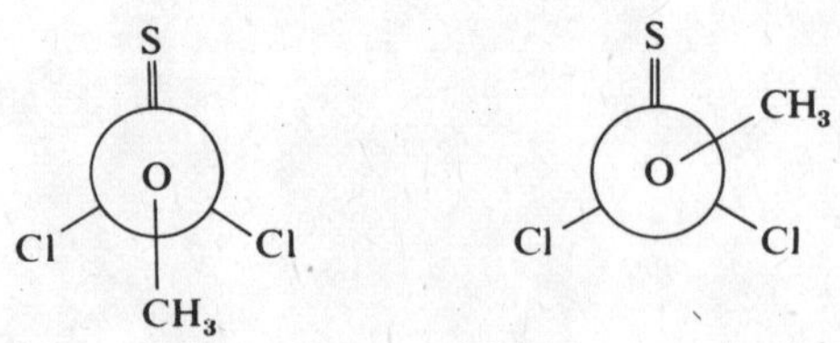

with a ratio of 1:4 at about room temperature by electron diffraction (Bezzubov and Naumov, 1976). In a vibrational spectroscopic study (Remizov, Kuramshin, Aganov and Butenko, 1973) the *gauche* conformer was found to predominate in all phases.

Oberhammer (1975) showed the electron diffraction data on $(CH_3O)_2P(S)Cl$ to be consistent with a mixture of the two rotational forms

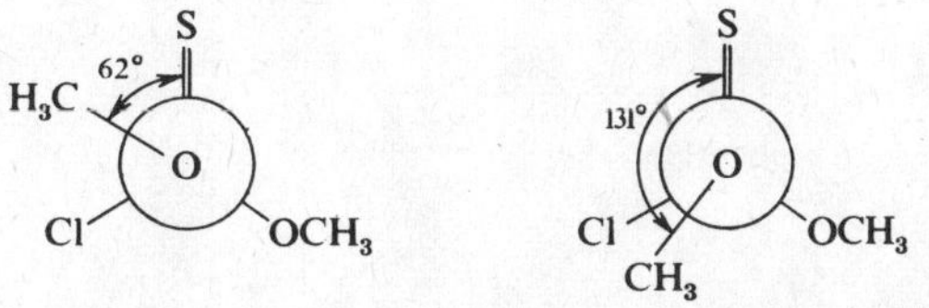

in about equal amounts. The vibrational spectra revealed also the presence of two forms, both having a symmetry plane (Nyquist and Muelder, 1968).

The electron diffraction investigation of *ethylenechloro trithiophosphate* (Naumov, Semashko, Zav'yalov, Cherkasov and Grishina, 1973) established a "half-chair" conformation

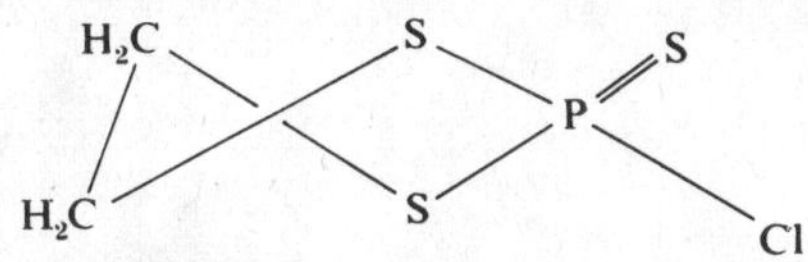

As determined in the same study, its oxygen analog takes up a similar form.

All the above-mentioned electron diffraction analyses produced geometrical parameters as well. Of these, however, only the P=S bond lengths are listed here:

$ClCH_2P(S)Cl_2$	188.8 (10) pm
$CH_3OP(S)Cl_2$	192.2(6) pm
$(CH_3O)_2P(S)Cl$	189.7 (3) pm
$\underline{SCH_2CH_2SP}(S)Cl$	188.5 (20) pm

The uncertainties indicated above (taken from different reports) may have different meaning.

1.4.1. X_3MY-TYPE MOLECULES

The most important geometrical parameters determined for X_3MY molecules are given in Figs. 7a, 7b, 7c. Here M may be phosphorus or arsenic, X halogen or methyl group, and Y may be sulphur, selenium, oxygen, BH_3 group, or a lone pair of electrons. The heavy-atom skeleton of these molecules has C_{3v} symmetry. The parameters collected in Figs. 7a, 7b, and 7c are reliably determined and have a well-defined physical meaning. A detailed critical discussion of these data has been given by Jacob and Samdal (1977).

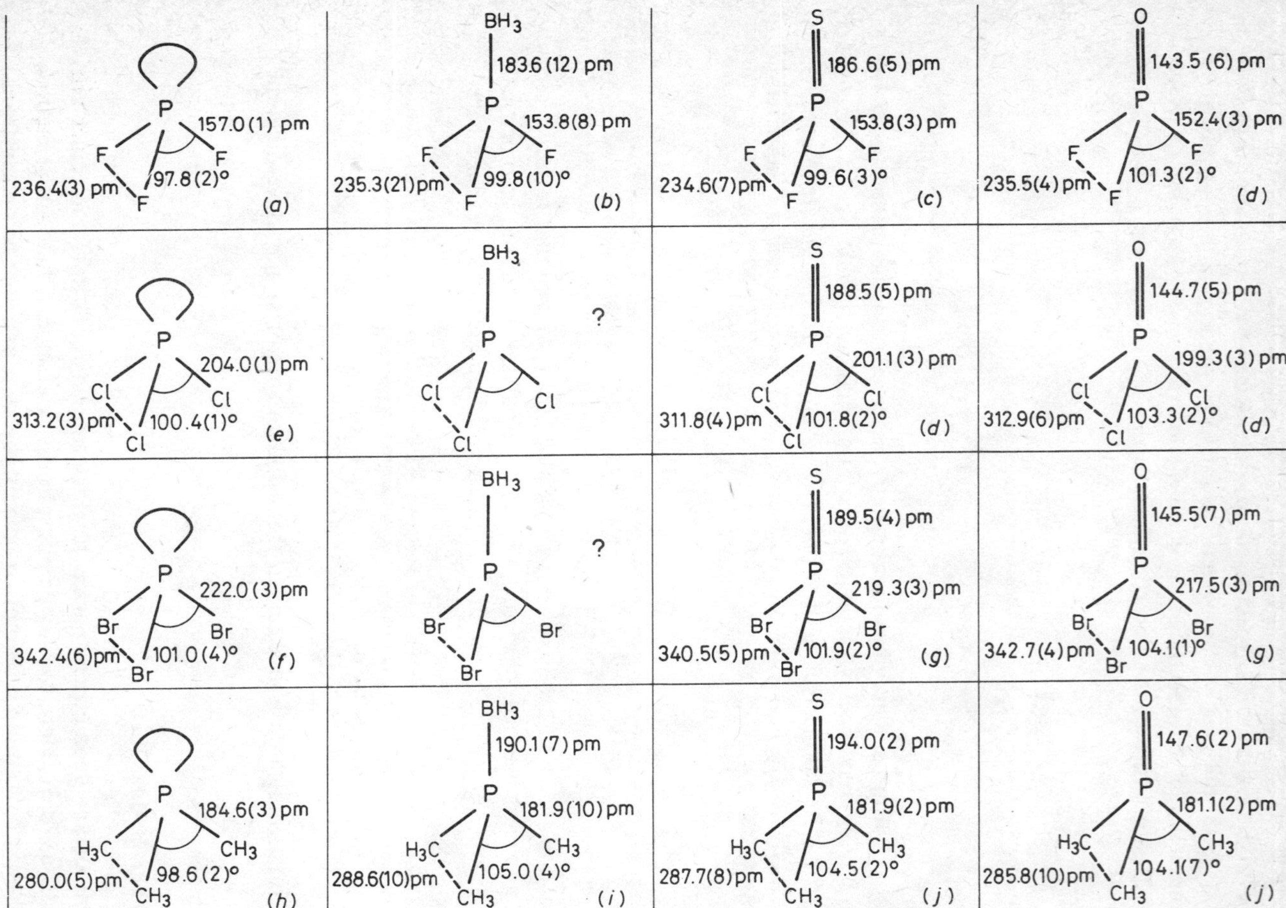

Fig. 7a. The molecular geometry of $X_3P{=}S$ and analogous compounds

(*a*) Morino, Kuchitsu and Moritani (1969)
(*b*) Kuczkowski and Lide (1967)
(*c*) Karakida and Kuchitsu (1976)
(*d*) Moritani, Kuchitsu and Morino (1971)
(*e*) Iwasaki and Hedberg (1962)
(*f*) Kuchitsu, Shibata, Yokozeki and Matsumura (1971)
(*g*) Jacob, Danielson and Samdal (1980)
(*h*) Bartell and Brockway (1960)
(*i*) Bryan and Kuczkowski (1972)
(*j*) Wilkins, Hagen, Hedberg, Shen and Hedberg (1975)

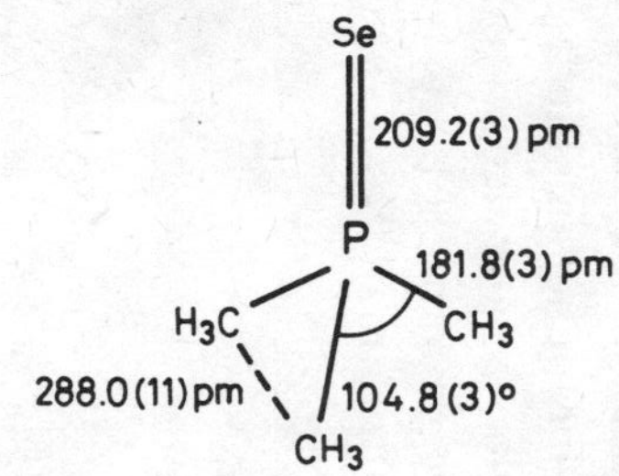

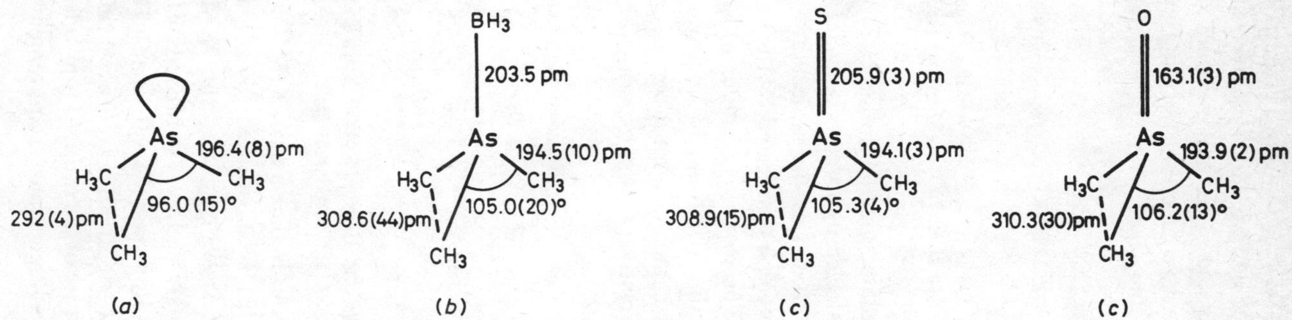

Fig. 7b. The molecular geometry of $(CH_3)_3P{=}Se$ (Jacob and Samdal, 1977)

Fig. 7c. The molecular geometry of $(CH_3)_3As{=}S$ and analogous compounds

(a) Jacob and Samdal (1977)
(b) Durig, Hudgens and Odom (1974)
(c) Wilkins, Hagen, Hedberg, Shen and Hedberg (1975)

The P=S, P=Se, and As=S bonds are essentially double bonds and the P=O and As=O bonds may be considered to be even stronger when their lengths are compared with those calculated by the expression of Schomaker and Stevenson (1941) applying the electronegativity corrections (Pauling, 1960). This is consistent with the changes in the bond stretching force constants of the methyl derivatives:

f(P=O)	803 N m^{-1}	Jacob and Samdal (1977)
f(P=S)	337 N m^{-1}	
f(P=Se)	235 N m^{-1}	
f(As=O)	568 N m^{-1}	Wilkins, Hagen, Hedberg, Shen and Hedberg (1975)
f(As=S)	316 N m^{-1}	

The variations in the calculated mean amplitudes of vibration are in agreement with the changes in the bond stretching force constants, as would be expected since the amplitude calculations are based on the molecular force field. At the same time, the relative bond strength is much less reflected by the bond vibrational amplitudes than by the force constants.* The mean amplitudes of vibration greatly depend on the masses of the atoms forming the bond in question. The pattern provided by the experimentally determined mean amplitudes of vibration may be even more ambiguous owing to the relatively large experimental error:

	Calculated	Experimental
l(P=O), pm	3.8 *(a)*	3.9(3) *(a)*
l(P=S), pm	4.8 *(a)*	3.6(4) *(a)*
l(P=Se), pm	5.3 *(a)*	4.4(4) *(b)*
l(As=O), pm	4.1 *(b)*	4.6(4) *(a)*
l(As=S), pm	4.7 *(b)*	5.2(4) *(a)*

(a) Wilkins, Hagen, Hedberg, Shen and Hedberg (1975)
(b) Jacob and Samdal (1977)

The geometrical variations in the series of X_3MY molecules will now be considered. As Y = lone electron pair (E) is "substituted" by a ligand, the X—M—X bond angles open and the M—X bonds shorten. For the $X_3ME \rightarrow X_3M—BH_3$ transition the geometrical changes are unambiguously predicted by the VSEPR model, as a lone pair is "substituted" by a bonding pair (M—B bond). However, when the lone pair is "substituted" by a double bond, there is no general rule for the geometrical changes in the VSEPR model. The nature of the central atom will determine whether the lone pair or the double bond has a larger space requirement. According to the geometrical

* Although generally used, it is not really the correct terminology, since the bond strength is related to the depth of the potential energy function (dissociation energy), while the bond stretching force constant is the second derivative of this function and thus expresses the rigidity or compliance of the bond rather than its strength.

variations demonstrated by the data in Figs. 7a, 7b, 7c, the lone pairs of phosphorus or arsenic as the central atom have larger space requirements and, accordingly, exercise stronger repulsions towards the neighbouring electron pairs than the P═S, P═Se, P═O, or As═S, As═O double bonds (or even triple bonds as the P═O and As═O may be considered). In the case of phosphorus as the central atom, this effect has already been stated in a discussion of the applicability of the VSEPR model for tetrahedral systems (Hargittai and Baranyi, 1977). As the X_3MS and X_3MO structures are compared from the point of view of the VSEPR model, two competing effects have to be considered. Oxygen as a ligand has a greater electronegativity than sulphur, thus larger X—M—X bond angles and shorter M—X bonds are expected in the X_3MO molecules than in the analogous X_3MS molecules. At the same time, the M═O bond has greater multiple-bond character than the M═S bond causing opposite effects to those described above. The experimentally determined geometries show that the effect of the electronegativity difference prevails.

Karakida and Kuchitsu (1976) established an empirical relationship between the X—M—X bond angles in X_3MY molecules on the one hand, and the covalent radii of the central atom M and ligand X and the electronegativity of ligand Y on the other,

$$\measuredangle X\text{—}M\text{—}X = 101.3 + 0.082 r_X - 0.088 r_M + \chi_Y$$

Earlier Konaka and Kimura (1973) gave an empirical relationship between the X—M—X bond angles of X_3MY molecules and the electronegativities of the atoms constituting these molecules,

$$\measuredangle X\text{—}M\text{—}X = 92 + 8\chi_M - 3\chi_X + \chi_Y$$

These relationships express well the importance of electronegativity in these structures. It is of considerable interest to remember the much earlier investigation of Robinson (1961) on the relationship between the P═O bond stretching frequency and the sum of the electronegativities of the ligands X, Y and Z in XYZPO molecules. Robinson (1961) set up a relative electronegativity scale for the ligands on the basis of the P═O bond stretching frequency variations.

The lengths of the P═Y bonds change considerably with the change of the electronegativities of the X ligand, as shown by the data in Fig. 7a. Thus, e.g. the P═S or P═O bonds are much shorter in the fluorine derivatives than in the chlorine derivatives. At the same time, the difference of r(P═S) and r(P═O) remains unchanged:

Fluorine derivatives	r(P═S) − r(P═O)	43 pm
Chlorine derivatives	r(P═S) − r(P═O)	44 pm

The shortening of the P═Y bond with increasing X ligand electronegativity is in accordance with the electronegativity rule of the VSEPR model. A quantitative description of this correlation is difficult, however, if the methyl group is one of the X ligand series. Figure 8 shows the lengths of the P═S and P═O bonds in X_3PY

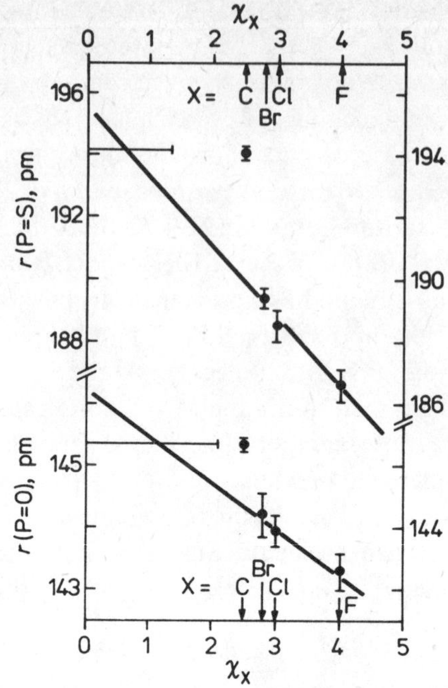

Fig. 8. The variations of the P=S and P=O bond lengths in S=PX_3 and O=PX_3 molecules with X ligand electronegativity (χ_X) after Jacob and Samdal (1977). The vertical bars indicate experimental uncertainties. The horizontal bars correspond to the methyl electronegativity intervals allowed by the experimental data on the halogen derivatives

molecules as a function of the X ligand electronegativities after Jacob and Samdal (1977). The points representing the data on the methyl derivatives deviate considerably from the least-squares lines comprising the data on the halogen derivatives. Thus the experimental bond lengths exaggerate the electronegativity difference between the methyl group and the other ligands. If one insists on these being in a linear relationship including the methyl ligand, the maximum values for methyl electronegativity could be 2.1 in $(CH_3)_3PO$ and 1.3 in $(CH_3)_3PS$, and even these can only be obtained if the optimal experimental error limits are considered.

The geometrical variations, or at least the direction of the changes observed from the data of Fig. 7c may well be accounted for by the VSEPR model. At the same time, an alternative, or at least complementary, interpretation is offered by consideration of the non-bonded atom–atom interactions. Jacob and Samdal (1977) consider these atom–atom interactions to be merely electrostatic. They estimate that changes of a few tenths of an electron in the atomic charges correspond to bond angle changes of several degrees. It is expected that the X halogen ligands as well as the Y ligands have negative charge and the methyl groups have positive charge. The repulsion between the methyl ligands and the attraction between the methyl and Y ligands both increase the C—M—C bond angle. This is indeed observed as the C—M—C bond angle opens considerably upon going from $(CH_3)_3ME$ to $(CH_3)_3MY$. On the other hand, the X ... X and X ... Y repulsions are competing effects from the point of view of the changes in the X—M—X bond angle in going from X_3ME to X_3MY.

1.4.2. HF_2PY-TYPE AND XF_2PS-TYPE MOLECULES

These asymmetric structures pose questions as interesting as do the X_3MY-type molecules. The microwave spectra of three isotopic species of thiophosphoryl difluoride hydride were analysed by Nave and Sheridan (1973). The molecule has C_s symmetry. The planar moment is determined by the position of the two fluorine atoms and, ignoring vibrational effects, 235.2 pm was obtained for the F ... F distance in all three isotopic molecules from the expression

$$I_a + I_c - I_b = m_F r_0^2(\text{F} \ldots \text{F})$$

The bond lengths and bond angles cited in Table 13 were determined by a least-squares fit ("r_0" parameters).

There are geometrical parameters for four HF_2PY molecules in Table 13 determined by microwave spectroscopy. The ligand Y is a lone electron pair, BH_3 group, sulphur or oxygen in the four molecules. The characteristic structural variations may be easily accounted for by the VSEPR model, as discussed for the X_3MY molecules. The difference between the two bond angles H—P—F and F—P—F, however, has to be commented upon. According to the VSEPR model two effects have to be considered here. One of them is the ligand electronegativity according to which ∡F—P—F is expected to be smaller than ∡H—P—F. On the other hand, the P—F bonds have a considerable amount of double bond character, as shown by the significant difference between the experimentally determined bond lengths and the one calculated for a single bond from the expression of Schomaker and Stevenson (1941),

TABLE 13
Geometrical parameters of HF_2PY molecules (Y = –, BH_3, S, O) from microwave spectroscopy

Parameter	HF_2P *(a)*	$HF_2P—BH_3$ *(b)*	HF_2PS *(c)*	HF_2PO *(d)*
r(P—Y), pm		183.2(9)	186.7(5)	143.7(10)
r(P—H), pm	141.2(6)	140.9(4)	139.2(5)	138.7(20)
r(P—F), pm	158.2(2)	155.2(6)	155.1(5)	153.9(5)
∡H—P—F (°)	96.3(5)	98.6(2)	100.3	101.9(20)
∡F—P—F (°)	99.0(2)	100.0(5)	98.6(2)	99.8(10)
r(F...F), pm	240.6 "r_0"	237.8 "r_0"	235.2 "r_0"	235.4 "r_0"

(a) Kuczkowski (1968)
(b) Pasinski and Kuczkowski (1971)
(c) Nave and Sheridan (1973)
(d) Centofanti and Kuczkowski (1968)

viz. 167 pm. According to the "multiple-bond effect" of the VSEPR model, the F—P—F bond angle is expected to be larger. The changes in the relationship of the two angles in the four structures indicate that the two competing effects are in a fine balance that can be easily changed depending on the nature of the Y ligand.

The structure of HF_2PS has also been determined by electron diffraction, together with those of other XF_2PS-type derivatives with X = Cl, Br, and I by Acha, Cromie and Rankin (1981). Only the P=S distances (r_g) are cited here:

HF_2PS	187.7(3) pm
ClF_2PS	186.5(8) pm
BrF_2PS	188.2(4) pm
IF_2PS	190.2(6) pm

The Cl, Br, I series clearly shows some electronegativity effect in the variations of the P=S bond lengths.

Boyd, Laurenson and Rankin (1981) reported the structure of HF_2PSe from a combined analysis of gas electron diffraction and dipolar couplings obtained for a solution in a nematic phase. The results include r_a(P=Se) = 202.6(4) pm.

1.5. OTHER STRUCTURES

The first reports on the existence of two isomers of difluorodisulphide, S_2F_2,

$S{=}SF_2$	FSSF
1,1-Difluorodisulphide	1,2-Difluorodisulphide

appeared in 1963 (microwave spectroscopy, Kuczkowski and Wilson, 1963; mass spectrometry, Kuczkowski, 1963; *cf.* also Seel, 1974). The molecular geometry of 1,1-difluorodisulphide was determined by microwave spectroscopy (Kuczkowski, 1964). The structure of 1,2-difluorodisulphide will be discussed with other disulphanes including HSSH, ClSSCl, and BrSSBr. Until very recently no other 1,1-disubstituted disulphide derivatives were known. Not long ago, however, 1,1-dichlorodisulphide, $S{=}SCl_2$, and 1,1-dibromodisulphide, $S{=}SBr_2$, have also been observed and their infrared spectra identified (Chadwick, Grzybowski and Long, 1978; Feuerhahn and Vahl, 1979). Furthermore, *ab initio* molecular orbital calculations have predicted $S{=}SH_2$ to be as stable with respect to HSSH as $S{=}SF_2$ is to FSSF (Hinchliffe, 1979). Thus it is hoped that the molecular geometries of further 1,1-disubstituted disulphides will eventually be determined.

The microwave spectra of three isotopic species of *1,1-difluorodisulphide* have been analysed (Kuczkowski, 1964). The length of the S=S bond was determined to be 186.0(15) pm, where the indicated error included both the errors in the rotational constants and the uncertainties in the analysis. Further discussion of the S=SF_2 structure will be found in the later section on the sulphoxides and related molecules with three-coordinated sulphur.

The geometry of the S=SO molecule was also determined by microwave spectroscopy (Tiemann, Hoeft, Lovas and Johnson, 1974). The data on three isotopic species have been analysed, $r_s = 188.248(10)$ pm.

The millimeter wavelength spectrum of *diatomic sulphur*, S_2, generated in high-temperature vapour, was recently investigated by Pickett and Boyd (1979). The r_0 internuclear distance of 189.1136 pm may be calculated from the B_0 rotational constant reported, 8831.8676 ± 0.006 MHz. It is pleasing to note that the result of an early visual electron diffraction study agrees within experimental error with this highly accurate parameter. Maxwell, Mosley and Hendricks (1936) determined 192(3) pm for S_2 in a vapour at 800°C. Similarly, diatomic *selenium* and *tellurium* were also investigated yielding 219(3) and 259(2) pm, respectively (Maxwell and Mosley, 1939).

Although there are significant differences in the S=S bond lengths of the S_2, SSO, SSF_2 series, the available bond stretching force constants show no regularities to correlate with them (490, 470, 500 N m^{-1} in that order, *cf.* Steudel, 1975).

The structure of the NS free radical has been investigated by various physical techniques, microwave spectroscopy, electron paramagnetic resonance, ultraviolet spectroscopy (see Uehara and Morino, 1969; and references therein). The best rotational constant, 23,156.01 MHz, comes from the pure rotational spectra. The corresponding internuclear distance is $r_0(^{14}N—^{32}S)$ 164.441 pm with an uncertainty probably only in the third decimal place. The equilibrium internuclear distance was determined in the SO free radical from the pure rotational spectra, $r_e(^{32}S—^{16}O)$ 148.108(5) pm (Amano, Hirota and Morino, 1967).

A complete substitution structure (r_s) was determined for *thioborine*, HBS, from millimeter-wave spectra (Pearson and McCormick, 1973). The rotational spectra of eight isotopic species were analysed, and the r_s parameters obtained in eight different ways. The r_s structures proved to be invariant to a great extent to the choice of the isotopic species combinations from which they had been deduced, as demonstrated by the bond lengths shown in Table 14. Although it was stated that thioborine was the first molecule to demonstrate a boron–sulphur double bond, such a bond had already been observed in an early electron diffraction analysis of *boron sulphide*, B_2S_3, by Akishin and Spiridonov (1959). A V-shape for the B_2S_3 molecules was established and other molecular shapes (*cf.* Fig. 9) could be excluded. The structure of boron sulphide will be discussed in more detail among the symmetrical sulphide molecules.

The structure of the ArOCS van der Waals molecule was investigated by rotational spectroscopy to examine the intermolecular potential between an atom and a linear molecule (Harris, Janda, Novick and Klemperer, 1975). The molecules subjected to molecular beam electric resonance spectroscopic investigation were identified by mass spectrometry (*cf.* Harris, Novick, Klemperer and Falconer, 1974; Dyke, Howard and Klemperer, 1972). The ArOCS molecules were generated by expansion of a 1 atm gas

TABLE 14
Bond lengths of thioborine in substitution structures derived eight different ways by Pearson and McCormick (1973)

Parent molecule	r_s(H—B) (pm)	r_s(B=S) (pm)
$H^{11}B^{32}S$	116.899	159.961
$H^{10}B^{32}S$	116.911	159.962
$H^{11}B^{34}S$	116.897	159.959
$H^{10}B^{34}S$	116.911	159.961
$D^{11}B^{32}S$	116.924	159.928
$D^{10}B^{32}S$	116.937	159.934
$D^{11}B^{34}S$	116.927	159.924
$D^{10}B^{34}S$	116.941	159.929
Mean	116.918	159.945
Range	0.04	0.04

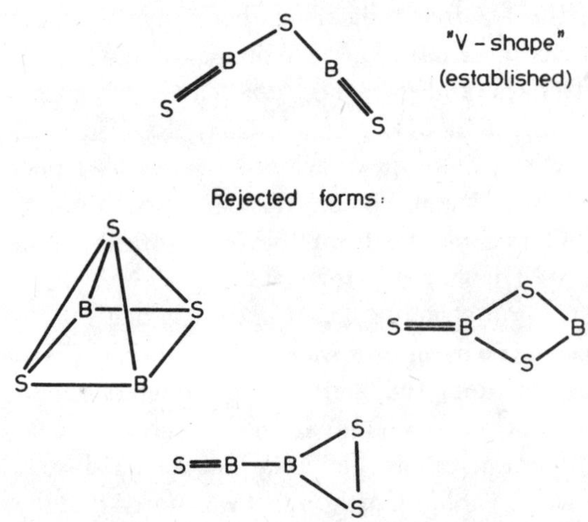

Fig. 9. Molecular models of boron sulphide, B_2S_3, considered in its electron diffraction analysis (Akishin and Spiridonov, 1959)

mixture of 99% Ar and 1% OCS. The OCS structure was assumed and the following distances were determined from the three observed rotational constants:

Ar—C	357.8(1) pm
Ar—O	360.1(1) pm
Ar—S	410.1 pm

Accordingly, the molecule is of T-shape and the Ar—C=S angle is 98.1°. A schematic

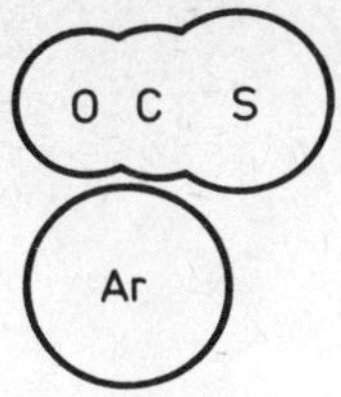

Fig. 10. Molecular model of the ArOCS van der Waals structure after Harris, Janda, Novick and Klemperer (1975)

van der Waals model of this molecule is shown in Fig. 10. It was suggested that the ArOCS molecule held together by weak forces, be considered as a donor–acceptor complex in which the interaction is formed between the highest energy occupied orbital of the argon atom and the lowest energy unoccupied orbital of the OCS molecule. It would be of considerable interest to know to what extent these weak interactions change the geometry of the OCS part of this van der Waals molecule as compared with the free OCS molecule. No information is available on this from the cited study, since the OCS geometry was assumed in the analysis. From general considerations, however, it may be possible to predict the *direction* of the expected structural changes. Both the electron pair repulsions and the atom–atom non-bonded repulsions tend to lengthen the bonds and to bend the molecule. As the OCS molecule itself is relatively rigid and the new interactions in the ArOCS complex are weak, very slight geometrical changes are to be expected.

Chapter 2

MOLECULES WITH TWO-COORDINATED SULPHUR

2.1. SYMMETRICAL SULPHIDES

Prior to the detailed discussion of the individual structures, the SX_2 sulphides whose molecular geometry has been determined in the vapour phase are listed. The analogous selenium and tellurium compounds investigated are also indicated.

SF_2	SCl_2	$SeCl_2$	$SeBr_2$
SH_2	SeH_2	TeH_2	
$S(BS)_2$			
$S(CN)_2$			
$S(CH_3)_2$	$S(CF_3)_2$	$Se(CH_3)_2$	$Se(CF_3)_2$
$S(SiH_3)_2$	$S(GeH_3)_2$	$Se(SiH_3)_2$	$Se(GeH_3)_2$
$S[N(CH_3)_2]_2$			
$S(C_6H_5)_2$	$S(C_5NH_4)_2$		

The simplest derivatives were studied primarily by microwave spectroscopy, while electron diffraction was used to elucidate the structure of the larger molecules. Both techniques have been employed to determine the structure of sulphur dichloride, sulphur dicyanide and dimethyl sulphide. The bond lengths and bond angles characterizing the average structure of dimethyl sulphide were determined by a combined analysis of electron diffraction and microwave spectroscopic data. Thus accurate structural information is available for the simplest compounds. However, while the bond length of sulphur difluoride is known to a hundredth of a picometer, and the tilt of the symmetry axis of the methyl group from the C—S bond has been unambiguously determined, we have only superficial information on the conformational properties of diphenyl sulphide or di-2-pyridyl sulphide. The interesting geometrical peculiarities and observed structural variations of the symmetrically substituted sulphides will be discussed not only in the present chapter but also in relation to other classes of compounds, e.g. sulphone molecules.

Sulphur difluoride, SF_2. This molecule has been unambiguously identified by mass spectrometry (Glemser, Heussner and Haas, 1963; Seel, Heinrich, Gombler and

TABLE 15
Molecular parameters of sulphur difluoride

Geometry	r(S—F), pm	∢F—S—F (°)
r_0 *(a)*	158.9	98.27
r_{av} *(b)*	159.208(8)	98.197(11)
r_e *(c)*	158.745(12)	98.048(13)

Rotational constants (MHz) *(c)*	A_e 26,858.871(50)	B_e 9259.153(19)	C_e 6886.114(15)

Vibrational frequencies	ν_1 (cm^{-1})	ν_2 (cm^{-1})	ν_3 (cm^{-1})
Estimated *(b)*	840 ± 20	357 ± 2	809 ± 10
Ne-matrix *(d)*	834.1(9.02 ± 0.05)	358	807.5(9.50 ± 0.05)
Ar-matrix *(d)*	831.9(8.92 ± 0.05)	357.5(1.85 ± 0.05)	804.1(9.38 ± 0.05)
N_2-matrix *(d)*	825.4	357.5	799.2
Gaseous *(e)*	838.5299		813.0413

(a) Johnson and Powell (1969)
(b) Kirchhoff, Johnson and Powell (1973)
(c) Endo, Saito, Hirota and Chikaraishi (1979)
(d) Haas and Willner (1978); Parenthesized values are the frequency differences observed by the isotopic substitution $^{32}S/^{34}S$
(e) Deroche, Bürger, Schulz and Willner (1981)

Budenz, 1969) and microwave spectroscopy (Johnson and Powell, 1969). The latter also established the C_{2v} symmetry of the molecule and determined the bond length and bond angle in terms of the r_0 parameters cited in Table 15. Extremely accurate average parameters, r_{av} (Table 15) were later communicated by Kirchhoff, Johnson and Powell (1973) from a more detailed study taking into account the effects of intramolecular motion. Finally, Endo, Saito, Hirota and Chikaraishi (1979) determined the equilibrium structure (r_e, Table 15) taking into account anharmonic corrections. In this study the microwave spectroscopic information available for the ground state was augmented by the observation and analysis of the rotational spectra in the first excited states of the three normal modes. The equilibrium rotational constants are also given in Table 15. The resulting moments of inertia still produce a small inertia defect, $\Delta_e = -66.2(20)$ amu pm^2, which is ascribed to an electronic contribution and to neglecting higher-order terms in the vibrational corrections. Since it was possible to derive the equilibrium parameters from three different combinations of the principal moments of inertia, a range of values resulted. The spread in the results obtained was given as the uncertainties in the mean values from the three combinations. The uncertainties due to experimental errors in the rotational constants were considered to be at least an order of magnitude smaller (Endo, Saito, Hirota and Chikaraishi, 1979). The difference between r_{av} and r_e of the S—F bond was ascribed mainly to the anharmonicity of the bond stretching.

Until very recently (Haas and Willner, 1978), there were no vibrational spectroscopic studies on sulphur difluoride. Kirchhoff, Johnson and Powell (1973) established a force field from the centrifugal distortion constants. Their estimated vibrational frequencies are shown in Table 15. As shown, ν_1 was presumed to lie at higher wave number than ν_3, in agreement with the vibrational data of CF_2, NF_2, OF_2 (for references, see Kirchhoff, Johnson and Powell, 1973), which differ, however, from those of SiF_2 (Hastie, Hauge and Margrave, 1969; Milligan and Jacox, 1968). In fact, considering vibrational data on molecules containing a sulphur–fluorine bond, Bliefert and Wanczek (1970) estimated the frequencies of SF_2 in the reverse order. In the microwave spectroscopic study of sulphur difluoride by Endo, Saito, Hirota and Chikaraishi (1979) the ν_1–ν_3 difference was determined to be 25.72 ± 0.33 cm^{-1} with ν_1 being definitely larger. A matrix-isolation infrared spectroscopic investigation by Haas and Willner (1978) reported frequencies consistent with the microwave spectroscopic data. The results of a recent gaseous infrared spectroscopic investigation by Deroche, Bürger, Schulz and Willner (1981) are also cited in Table 15. Another recent study on selenium difluoride by Haas and Willner (1979) gave $\nu_1 = 702.0(697.2)$ and $\nu_3 = 675.0(669.8)$ cm^{-1} for $^{76}SeF_2$($^{82}SeF_2$). Comparing the force fields of sulphur difluoride and silicon difluoride, Haas and Willner (1978) noted that while the other internal force constants are similar, the $f_{r\alpha}$ constant in SiF_2 is three times larger than that in SF_2. This finding indicates that there is a greater degree of angular deformation induced by changing the bond length in the silicon derivative than in sulphur difluoride.

Sulphur dichloride, SCl_2. The geometry of this compound was determined by two independent microwave spectroscopic studies of Murray, Little, Williams and Weatherly (1976) and Davis and Gerry (1977), and also by an electron diffraction investigation of Morino, Murata, Ito and Nakamura (1962). The electron diffraction paper reported the following parameters:

r_g(S—Cl)	200.6(4) pm	l(S—Cl)	5.3(4) pm
r_g(Cl...Cl)	314.3(9) pm	l(Cl...Cl)	10.2(7) pm
∢Cl—S—Cl	103.0(4)°		

Cyvin (1968) calculated the following mean amplitudes of vibration for the $^{35}Cl_2S$ molecule at 298 K:

l(S—Cl) 4.87 pm and l(Cl...Cl) 9.30 pm

Of the two available microwave spectroscopic studies, the one by Davis and Gerry (1977) seems to be the more complete. They examined the rotational spectrum of the $^{32}S^{35}Cl_2$ molecule in the $\nu_2 = 1$ excited vibrational state and the rotational spectrum of the ground state $^{32}S^{35}Cl^{37}Cl$ molecule. The collected data enabled them to determine three types of structures: r_0 effective structure, r_s partial substitution structure and r_{av} average structure. As only two of the three principal moments of inertia are necessary to determine the two independent geometric parameters, the choice of the principal moments of inertia influenced the structural parameters obtained. The results pertaining to the $^{32}S^{35}Cl_2$ molecule are cited in Table 16. The partial substitution

TABLE 16
The bond length and bond angle of sulphur dichloride from the microwave spectroscopic study of Davis and Gerry (1977)

		r(S—Cl), pm	∢Cl—S—Cl (°)
Ground vibrational state			
I_a, I_b	r_0	201.249	102.808
I_a, I_c		201.358	102.858
I_b, I_c		201.599	102.599
Mean		201.402	102.742
I_a, I_b	r_s	201.376	102.662
I_a, I_c		201.350	102.699
I_b, I_c		201.511	102.564
Mean		201.412	102.642
I_a, I_b	r_{av}	201.524	102.730
I_a, I_c		201.524	102.730
I_b, I_c		201.526	102.729
Mean		201.525	102.730
Excited vibrational state ($v_2 = 1$)			
I_a, I_b	r_{av}	201.537	102.931
I_a, I_c		201.538	102.931
I_b, I_c		201.538	102.930
Mean		201.538	102.931

structure was obtained from the substitution coordinates of the chlorine atom and the $\Sigma m_i b_i$ condition. The average structures were produced taking into account the effects of the molecular vibration. It can be clearly seen that the parameters depend to a decreasing degree on the choice of the particular pair of principal moments of inertia used in the calculation in the order $r_0 \rightarrow r_s \rightarrow r_z$. It can be noted that bond lengths are essentially the same in the $v_2 = 1$ excited vibrational state and in the ground state. On the other hand, the bond angle slightly opens as a consequence of excitation of the bending vibration.

Selenium dichloride, $SeCl_2$ and *selenium dibromide*, $SeBr_2$. Visual electron diffraction studies by Akishin, Spiridonov and Mishulina (1960) yielded the following bond lengths:

r(Se—Cl) 218(2) pm
r(Se—Br) 232(2) pm

In fact, the samples studied were tetrahalides. The analysis of the data revealed, however, that the electron scattering originated from dihalide and halogen molecules corresponding to the thermal reaction

$$SeX_4 \rightarrow SeX_2 + X_2$$

TABLE 17
Equilibrium geometrical parameters of H_2S, H_2Se and H_2Te

E = S, Se, Te	r_e(E—H), pm	∢H—E—H (°)
H_2S *(a)*	133.56	92.12
H_2Se *(b)*	146.0	90.57
H_2Te *(c)*	165.8	90.25

(a) Edwards, Moncur and Snyder (1967); Burrus and Gordy (1953)
(b) Hill and Edwards (1965); Jache, Moser and Gordy (1956)
(c) Moncur, Willson and Edwards (1974)

Hydrogen sulphide, H_2S, *hydrogen selenide*, H_2Se, and *hydrogen telluride*, H_2Te. The parameters of the H_2S and H_2Se equilibrium structures were determined by a combined analysis of microwave and infrared spectroscopic data. The equilibrium structure of H_2Te originated from a rotational–vibrational infrared spectroscopic study. The bond lengths and bond angles of all the three molecules are compiled in Table 17. The geometry of hydrogen sulphide in an excited electronic state (Gallo and Innes, 1974) and that of the hydrogen sulphide positive ion, H_2S^+ (Duxbury, Horani and Rostas, 1972) were determined from ultraviolet spectra.

Boron sulphide, B_2S_3. The B═S terminal bond of this molecule has already been mentioned. In the electron diffraction investigation a rotating sector was employed in the experiments but the electron scattering intensities were estimated visually (Akishin and Spiridonov, 1959). Boron oxide was studied in a similar way at about the same time by the same authors (Akishin and Spiridonov, 1960). Later, more advanced electron diffraction studies of boron oxide (Ezhov, Tolmachev, Spiridonov and Rambidi, 1968; Ezhov, Tolmachev and Rambidi, 1970) established a B—O—B bond angle of 132° which differed significantly from the result of the early study (95°). There has been no new investigation of the boron sulphide structure, but some arguments are presented below by which we will try to show that the early study reliably determined the B—S—B angle as $96 \pm 5°$. The fact that the visual electron diffraction technique could yield such an inaccurate bond angle for the B_2O_3 molecule can probably be explained by the difficulties caused by the relatively small electron scattering power of the oxygen atom (compared to the sulphur atom with twice as large atomic number). The experimental radial distribution curve of boron sulphide is reproduced in Fig. 11. There are three distinct maxima on this curve; of these the first originates from the boron–sulphur bonds and the last two may be assigned to the sulphur. . .sulphur non-bonded interactions. The assignment of the sulphur. . .sulphur contributions is greatly helped by the relatively large scattering power of the sulphur atoms. The angle of 96° determined in the S. . .S. . .S triangle can be considered to be the B—S—B bond angle to the extent to which the SBS chains are supposed to be linear. An up-to-date electron diffraction reinvestigation of boron sulphide would be desirable in any case, since

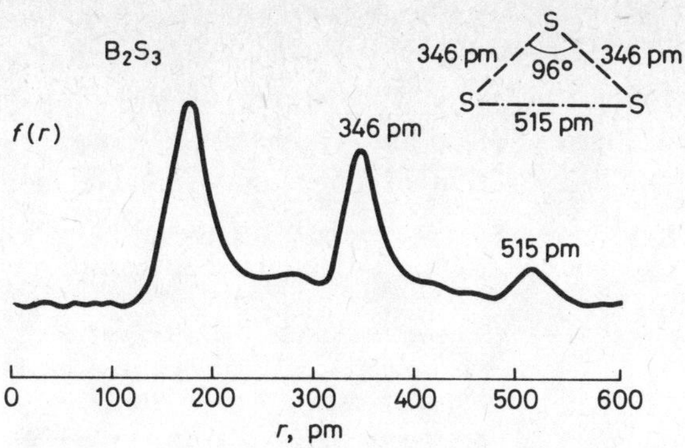

Fig. 11. The electron diffraction radial distribution of boron sulphide after Akishin and Spiridonov (1959) and the assignment of S. . .S interactions

TABLE 18

Mean amplitudes of vibration (*l*, pm) of boron sulphide and sulphur dicyanide at room temperature from spectroscopic calculations

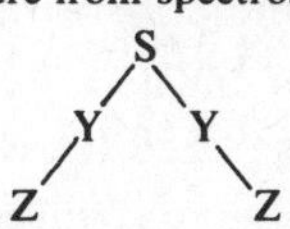

Atomic pair	B_2S_3 *(a)*	$S(CN)_2$ *(a)*	$S(CN)_2$ *(b)*
S—Y	5.51	4.68	4.55
Y—Z	5.03	3.68	3.49
Y. . .Y	8.33	6.73	9.87
Z. . .Z	8.78	7.91	18.10
Y. . .Z	8.60	7.28	12.60
S. . .Z	6.31	6.03	4.84

(a) Nagarajan, Durig and Perumal (1969)
(b) Brunvoll (1967)

many interesting structural details could be determined. This is, however, a relatively non-volatile compound necessitating high-temperature electron scattering experiments. The shrinkage effect expected at the high-temperature conditions may then hinder the determination of the linearity of the SBS chain or a deviation from it. Spectroscopic calculations by Nagarajan, Durig and Perumal (1969) have indicated that the boron sulphide molecules are relatively rigid at room temperature. The calculated mean amplitudes of vibration are cited in Table 18. Unfortunately, no calculations for the molecule at higher temperatures have been communicated.

Sulphur dicyanide, $S(CN)_2$. The molecular structure has been investigated both in the gaseous (Arnold and Rogowski, 1965; Pierce, Nelson and Thomas, 1965) and crystalline phases (Emerson, 1966). The results of the up-to-date microwave spectroscopic study of Pierce, Nelson and Thomas (1965) and those of the crystal-phase x-ray diffraction investigation of Emerson (1966) are in agreement if the relatively large uncertainties of the x-ray diffraction data are taken into account. There is no indication of any striking structural difference between the two phases:

	Vapour phase microwave spectroscopy Pierce, Nelson and Thomas (1965)	Crystalline phase x-ray diffraction Emerson (1966)
C—S (pm)	170.1(2)	173(2)
C≡N (pm)	115.7(2)	112(2)
C—S—C (°)	98.4(2)	95.6(8)
S—C≡N (°)	174.8*	177(1)

The C—S bonds are considerably shorter than the expected single bond on the basis of covalent radii and electronegativity difference.

The microwave spectroscopic investigation by Pierce, Nelson and Thomas (1965) established the planarity and C_{2v} symmetry of the sulphur dicyanide molecule. The SCN chain was found to be non-linear both in the vapour and crystalline phases. The vapour-phase data are, of course, decisive in the determination of the structure of the free molecules. Some effects, however, were ignored in the analysis, and it was felt that further evidence was required on the configuration of the chain. On the basis of comparative *ab initio* molecular orbital calculations, Rosmus, Stafast and Bock (1975) found the structure with the bent SCN chain to be slightly more stable than the one with the linear chain. Figure 12 demonstrates the electron density distributions in two projections after Rosmus, Stafast and Bock (1975) and the non-linearity of the SCN chains is well seen. The canonical representation of the electron density distribution shows another interesting peculiarity. There is some electron density between the two carbon atoms which may indicate the presence of some fractional bond. It may be significant that the carbon . . . carbon distance in sulphur dicyanide is much shorter than, for example, that in dimethyl sulphide (see later):

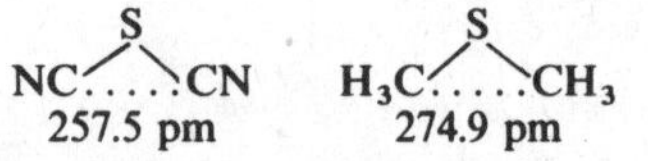

Pierce, Nelson and Thomas (1965), and also Nayar and Aruldhas (1968) reported estimated force fields for sulphur dicyanide. The S—C bond stretching force constant is

* Pierce, Nelson and Thomas (1965) give the angle between the directions of the two C≡N bonds (*viz.* 108.4 ± 0.3°). This is, of course, an unambiguous characterization, while in addition to the S—C≡N bond angle used in the present discussion, the direction of the deviation from linearity must also be indicated. The C≡N bonds are tilted towards the lone electron pairs of the sulphur atom (*cf.* Fig. 12).

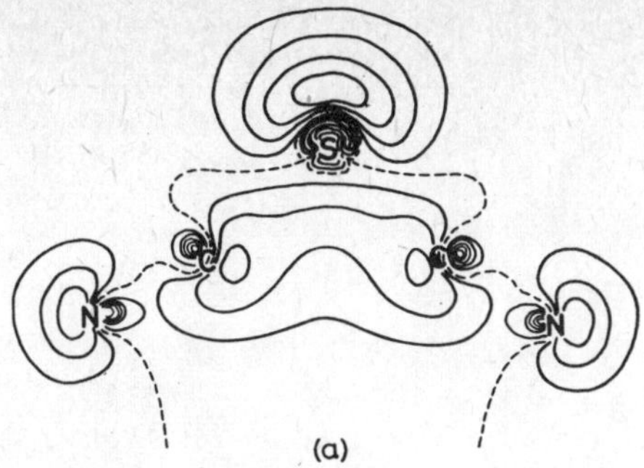

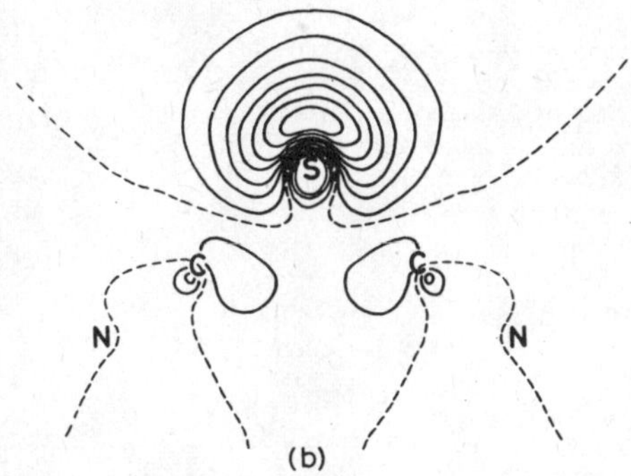

Fig. 12. Contour diagrams of the (a) canonical and (b) localized in-plane sulphur lone electron pair of sulphur dicyanide from *ab initio* calculations after Rosmus, Stafast and Bock (1975). It is seen that the S—C≡N chain is not linear. Furthermore, the electron density between the two carbon atoms indicates the formation of a fractional bond

430 N m^{-1} according to the latter authors. The C≡N bond stretching force constants of a series of cyanide molecules are given in Table 19 together with the bond lengths. According to Nayar and Aruldhas (1968) the smaller force constant in $S(CN)_2$ indicates a decreased C≡N bond order in this molecule. At the same time, the C≡N bond in $S(CN)_2$ is relatively short. The structure of sulphur dicyanide certainly calls for an electron diffraction determination, which would be expected to yield some good mean vibrational amplitudes in addition to the geometry. Calculated amplitudes are cited in Table 18. The two sets of results agree well for the bonds, but differ considerably for the non-bonded interactions. One of the vibrational parameters of great interest is, of course, l(C . . . C), since in the case of partial bonding, the carbon–carbon non-bonded interaction would be expected to be relatively rigid.

Dimethyl sulphide, $S(CH_3)_2$. Pierce and Hayashi (1961) analysed the rotational spectra of five isotopic species in a microwave spectroscopic study. In the determination of the geometrical parameters they utilized one single assumption, *viz.* $r(C—H_{as}) = r(C—H_s)$, where the meaning of the subscripts *as* and *s* is shown in the molecular model of Fig. 13. The parameters determined correspond essentially to a

TABLE 19

C≡N bond lengths and bond stretching force constants

Compound	r(C≡N), pm			$f(C\equiv N) \times 10^{-2}$, $N\,m^{-1}$
HCN	115.321(5)	r_e	*(a)*	18.7 *(f)*
FCN	115.9(3)	r_s	*(b)*	17.9 *(f)*
ClCN	116.0(7)	r_e	*(c)*	
BrCN	115.8(3)	r_s	*(b)*	18.5 *(f)*
CH_3CN	115.9(2)	r_g	*(d)*	17.9 *(g)*
$S(CN)_2$	115.7(2)		*(e)*	16.5 *(h)*

(a) Winnewisser, Maki and Johnson (1971)
(b) Tyler and Sheridan (1963)
(c) Lafferty, Lide and Toth (1965)
(d) Karakida, Fukuyama and Kuchitsu (1974)
(e) Pierce, Nelson and Thomas (1965)
(f) Orville-Thomas (1965)
(g) Krishna Pillai and Cleveland (1960)
(h) Nayar and Aruldhas (1968)

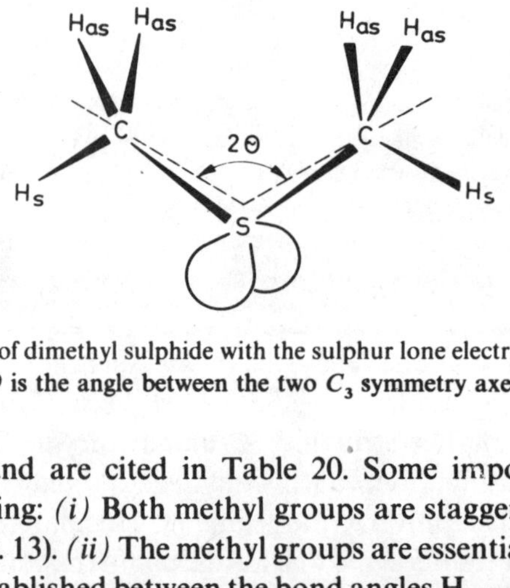

Fig. 13. The molecular model of dimethyl sulphide with the sulphur lone electron pairs and the notation of the hydrogen atoms. 2 Θ is the angle between the two C_3 symmetry axes of the methyl groups

substitution structure and are cited in Table 20. Some important features of this geometry are the following: *(i)* Both methyl groups are staggered with respect to the other C—S bond (*cf.* Fig. 13). *(ii)* The methyl groups are essentially symmetric. A mere 4′ difference could be established between the bond angles H_{as}—C—H_{as} and H_{as}—C—H_s, and this difference is much smaller than the 20′ uncertainty of the bond angles themselves. It should be indicated that the influence of zero-point vibrations has been taken into account in the error estimates. *(iii)* The symmetry axes of the methyl groups do not coincide with the axes of the C—S bonds. The angle between the two symmetry axes (2Θ) is 5°35′ larger than the C—S—C bond angle. Accordingly, the tilt of each methyl group is 2°47′. The methyl groups are tilted towards the lone electron pairs of the sulphur atom (*cf.* Fig. 13). The methyl tilt in various molecules will be compared in a subsequent section.

TABLE 20
Parameters of substitution structures of dimethyl sulphide from Pierce and Hayashi (1961) and dimethyl selenide from Pandey and Dreizler (1977)

E=S or Se	$(CH_3)_2S$	$(CH_3)_2Se$
r(E—C), pm	180.2(2)	194.5(0.4)
r(C—H_s), pm	[=r(C—H_{as})]*	108.8(13)
r(C—H_{as}), pm	109.1(5)	109.6(4)
∢C—E—C	98°52′±10′	96°19′±5′
∢E—C—H_s	106°38′	105°0′±40′
∢E—C—H_{as}	110°45′	110°20′±15′
∢H_{as}—C—H_{as}	109°32′±20′	109°52′±30′
∢H_{as}—C—H_s	109°36′±20′	110°37′±30′
2Θ	104°22′±15′	101°34′±1°40′
Angle of tilt	2°47′±30′	2°38′±50′

* Assumed

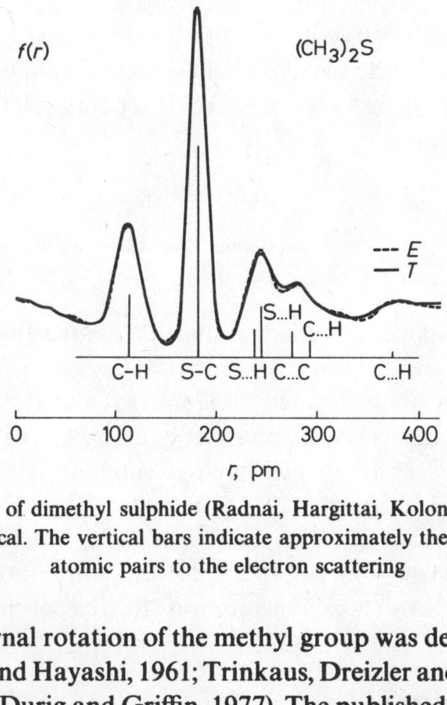

Fig. 14. Radial distribution of dimethyl sulphide (Radnai, Hargittai, Kolonits and Gregory, 1975). *E*—Experimental, *T*—theoretical. The vertical bars indicate approximately the relative contributions of the atomic pairs to the electron scattering

The barrier to internal rotation of the methyl group was determined by microwave spectroscopy (Pierce and Hayashi, 1961; Trinkaus, Dreizler and Rudolph, 1973) and by Raman spectroscopy (Durig and Griffin, 1977). The published barriers are around 8–9 kJ mol^{-1}.

The structure of dimethyl sulphide was also determined by two independent electron diffraction studies (Radnai, Hargittai, Kolonits and Gregory, 1975; Iijima, Tsuchiya and Kimura, 1977). Radial distribution curves for dimethyl sulphide are shown in Fig. 14. The geometrical parameters determined by electron diffraction are

TABLE 21
Average structure of dimethyl sulphide

Parameter	Electron diffraction and microwave spectroscopy *(a)*	Electron diffraction *(b)*
r_g(S—C), pm	180.7(2)	181.1(4)
r_g(C—H), pm	111.6(3)	
$\measuredangle_{av}$C—S—C (°)	99.05(4)	99.2(6) *(d)*
$\measuredangle$H—C—H (°)	109.3(5)	
Angle of tilt	*(c)*	*(c)*

(a) Iijima, Tsuchiya and Kimura (1977)
(b) Radnai, Hargittai, Kolonits and Gregory (1975)
(c) Assumed from the microwave spectroscopic study by Pierce and Hayashi (1961)
(d) From r_a interatomic distances

given in Table 21. In their analysis, Iijima, Tsuchiya and Kimura (1977) have combined their own electron diffraction data with the rotational spectroscopic data of Pierce and Hayashi (1961). Therefore, of the two electron diffraction studies the results of Iijima, Tsuchiya and Kimura (1977) are to be preferred. It is of interest to compare the length of the C—S bond in various representations:

r_s 180.2(2) pm (Pierce and Hayashi, 1961)
r^0_α (r_z) 180.5(2) pm
r_g 180.7(2) pm } (Iijima, Tsuchiya and Kimura, 1977)

The mean vibrational amplitudes were determined from the electron diffraction data only in the study by Radnai, Hargittai, Kolonits and Gregory (1975). They are cited in Table 22, together with those calculated by Gebhardt and Cyvin (1972).

As might be expected and as is shown unambiguously by the parameters in Tables 20 and 21, the C—S—C skeleton of dimethyl sulphide is highly bent. Hence a publication (Hamada and Morishita, 1978) concluding that this molecule as well as several others, CH_3—X—CH_3, X = O, S, Se, Te, have D_{3d} symmetry with a linear C—X—C skeleton, invited strong criticism (McKean and Duncan, 1979) based on a wide range of spectroscopic evidence, in addition to the results of the geometry determinations.

Dimethyl selenide, $Se(CH_3)_2$. The molecular geometry was determined by microwave spectroscopy (Beecher, 1966; Pandey and Dreizler, 1977). A substitution structure is given in Table 20. The C_3 symmetry axis of the methyl group does not coincide with the axis of the C—Se bond, and the tilt is nearly 3°. The C—H bond in the *anti* position to the non-adjacent C—Se bond is somewhat shorter than the other C—H bonds.

TABLE 22
Mean amplitudes of vibration (pm) of dimethyl sulphide

Atomic pair	Calculated *(a)* 0 K	Calculated *(a)* 298 K	Observed *(b)*
C—H	7.83	7.83	9.0(3)
C—S	4.80	4.94	5.0(2)
C...C	7.76	9.56	9.4(10)
C...H_s	10.68	11.66	}15.3(23)
C...H_{as}	16.62	21.58	
S...H_s	10.62	10.72	}11.1(5)
S...H_{as}	10.99	11.11	

(a) Gebhardt and Cyvin (1972)

(b) Electron diffraction; Radnai, Hargittai, Kolonits and Gregory (1975). The experimental electron diffraction data could be better approximated by a model in which the methyl group was tilted than by the model in which the symmetry axis of the methyl group and the C—S bond coincided. The results cited here refer to the former model. However, only mean values could be determined for the distances which differed from each other only because of the tilt

For a long time only visual electron diffraction data (Bowen, 1954a) were available for the structure of *bis(trifluoromethyl)monosulphide*, $S(CF_3)_2$. A large and uncertain C—S—C bond angle of $106 \pm 3°$ was not consistent with the structural data on related molecules. Oberhammer, Gombler and Willner (1981) recently reinvestigated this substance and communicated r_α^0(C—S) 181.9(3) pm and $\measuredangle_\alpha$C—S—C 97.3(8)°.

Bis(trifluoromethyl)monoselenide, $Se(CF_3)_2$. The earlier visual electron diffraction study (Bowen, 1954a) was followed by a sector-microphotometer determination (Marsden and Sheldrick, 1971a). Two solutions were found relating to the conformational properties of this molecule. They are illustrated by the forms (A) and (B) of Fig. 15. Both forms have C_2 symmetry, although, form (A) may be considered to be of slightly distorted C_{2v} symmetry. It was not decided whether both forms may be present in the vapour, or one of the forms was merely a mathematically possible solution. This conformational analysis was apparently hindered by the lack of additional information on the mean vibrational amplitudes for the rotation-dependent interactions. The contribution to the total electron scattering from these interactions C...F and F...F is relatively small. The extensively used assumptions may have predetermined, to some extent, the conformational findings. The fact that the longest C...F distances, i.e. those corresponding to an almost *anti* arrangement had larger l values in the refinements than the shorter C...F interactions (*gauche*-type) is particularly suspect. There is also a question as to whether the two rather similar forms could be substituted by one with larger amplitudes of vibration associated with the rotation-dependent interactions. The uncertainties in the determination of the conformational properties of bis(trifluoromethyl)monoselenide did not hinder an

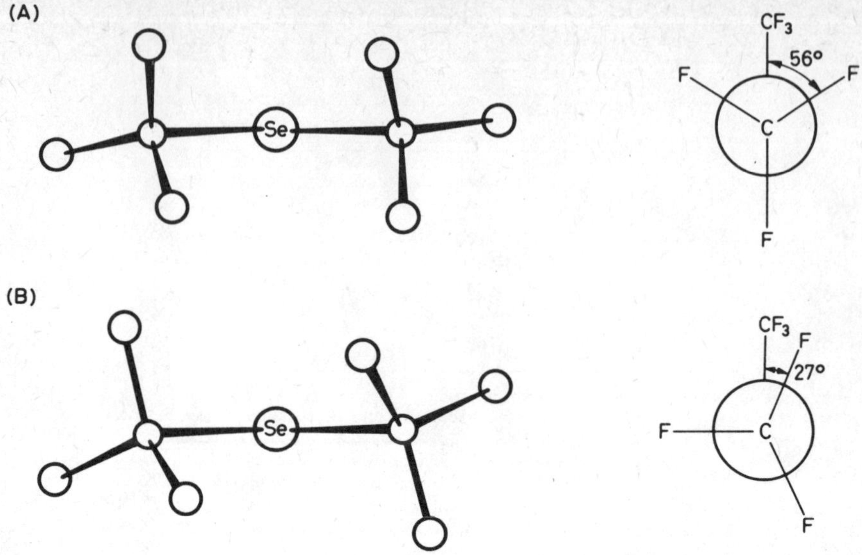

Fig. 15. Two models for rotational isomers of bis(trifluoromethyl)monoselenide after Marsden and Sheldrick (1971a)

accurate determination of the geometry of the F_3CSe part characterized by the following parameters (r_a):

r(C—Se) 197.8(9) pm r(C—F) 133.3(4) pm
∢F—C—F 108.8(6)°

The bond angle C—Se—C has proved to be sensitive to the choice of the conformational model. The large error limit in ∢C—Se—C 95.5±3.5° seems to be a realistic estimate. The angle here is the mean of the two obtained for the two minima. The electron scattering data were found to be consistent with a small tilt of the trifluoromethyl group towards the lone electron pairs of the selenium atom.

The liquid-phase Raman spectra of bis(trifluoromethyl)monosulphide are interpreted on the basis of a C_{2v} symmetry molecular model (Carter, Shiow-Chyn Wang and Shreeve, 1973).

Disilyl sulphide, $S(SiH_3)_2$ and *digermyl sulphide*, $S(GeH_3)_2$. The electron diffraction studies by Almenningen, Hedberg and Seip (1963) and Glidewell, Rankin, Robiette, Sheldrick, Beagley and Cradock (1970), respectively, yielded bond lengths, bond angles and some selected mean vibrational amplitudes. The most important geometrical parameters are shown in Table 23. It was also possible to collect a limited amount of information on the relative orientation of the silyl groups and on the barrier to internal rotation. For both molecules the experimental data are consistent with a C_{2v} symmetry model and also with the Si—H bonds (Ge—H bonds) staggering the other, i.e. non-

TABLE 23
Geometrical parameters of disilyl sulphide and digermyl sulphide and analogous selenides

M = Si or Ge	$(SiH_3)_2S$		$(GeH_3)_2S$
	(a1)	*(b)*	*(a2)*
r(M—S), pm	213.6(2)	212.9(3)	220.9(4)
r(M—H), pm	149.4(8)	151.0(10)	151.2(16)
∢M—S—M (°)	97.4(7)	98.35(5)	98.9(1)
∢H—M—H (°)	110.2(35)	111(1)	110.6(20)

M = Si or Ge	$(SiH_3)_2Se$ *(a3)*	$(GeH_3)_2Se$ *(a4)*
r(M—Se), pm	227.4(4)	234.4(3)
r(M—H), pm	151.6(5)	156.3(15)
∢M—Se—M (°)	96.6(7)	94.6(5)

(a) Electron diffraction, r_a parameters
(a1) Almenningen, Hedberg and Seip (1963)
(a2) Glidewell, Rankin, Robiette, Sheldrick, Beagley and Cradock (1970)
(a3) Almenningen, Fernholt and Seip (1968)
(a4) Murdoch, Rankin and Glidewell (1971)
(b) Microwave spectroscopy, r_0 parameters; Dössel and Sutter (1978)

adjacent Si—S (Ge—S) bond. The mean amplitudes of vibration for the Si...H interactions were determined somewhat more accurately than those for the Ge...H interactions. The relatively large l(Si...H) values indicate that the internal rotation of the silyl groups is less hindered in disilyl sulphide than that of the methyl groups in dimethyl ether (Kasai and Myers, 1959). Both electron diffraction analyses were performed under the assumption that the symmetry axes of the MH_3 groups (M = Si or Ge) coincide with the axes of the M—S bonds. A recent microwave spectroscopic study of disilyl sulphide by Dössel and Sutter (1978) showed a silyl tilt of 1.65°. The other geometrical parameters from this study are cited in Table 23. The very small uncertainties in the parameters presumably do not reflect the possible consequences of the assumptions utilized in the structure analysis. It is hoped that an investigation of isotopic species will eventually make an accurate geometry available. The microwave spectroscopic study of Dössel and Sutter (1978) yielded also the barrier to internal rotation of the silyl group, 2.18 kJ mol^{-1}, being indeed very much smaller than that of the methyl group in dimethyl ether.

Disilyl selenide, $Se(SiH_3)_2$ and *digermyl selenide*, $Se(GeH_3)_2$. The molecular structures of these two compounds were determined by Almenningen, Fernholt and Seip (1968) and Murdoch, Rankin and Glidewell (1971), respectively, using electron

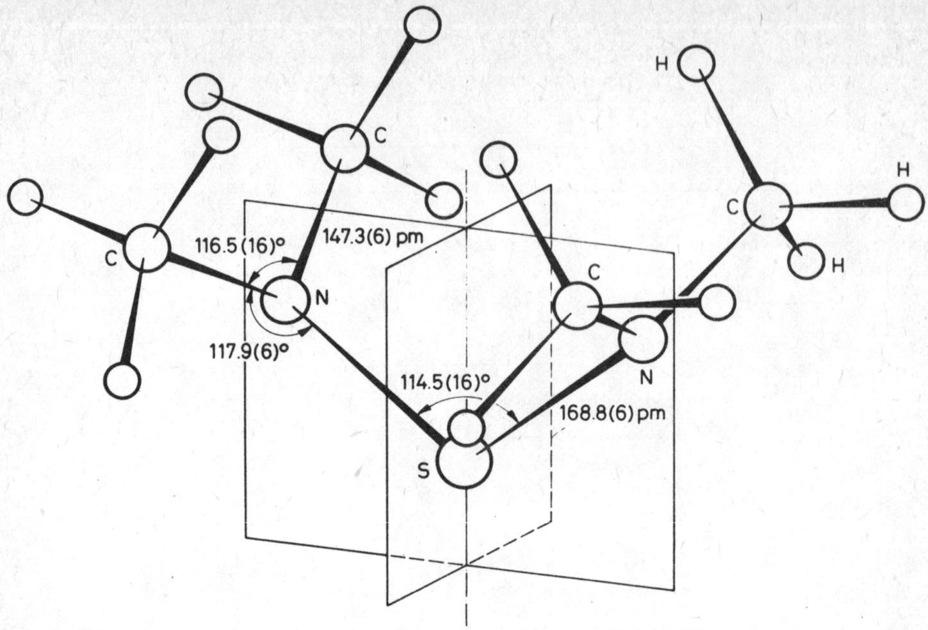

Fig. 16. The prevailing conformer of thiobis(dimethylamine) with the bond lengths and bond angles of the heavy-atom skeleton (Hargittai and Hargittai, 1973a)

diffraction. The analyses were similar to those on the sulphur analogues. The most important geometrical parameters are given in Table 23.

N,N'-Thiobis(dimethylamine), $(CH_3)_2NSN(CH_3)_2$. The bond lengths and bond angles resulting from an electron diffraction study (Hargittai and Hargittai, 1973a) are given in Fig. 16. The molecular conformation shown in this Figure was found to predominate in the vapour at slightly above room temperature. This form is also represented below by one of the Newman projections **(I)** representing the view along one of the two N—S bonds.

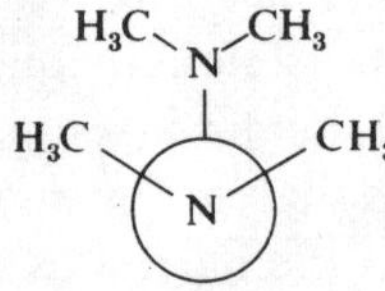

I

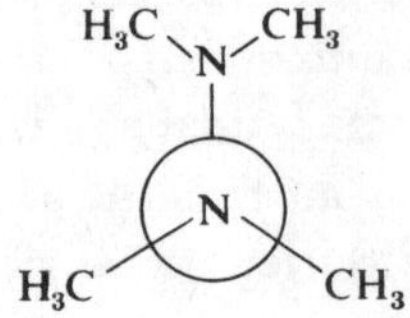

II

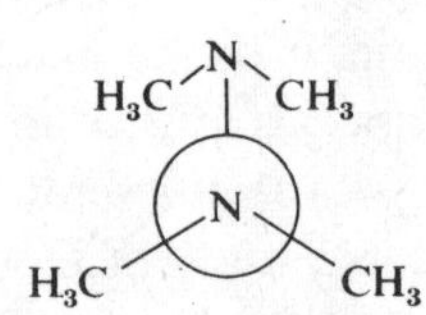

III

The experimental data indicate that a smaller fraction of the molecules may be present in either or both forms, **(II)** and **(III)**. In these forms one or both methyl groups are rotated by 180° with respect to those in form **(I)**, or a corresponding inversion of the nitrogen bond configuration takes place.

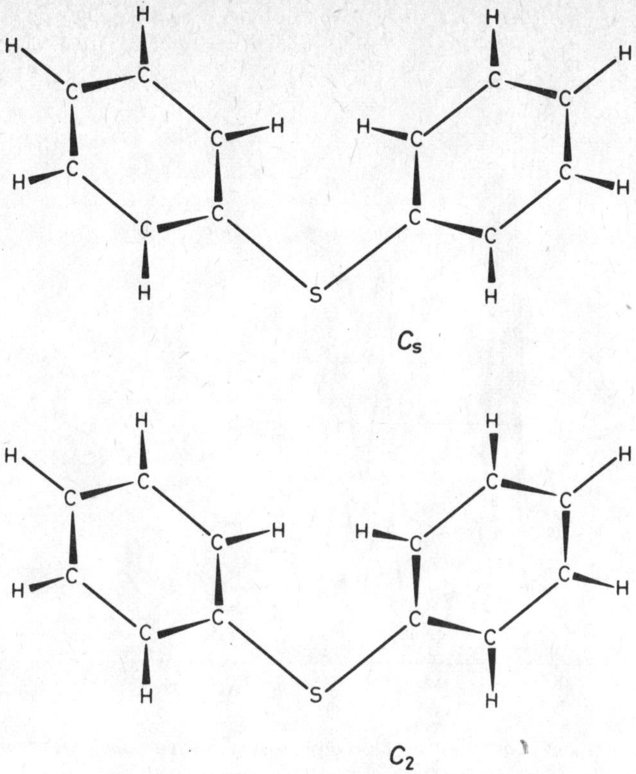

Fig. 17. Two models of rotational isomers of diphenyl sulphide

Diphenyl sulphide, $S(C_6H_5)_2$. According to an electron diffraction analysis (Rozsondai, Moore, Gregory and Hargittai, 1977), the presence of at least two conformers is in agreement with the experimental data and further forms may not be excluded. The two forms referred to have C_2 and C_s symmetry and are shown in Fig. 17. An accurate electron diffraction determination of the conformational properties of diphenyl sulphide is hindered by the relatively small contribution from the rotation-dependent interactions to the total electron scattering and by the presumed large-amplitude torsional motion of the phenyl groups around the C—S bonds. Nagel, Steiger, Fruwert and Geiseler (1975) studied the conformational behaviour of diphenyl sulphide by vibrational spectroscopy. They interpreted their data with a model in which the two phenyl groups did not have equivalent orientations. The conformational behaviour of diphenyl sulphide and analogous molecules has also been investigated by semi-empirical calculations (see, e.g. Galasso, De Alti and Bigotto, 1971; Smeyers and Hernandez-Laguna, 1978).

The experimental radial distribution is shown in Fig. 18 together with the respective curves for the sulphoxide and sulphone analogs of diphenyl sulphide (their structures will be discussed later). The determination of the geometrical parameters (r_a) shown

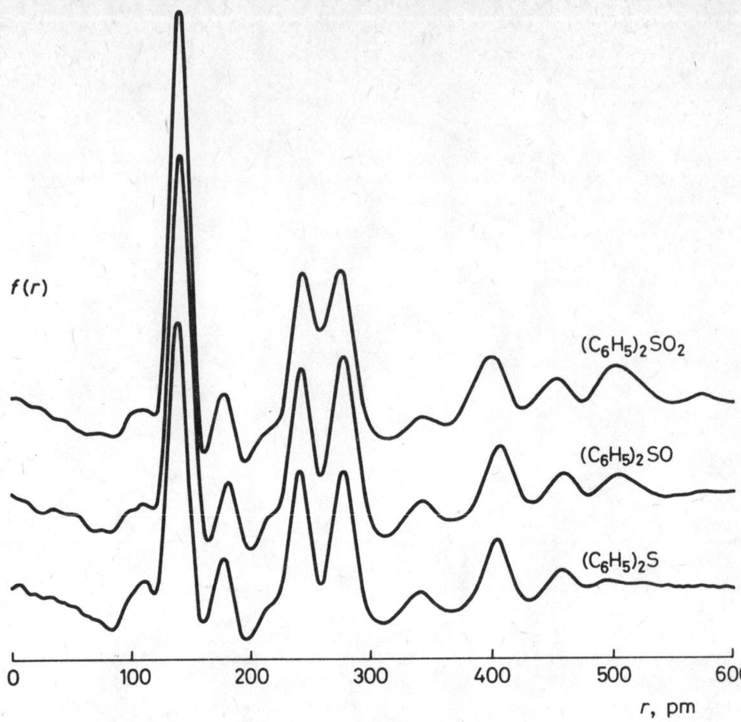

Fig. 18. Radial distributions of diphenyl sulphide, diphenyl sulphoxide and diphenyl sulphone (Rozsondai, Moore, Gregory and Hargittai, 1977; 1979)

below is not influenced by the uncertainty in the conformational analysis. The bond vibrational amplitudes are also given here:

r(S—C)	177.1(5) pm	l(S—C)	4.9(4) pm
r(C—C), mean	139.9(3) pm	l(C—C)	4.6(2) pm
r(C—H), mean	108.9(8) pm	l(C—H)	8.1(8) pm
∢C—S—C	103.7(13)°		

The benzene rings were found to be undistorted within the experimental errors (Rozsondai, Schultz and Hargittai, 1981).

Di-2-pyridyl sulphide, $S(C_5NH_4)_2$. The conformational properties of this molecule are apparently also very complex. Although they have been repeatedly investigated using various techniques (Chachaty, Pappalardo and Scarlata, 1976; Galasso, Pappalardo and Scarlata, 1977; Bigotto, Galasso, Pappalardo and Scarlata, 1978), only a rather limited and tentative pattern emerges. The radial distribution curve obtained from an electron diffraction experiment with 140 °C nozzle temperature is shown in Fig. 19. As might be anticipated, this curve very much resembles the radial

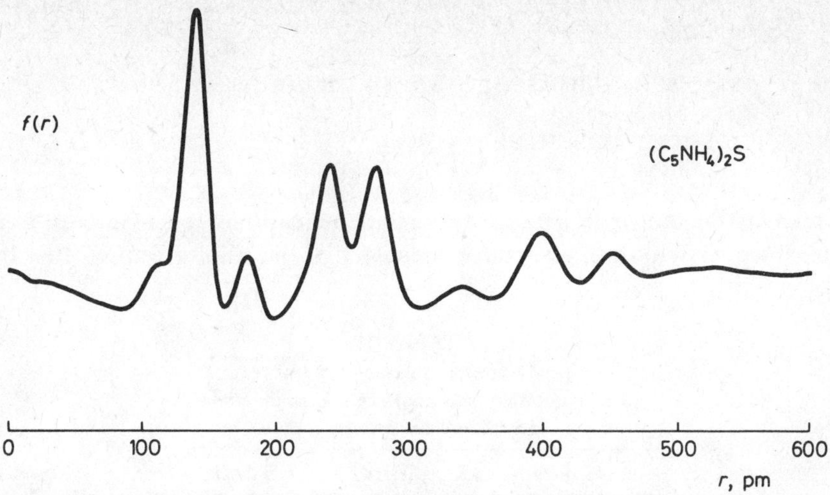

Fig. 19. Radial distribution of di-2-pyridyl sulphide (Rozsondai, Hargittai and Pappalardo, 1979)

distribution of diphenyl sulphide (Fig. 18). In the electron diffraction analysis by Rozsondai, Hargittai and Pappalardo (1979), the experimental data could be well approximated by either C_2 or C_s symmetry models, or even better by one without any symmetry ($\varphi_1 = 23°$ and $\varphi_2 = 67°$ for the angles of rotation around the C—S bonds). There was no unambiguous solution, however. The data were certainly consistent with the notion of the rings performing large-amplitude torsional motion or a rapid interconversion of various forms, and with the broad potential energy minima revealed by quantum chemical calculations (Chachaty, Pappalardo and Scarlata, 1976; Galasso, De Alti and Bigotto, 1971; Smeyers and Hernandez-Laguna, 1978). The C—S bond length and C—S—C bond angle determined by electron diffraction were 178.6(4) pm and 104.4(7)°, respectively.

2.2. MIXED SULPHIDES

The molecular geometry of the following unsymmetrically substituted sulphides has been determined in the vapour phase (excluding sulphanes to be discussed in the next section):

CH_3SH CH_3SeH CF_3SH
$ClCH_2SCH_3$ CH_3SCH_2CN CH_3SCl CCl_3SCl CF_3SCl CF_3SF
CH_3CH_2SH CH_3CH_2SeH $CH_3CH_2CH_2SH$
$HSCH_2CH_2SH$ $HOCH_2CH_2SH$
$CH_3SCH_2CH_3$
C_6H_5SCl $C_6H_5SCH_3$ C_6H_5SeBr $C_6H_5SeCH_3$
$CH_2{=}CHSH$ $CH_2{=}CHSCH_3$ $CH_2{=}C{=}CHSCH_3$ $C_6H_5SCH{=}CH_2$
$HC{\equiv}CSCH_3$ $CH_3SC{\equiv}CSCH_3$

CH_3SCN CH_3CH_2SCN CF_3SeCN
HC(O)SH HC(S)SH
$S{=}C(SCH_3)_2$ $O{=}C(SCH_3)_2$ $H_2C{=}C(SCH_3)_2$
$(CH_3)_2BSCH_3$ $CH_3B(SCH_3)_2$ $B(SCH_3)_3$ $B(SeCH_3)_3$
$(CF_3S)_3N$
$CH_3SP(O)Cl_2$ $P(SCH_3)_3$
$(CF_3S)_3CC(SCF_3)_3$

As many of the molecules appearing here are mercaptans, it is of interest to refer to an extensive review on the structural chemistry of the thiol group written by Paul (1974).

TABLE 24
Geometrical parameters (r_0)
of methane thiol and methane selenol determined
by microwave spectroscopy

E=S or Se	CH_3SH *(a)*	CH_3SeH *(b)*
r(E—H), pm	133.5(10)	147.3(20)
r(C—E), pm	181.4(5)	195.9(10)
r(C—H), pm	109.2(10)	108.8(10)
∢C—E—H (°)	96.5(5)	95.5(20)
∢H—C—H (°)	109.8(5)	110(10)

(a) Kojima (1960)
(b) Thomas (1973)

Methane thiol, CH_3SH. Several microwave spectroscopic studies have dealt with the molecular structure (Kojima and Nishikawa, 1957; Kojima, 1960; for further references see Landolt–Börnstein, 1976). The geometrical parameters are listed in Table 24. The tilt of the methyl group was determined to be 2.2°. This result was later confirmed by *ab initio* calculations which gave 2.5–2.9° (Kao, 1978). The barrier to internal rotation of the methyl group was found to be 5.30 kJ mol^{-1} from the microwave spectra. The *ab initio* calculations were in agreement with the experimental data as regards the conformation with respect to rotation around the C—S bond. The staggered form was found to be more stable than the eclipsed conformer by 4–6 kJ mol^{-1}.

Methane selenol, CH_3SeH. The bond lengths and bond angles (r_0) determined by microwave spectroscopy (Thomas, 1973) are given in Table 24. The conformation is staggered, the methyl group is tilted towards the lone electron pairs of the selenium atom by 1.5(10)°.

Trifluoromethane thiol, CF_3SH. Following an earlier visual electron diffraction study (Bowen, 1954b), the molecular structure of this compound was determined more accurately by an up-to-date electron diffraction investigation of Marsden (1974), assuming C_{3v} symmetry for the CF_3S fragment and that the S—H bond was staggered with respect to the CF_3 group. Furthermore, assuming several vibrational amplitudes and 133 pm for r(S—H) and 96° for ∢C—S—H, the following parameters were determined:

r_a(C—S)	180.0(5) pm	l(C—S)	5.8(8) pm
r_a(C—F)	133.5(3) pm	l(C—F)	4.1(5) pm
∢S—C—F 111.2(3)°			

Matrix-isolation infrared spectra and a normal coordinate analysis of trifluoromethane thiol were reported by Perttilä (1976). It should be noted that the electron diffraction geometrical parameters were cited in this study as if the mean amplitudes of vibration were the error limits of the interatomic distances.

Monochlorodimethyl sulphide, $ClCH_2SCH_3$. The results of two parallel electron diffraction analyses (in Leiden and Kazan') were jointly reported (Naumov, Garaeva, Geise and Mijlhoff, 1979). The most interesting observation is the shortening of the C—S bond in the chlorine-substituted half of the molecule and its lengthening in the other half when compared with that in dimethyl sulphide (see above). The most important geometrical parameters (r_a) are listed below from the two investigations, the numbers in parentheses being the least-squares standard deviations multiplied by a factor of three.

	Leiden	Kazan'
r(S—CH_2Cl) (pm)	178.5(5)	178.5(5)
r(S—CH_3) (pm)	182.7(5)	182.4(5)
r(C—Cl) (pm)	176.3(6)	176.6(6)
∢C—S—C (°)	98.9(12)	100.9(12)
∢S—C—Cl (°)	114.6(6)	114.6(9)
∢C—S—C—Cl (°)	59(8)	49(10)

The conformation around the ClH_2C—S bond was unambiguously determined to be *gauche*, in agreement with other studies (for references, see the cited paper).

A microwave spectroscopic study of *(methylthio)acetonitrile*, CH_3SCH_2CN, by Kewley (1978) proved the *gauche* conformer to be the most stable rotational form for this compound also. The observed rotational constants were consistent with ∢C—S—C $104 \pm 2°$ and an angle of rotation of $66 \pm 5°$. The data did not indicate any appreciable tilt.

Methane sulphenyl chloride, CH_3SCl. So far three reports on this structure have appeared with increasingly accurate microwave spectroscopic results based on increasing numbers of isotopic species (Guarnieri, 1968; 1970; Guarnieri, Charpentier and Kück, 1973). The r_0 parameters shown in Fig. 20a were obtained from a fitting procedure using the data on five isotopic species and an assumption of C_{3v} symmetry for the methyl group. The uncertainties parenthesized in Fig. 20a characterize merely the goodness of the fit. Comparison of the r_0(S—Cl) parameter with the one obtained from the available coordinates of sulphur and chlorine, *viz.* r_s(S—Cl) = 203.02(7) pm

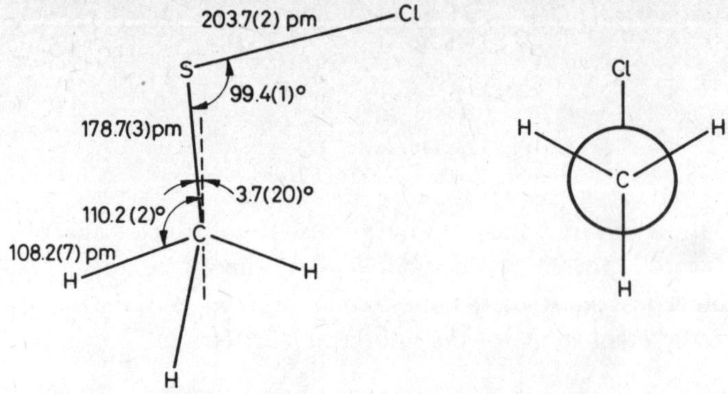

Fig. 20a. Conformation and geometrical parameters of methane sulphenyl chloride from microwave spectroscopy (Guarnieri, Charpentier and Kück, 1973)

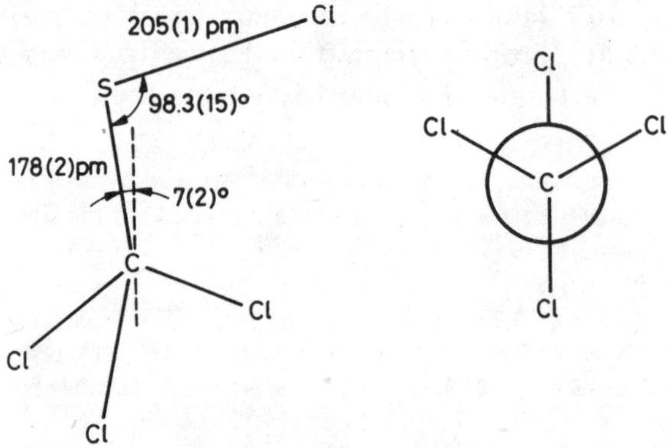

Fig. 20b. Conformation and geometrical parameters of trichloromethane sulphenyl chloride from electron diffraction (Alekseev and Velichko, 1967)

shows a difference of 0.7 pm. This indicates more realistic uncertainties of the r_0 parameters. A staggered conformation was established. The methyl group is slightly tilted towards the lone electron pairs of the sulphur atom.

Trichloromethane sulphenyl chloride, CCl_3SCl. Very few details have been communicated about the electron diffraction analysis of this compound (Alekseev and Velichko, 1967). The geometrical parameters are shown in Fig. 20b. The conformation is staggered, and the CCl_3 group is tilted towards the sulphur lone electron pairs.

The structures of *trifluoromethane sulphenyl fluoride*, CF_3SF, and *trifluoromethane sulphenyl chloride*, CF_3SCl, have been studied by modern electron diffraction techniques (Oberhammer, Gombler and Willner, 1981). Some important geometrical parameters are shown in Figs. 20c and 20d together with the staggered conformation and the trifluoromethyl tilt is also indicated.

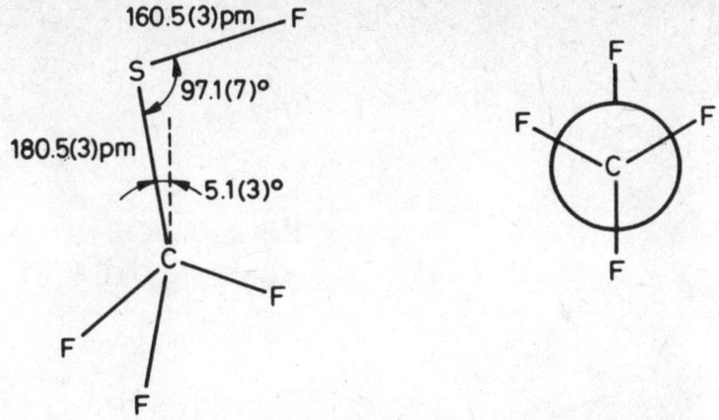

Fig. 20c. Conformation and geometrical parameters of trifluoromethane sulphenyl flouride from electron diffraction (Oberhammer, Gombler and Willner, 1981)

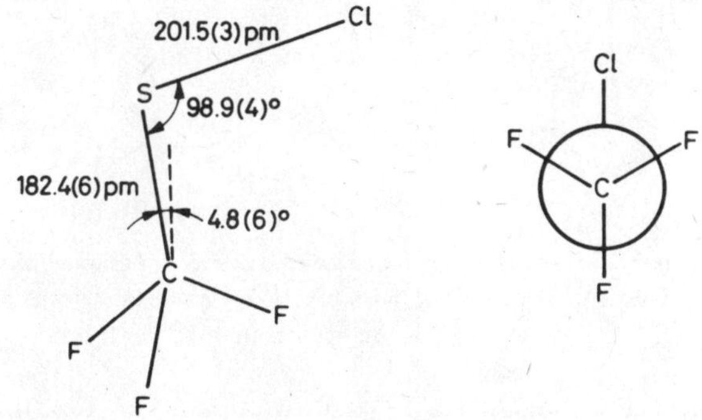

Fig. 20d. Conformation and geometrical parameters of trifluoromethane sulphenyl chloride from electron diffraction (Oberhammer, Gombler and Willner, 1981)

Ethane thiol, CH_3CH_2SH. Two independent microwave spectroscopic investigations yielded consistent results on the conformation and geometry (Hayashi, Imaishi and Kuwada, 1974; Nakagawa, Kuwada and Hayashi, 1976; Schmidt and Quade, 1975). The results from Hayashi's laboratory are cited here in more detail. They are based on the spectra from 10 isotopic species. *Gauche* and *anti* forms are formed by rotation around the C—S bond in equilibrium in the vapour of ethane thiol. Essentially r_s structures have been determined for both conformers. Some important parameters are shown in Fig. 21. The increased uncertainties originate from the influence of zero-point vibrations and from some of the atoms being located near one of the principal axes of inertia. In the two conformers almost all bond lengths and bond angles differ only within the error limits. The notable exceptions are the carbon bond angles in the methylene groups and especially the S—C—C bond angle (*cf.* Fig. 21). This difference was attributed to a tilt of the axis of internal rotation from the C—S bond. The angle between the rotational axis and the C—C bond was assumed to be the same for both isomers and is shown in square brackets.

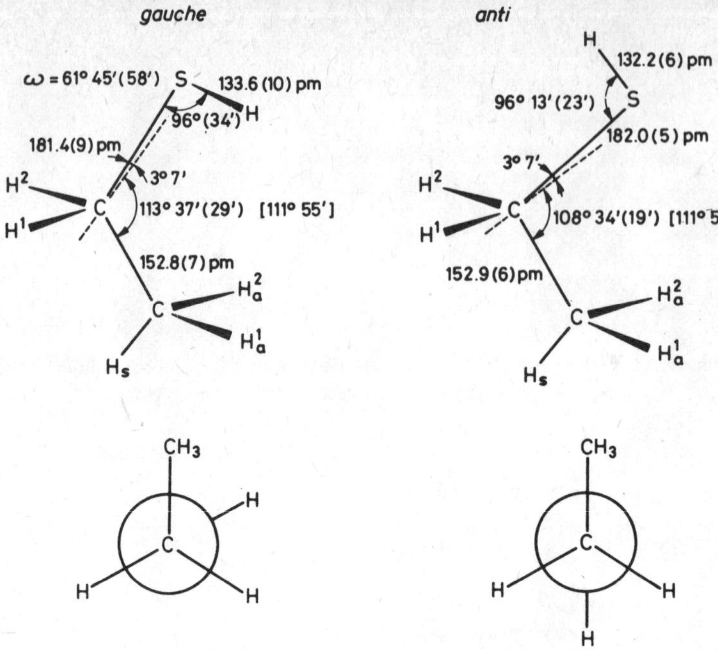

Fig. 21. The *gauche* and *anti* conformers and geometrical parameters of ethane thiol from the microwave spectroscopic studies of Hayashi, Imaishi and Kuwada (1974) and Nakagawa, Kuwada and Hayashi (1976). ω is the angle of rotation around the C—S bond

According to the relative intensity measurements of the microwave spectra (Schmidt and Quade, 1975), the *gauche* form is more stable than the *anti* by 1.7 ± 0.2 kJ mol^{-1}, thus corresponding to an *anti*/*gauche* relative abundance of 1:4 at room temperature. The barrier to the internal rotation of the mercapto group was determined to be 5.42 and 5.24 kJ mol^{-1} in the two microwave spectroscopic studies (Nakagawa, Kuwada and Hayashi, 1976; and Schmidt and Quade, 1975, respectively) and 5.65 kJ mol^{-1} in a far-infrared spectroscopic study by Manocha, Fateley and Shimanouchi (1973). The potential curve of the mercapto rotation in ethane thiol is illustrated in Fig. 22 after Nakagawa, Kuwada and Hayashi (1976).

Ethane selenol, CH_3CH_2SeH. Durig and Bucy (1977) analysed the microwave spectra of four isotopic species to determine the geometries of the *gauche* and *anti* conformers both present in the vapour phase. The ethyl group was assumed to have the staggered conformation. Further assumptions included r(C—H) = 109.3 pm and ∢C—C—H = 109.7° for the methylene group. The following bond lengths and bond angles were obtained:

Bond lengths:	r_0(C—C)	154.3(0) [5] pm
	r_0(C—Se)	195.8(1) [5] pm
	r_s(Se—H)	144.4(6) pm

Bond angles:	for *gauche*	for *anti*
∢C—Se—H	93.9(3) [10]°	93.3(3) [10]°
∢Se—C—C	113.13(1) [100]°	108.83(9) [100]°

The parenthesized uncertainties, as usual, refer to the last digit of the parameter. The first numbers are the standard deviations from the least-squares fitting and are too small to be realistic. Durig and Bucy (1977) estimated the values in the square brackets taking into account the assumptions used in the structure analysis. The discrepancy between the Se—C—C bond angles in the two conformers is again a consequence of a tilt of the rotational axis. The angle of the axis with the C—C bond is 111.6(10)° for both isomers. The angle of rotation of the SeH group was determined to be 62.02(13) [100]° in the *gauche* form*. The potential curve describing the rotation of the SeH group in ethane selenol is shown in Fig. 22 after Durig and Bucy (1977). Their microwave spectroscopic results were augmented by infrared and Raman spectroscopic investigations of the intramolecular motion of ethane selenol.

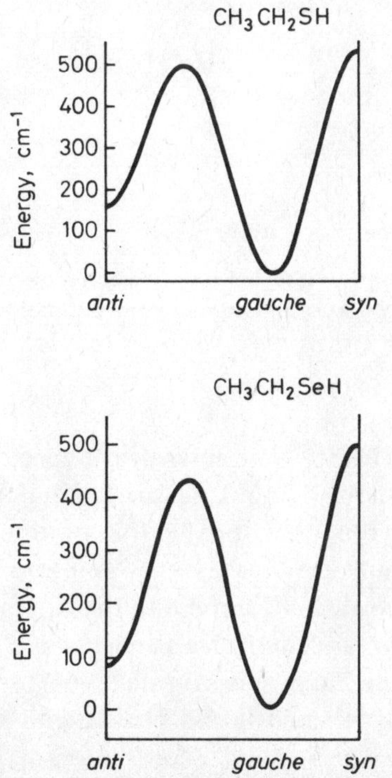

Fig. 22. Potential energy curves of internal rotation about the C—S bond in ethane thiol and about the C—Se bond in ethane selenol after Nakagawa, Kuwada and Hayashi (1976) and Durig and Bucy (1977), respectively

* The papers discussing the conformational properties of ethane thiol and ethane selenol give different definitions of the angles of rotation. Here the angle of rotation is 0° when the S—H (or Se—H) bond eclipses the C—C bond (*syn* form). Accordingly, the angle of rotation of the *anti* form is 180°.

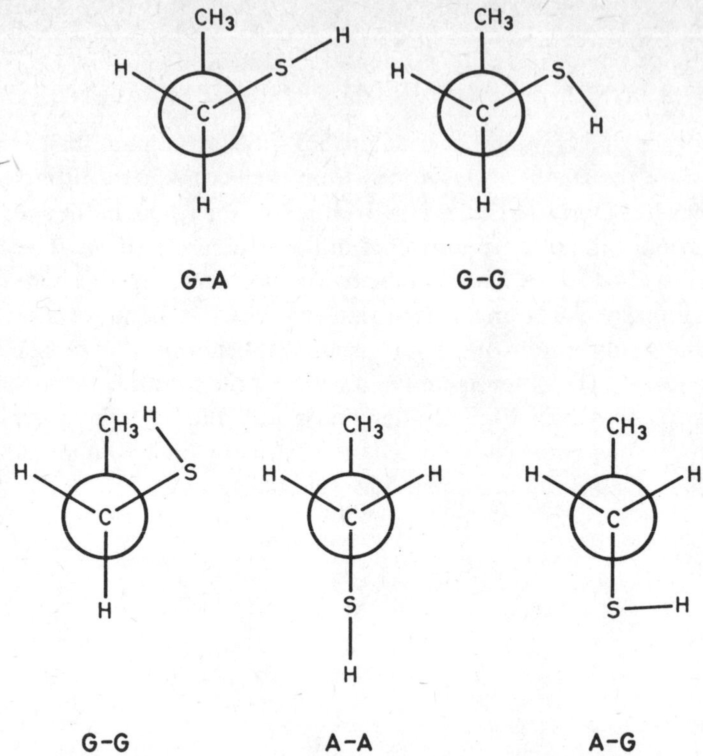

Fig. 23. Five rotational isomers of propane thiol shown by Newman projections representing views along the central C—C bond. Notation: the first and second letters refer to the rotational forms about the central C—C bond and the C—S bond, respectively (Ohashi, Ohnishi, Tagui, Sakaizumi and Yamaguchi, 1977)

Propane thiol, $CH_3CH_2CH_2SH$. The microwave spectra of two isotopic species were analysed by Ohashi, Ohnishi, Tagui, Sakaizumi and Yamaguchi (1977). Earlier Hayashi, Shiro and Murata (1966) investigated the rotational isomerism by means of the vibrational spectra. There are two axes of rotation in the C—C—C—S—H skeleton, *viz.* around the central C—C bond and the C—S bond. Possible rotational isomers involving the *gauche* and *anti* arrangements are shown in Fig. 23 by the Newman projections. Hayashi, Shiro and Murata (1966) found the **A–A** conformer to be the most stable in the vapour and liquid phases with some amount of the **G–A** conformer also being present. The first and second letters of the notation used here refer to the rotational forms about the C—C and C—S bonds, respectively, **A** = *anti* and **G** = *gauche*. The **A–G** form was also shown to be present by means of rotational spectroscopy (Ohashi, Ohnishi, Tagui, Sakaizumi and Yamaguchi, 1977). This was determined by the assumption of reasonable bond lengths and bond angles and the resulting calculated rotational constants for this conformer approximated best the experimental ones. It was not stated how sensitive the rotational constants were to changes in the geometrical parameters. Only the *A* rotational constant differed considerably from the experimental one for the **A–A** conformer which, according to the vibrational spectroscopic study, was preferred. However, since the assumed geometrical parameters were indeed very reasonable, the presence of the **A–G** form seems to be established. Furthermore, the **A–G** form is consistent with the *gauche* conformer being the most stable with respect to rotation about the C—S bond in ethane thiol (Nakagawa, Kuwada and Hayashi, 1976).

The conformational behaviour of *prop-2-ene-1-thiol*,

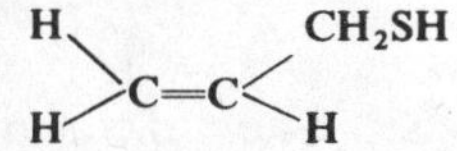

in the vapour phase was studied by microwave spectroscopy (Sastry, Dass, Brooks and Bhaumik, 1969). Assuming reasonable geometrical parameters, the form shown here by Newman projections representing views along the (*a*) C—C bond and (*b*) C—S bond after Paul (1974)

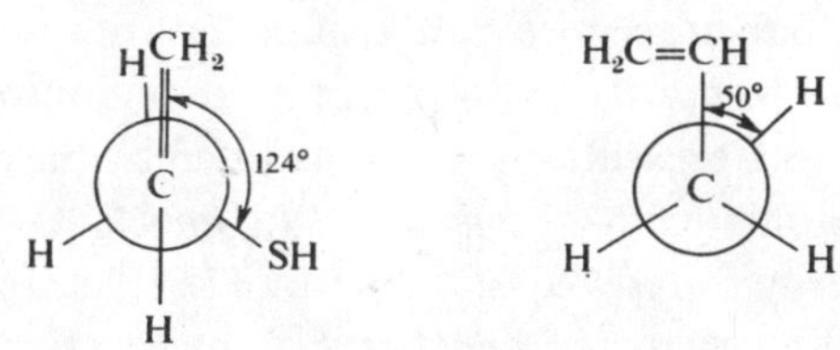

(*a*) (*b*)

appeared to approximate best the experimental data. The thiol group is *gauche* with respect to the C=C double bond, and the S—H bond is *gauche* with respect to the C—C single bond.

Electron diffraction and infrared spectroscopic investigations of *2-furanmethane thiol*,

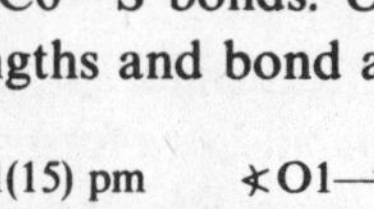

established a non-planar conformation (Schultz, Fellegvári, Kolonits, Kiss, Pete and Bánki, 1978). The angle of rotation was found to be 39(4)° with 0° defining the *anti* orientation of the O1—C2 and C6—S bonds. Utilizing assumptions for the furan geometry, the following bond lengths and bond angles have been determined:

r_a(C2—C6)	148.1(15) pm	∢O1—C2—C6	114.4(11)°
r_a(C6—S)	183.4(5) pm	∢C2—C6—S	112.5(8)°

According to microwave spectroscopic data (*cf.* Bastiansen, Seip and Boggs, 1971) on 2-propane thiol, $(CH_3)_2CHSH$ there is an equilibrium of two conformers in the vapour phase. The mutual orientation of the C—H and S—H bonds are *gauche* and *anti* in the two forms with respect to rotation about the C—S bond:

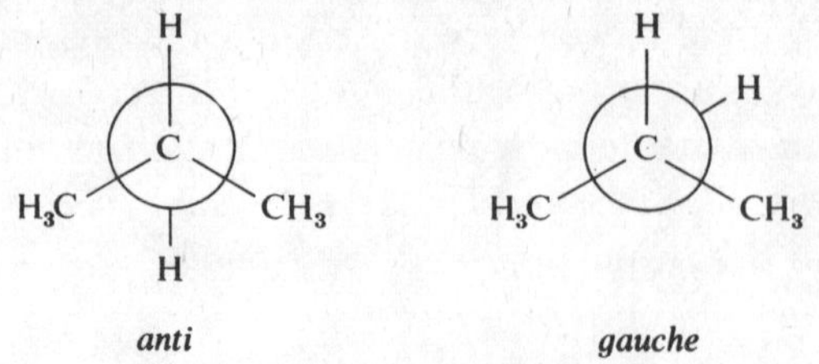

Ethane-1,2-dithiol, $HSCH_2CH_2SH$. The molecular geometry and conformational behaviour in the vapour phase were determined by electron diffraction (Schultz and Hargittai, 1973). The length of the S—C bond is 181.9(6) pm, the bond angles C—C—S and C—S—H being 112.0(13)° and 90(5)°, respectively. The position of the thiol hydrogen was not accurately determined in this work, and little could be said about the rotation about the C—S bond. On the other hand, electron diffraction yielded reliable data on the conformational properties concerning the rotation about the carbon–carbon bond. This is due primarily to the large relative contribution of the S. . .S interactions to the total electron scattering, as illustrated by the radial distributions of Fig. 24a. The relative abundance 62% *anti* and 38% *gauche* established corresponds approximately to an energy difference of 3.3 kJ mol^{-1} (the nozzle temperature was 70 °C in the electron diffraction experiment). It is of interest to compare the results of the vapour phase electron diffraction study with the conclusions from the vapour phase and crystalline phase vibrational spectroscopic investigation by Hayashi, Shiro, Oshima and Murata (1965). It was obvious from the vapour phase spectra of ethane-1,2-dithiol that both the *gauche* and *anti* forms were present. As the sample was crystallized, however, the spectral lines characterizing the *gauche* form disappeared. According to the infrared and Raman spectra of the crystalline sample, the molecule has a centre of symmetry. Hence in the crystal phase, either all three possible axes of rotation have an *anti* arrangement of bonds around them, or are *anti* around the central bond and *gauche* around the two other axes. These two conformations are designated as *aaa* and *gag*, respectively, and are shown in Fig. 24b. The form that is present in the vapour phase, but disappears upon crystallization, is *gauche* with respect to the rotational isomerism around the central bond. No conclusion could be made, however, from the vibrational spectra concerning the rotational behaviour around the C—S bond in this conformer. The normal coordinate analysis was consistent with *anti* arrangement around the C—S bond (form *aga*, Fig. 24b). Hayashi, Shiro, Oshima and Murata (1965) concluded that the *aaa* and *aga* forms are present in the vapour phase, whereas the crystal phase consists of the *aaa* form exclusively. The *anti* form was found to be more stable than the *gauche* form by 2.65 kJ mol^{-1}, according to the vibrational spectra. Thus there was satisfactory agreement between the electron diffraction and vibrational spectroscopic results. The latter also contained estimates for the rotational

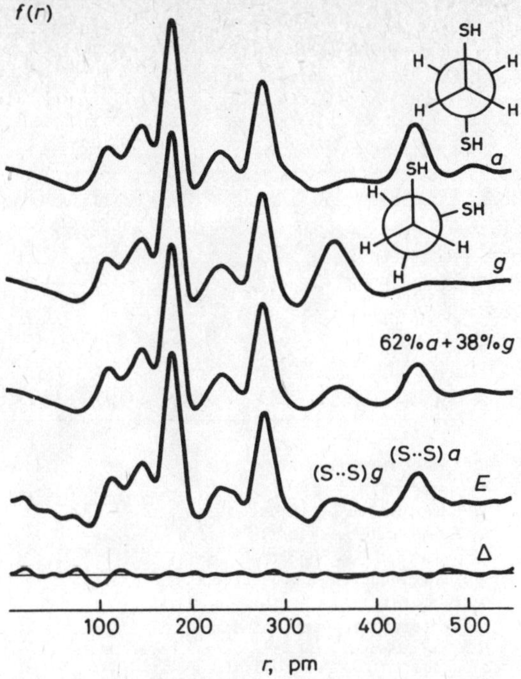

Fig. 24a. Radial distribution of ethane-1,2-dithiol; E — experimental; a, g, and 62%a + 38%g theoretical calculated for the *anti* and *gauche* conformers and their mixture, respectively; Δ — difference curve, experimental minus theoretical calculated for the mixture of two forms (Schultz and Hargittai, 1973)

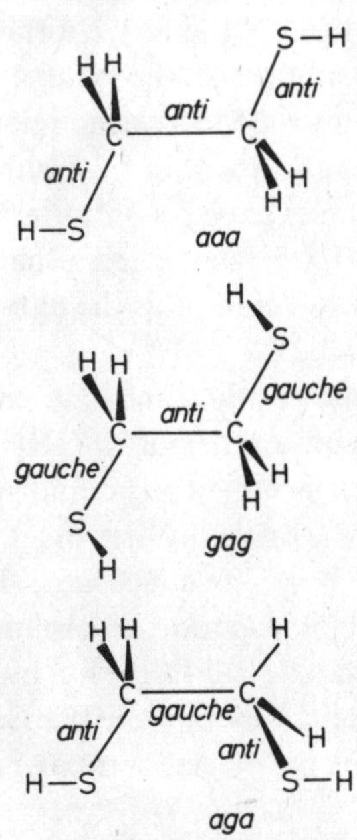

Fig. 24b. Three rotational isomers of ethane-1,2-dithiol considering all three axes of rotation

TABLE 25
The *anti–gauche* energy difference (ΔE) and the substituent mass (m_X) in 1,2-disubstituted ethane derivatives

Compound	m_X	ΔE, kJ mol^{-1}	
$BrCH_2CH_2Br$	80	9	*cf.* (*a*, *b*)
$ClCH_2CH_2Cl$	35	4	*cf.* (*c*)
$HSCH_2CH_2SH$	33	3	*cf.* (*d*)
$NCCH_2CH_2CN$	26	6	*cf.* (*e*)
FCH_2CH_2F	19	−2–−8	*cf.* (*a*, *f*, *g*)
$HOCH_2CH_2OH$	17	−4	*cf.* (*h*)
$H_2NCH_2CH_2NH_2$	16	−7	*cf.* (*i*)
$H_3CCH_2CH_2CH_3$	15	2	*cf.* (*j*)

(*a*) Brunvoll (1962)
(*b*) Fernholt and Kveseth (1978)
(*c*) Kveseth (1975)
(*d*) Schultz and Hargittai (1973)
(*e*) Fernholt and Kveseth (1979)
(*f*) van Schaick, Geise, Mijlhoff and Renes (1973)
(*g*) Harris, Holtzclaw and Kalasinsky (1977)
(*h*) Fernholt (1978)
(*i*) Yokozeki and Kuchitsu (1971b)
(*j*) Ohashi, Ohnishi, Tagui, Sakaizumi and Yamaguchi (1977)

barriers around the C—C and C—S bonds, *viz.* 20 kJ mol^{-1} and 7.1 kJ mol^{-1}, respectively.

The conformational properties of ethane-1,2-dithiol (with respect to the rotation about the central bond) are similar to those of 1,2-dichloroethane. Schultz and Hargittai (1973) called attention to the increasing relative amount of the *gauche* form with the decreasing mass of the substituent in 1,2-disubstituted ethane derivatives. The mass of the substituent, of course, may be only one of the factors determining the conformational behaviour of 1,2-disubstituted ethane derivatives. Nevertheless, a certain correlation pattern seems to emerge as the data on a series of analogous ethane derivatives are compared in Table 25.

Vibrational spectroscopic studies and some semi-empirical calculations were made by Ohsaku and Murata (1978) on a series of $RSCH_2CH_2SR$ molecules (R is, e.g., H, methyl, ethyl, etc.) in order to determine their conformational properties. The results are consistent with the *gauche* form being the most stable as far as the rotational isomerism about the central bond is concerned. In addition, further CNDO/2 calculations were carried out for a series of organo-sulphur molecules including methane thiol, ethane thiol, ethane-1,2-dithiol, etc. by Ohsaku, Bingo, Sugikawa and Murata (1979). Various parametrizations are discussed by these authors as to their applicability for the determination of geometrical parameters and conformational analysis.

Hayashi, Shiro, Murakami and Murata (1965) investigated the vibrational spectra of 2-chloroethane thiol and 2-bromoethane thiol. Both molecules were found to be in the *anti* form

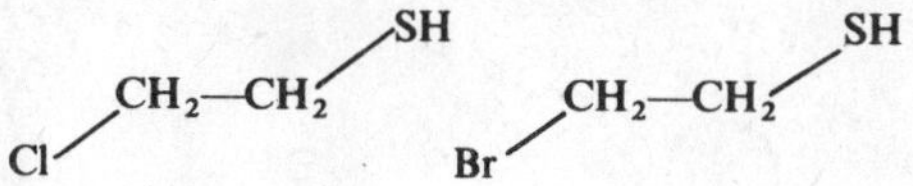

in all phases. Thus no intramolecular S—H...halogen hydrogen bond is formed which would stabilize a *gauche* form. At the same time the *gauche* conformer of 2-chloroethanol (ethylene chlorohydrin) is more stable than its *anti* form by 10 kJ mol^{-1} possibly as a consequence of the formation of the intramolecular O—H...Cl hydrogen bond (Almenningen, Fernholt and Kveseth, 1977).

2-Mercaptoethanol, $HOCH_2CH_2SH$. The microwave spectroscopic investigation by Sung and Harmony (1977) provided the first direct geometrical evidence for the participation of sulphur in an intramolecular hydrogen bond in the vapour phase. The microwave spectra of five isotopic species have been analysed. While it was not possible to derive a complete structure accurately, several r_s internuclear distances could be determined and for some other parameters r_0 values were calculated utilizing reasonable assumptions. Some of the most important bond lengths, non-bond distances and bond angles are the following:

r_s(S—H)	134.0(22) pm	r_0(C—C)	153.8(10) pm
r_s(S...H)	256.5(3) pm	r_0(C—O)	142.3(10) pm
r_s(H...H)	302.6(28) pm	∢C—C—S	113.5(10)°
r_0(O—H)	101.5(15) pm	∢C—C—O	111.4(10)°
r_0(S—H)	134.0(15) pm	∢C—O—H	104.4(15)°
r_0(C—S)	180.6(10) pm	∢C—S—H	96.5(20)°

There is very good consistency between the r_0 and r_s parameters. The most interesting result is the unambiguously determined *ggg* conformation shown in Fig. 25. This rotational form relative to the central bond is similar to those found in related molecules as demonstrated by the angles of rotation:

2-Mercaptoethanol	58.3(10)° (Sung and Harmony, 1977)
Aminoethanol	55.4(20)°, microwave spectroscopy (Penn and Curl, 1971)
Chloroethanol	63.2(6)°, microwave spectroscopy (Azrak and Wilson, 1970);
	62.4(6)°, electron diffraction (Almenningen, Fernholt and Kveseth, 1977)*

In the *ggg* conformer of 2-mercaptoethanol the hydroxyl hydrogen gets close to the sulphur atom indicating the formation of an S...H—O intramolecular hydrogen bond. The S...H distance is considerably smaller than the sum of the van der Waals radii (305 pm). Sung and Harmony (1977) pointed out the relatively long O—H bond

* Chloroethanol has been investigated by electron diffraction at five different temperatures. For comparison with the microwave spectroscopic results, the angle of rotation determined at the lowest temperature (37 °C) was selected.

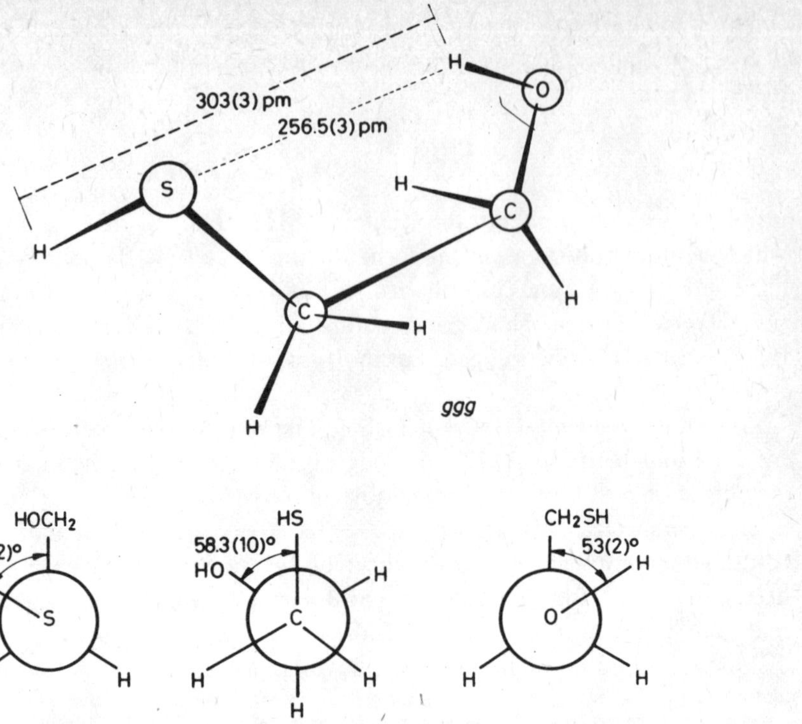

Fig. 25. The rotational isomerism of 2-mercaptoethanol from microwave spectroscopy by Sung and Harmony (1977). There is *gauche* orientation about all three possible axes of rotation. The sulphur atom participates in an intramolecular hydrogen bond

and relatively small C—O—H bond angle as further evidence for the S. . .H—O hydrogen bond. At the same time no particular change has been observed in the sulphur bond configuration, neither did the microwave spectra provide any evidence for any conformation in which a S—H. . .O hydrogen bond would be formed.

Cyclohexane thiol, $C_6H_{11}SH$. Bastiansen and Hassel (1946) established the equatorial orientation of the thiol group from electron diffraction data

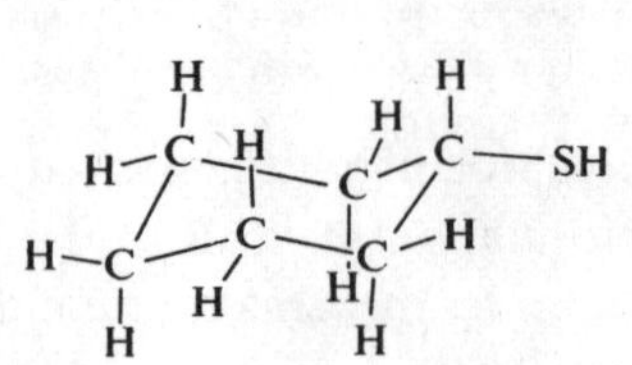

Although suggestions for the axial orientation appeared (Chiurdoglu, Reisse and Rogier, 1961), the equatorial orientation of the thiol group in cyclohexane thiol was further confirmed by N.M.R. spectroscopy (Eliel and Thill, 1963).

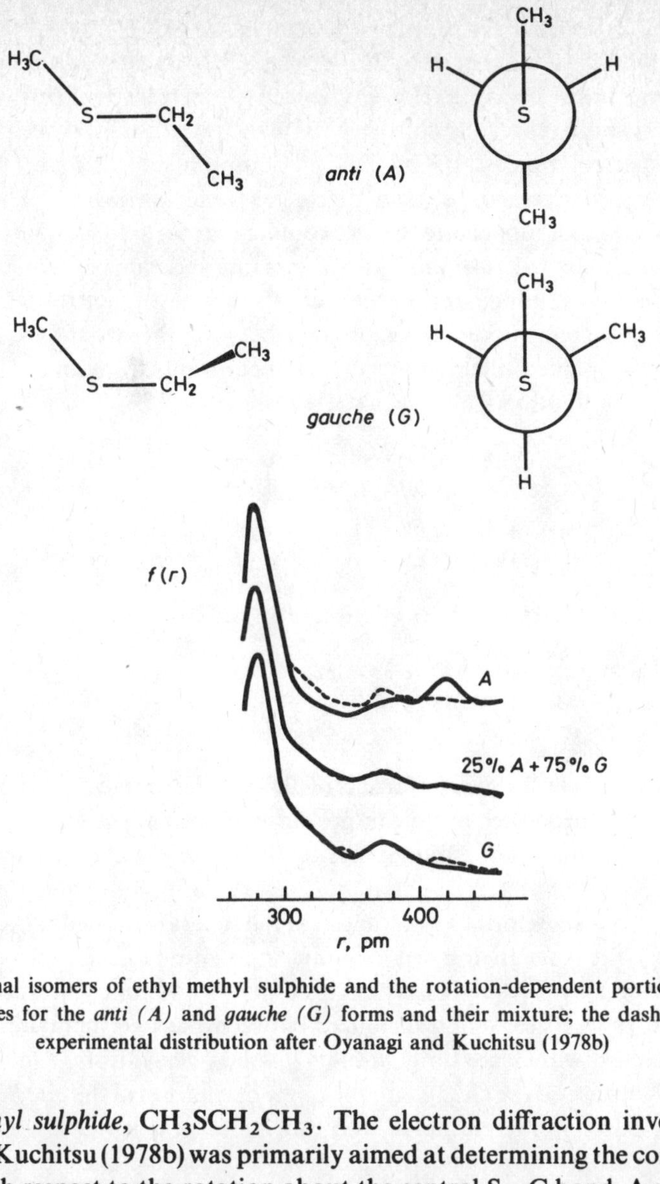

Fig. 26. Rotational isomers of ethyl methyl sulphide and the rotation-dependent portions of its radial distribution curves for the *anti (A)* and *gauche (G)* forms and their mixture; the dashed lines are the experimental distribution after Oyanagi and Kuchitsu (1978b)

Ethyl methyl sulphide, $CH_3SCH_2CH_3$. The electron diffraction investigation by Oyanagi and Kuchitsu (1978b) was primarily aimed at determining the conformational properties with respect to the rotation about the central S—C bond. A parallel study (Oyanagi and Kuchitsu, 1978a) offered an opportunity to make a comparison with the structure of the ether analog. According to the vibrational spectra, the *gauche* and *anti* forms of ethyl methyl sulphide have about the same energy (Sakakibara, Matsuura, Harada and Shimanouchi, 1977; Nogami, Sugeta and Miyazawa, 1975; Ohsaku, Shiro and Murata, 1973). The rotational forms are shown in Fig. 26. The experimental electron diffraction data could be best approximated by a mixture of 75% *gauche* and 25% *anti* forms. The conformation-dependent portion of the radial distribution is shown also in Fig. 26 after Oyanagi and Kuchitsu (1978b). The uncertainty given for the relative abundance of the two conformers, ±15%, was established by careful error analysis taking into account the influence of numerous error sources. The conformational findings are fully consistent with the results of a low-frequency infrared and Raman spectroscopic study by Durig, Compton and Jalilian (1979) showing almost no energy difference between the *gauche* and *anti* forms. The geometrical parameters determined appear to be as would be expected. The mean C—S bond length was found to be 181.3(4) pm. This value is intermediate between the C—S bond lengths for dimethyl sulphide and for ethane-1,2-dithiol. In fact it was even possible, although with large uncertainty, to distinguish between the two different C—S bonds in ethyl methyl sulphide utilizing the rotational constants from an earlier microwave spectroscopic investigation (Imaishi and Hayashi, 1974):

	Ethyl methyl sulphide	Dimethyl sulphide	Ethane-1,2-dithiol
$r(H_3C—S)$ (pm)	180.6(27)	180.7(2)	
$r(H_2C—S)$ (pm)	181.8(27)		181.9(6)

Further geometrical parameters of ethyl methyl sulphide include

r(C—C)	153.6(8) pm	∢C—S—C	97.1(11)°
r(C—H)	111.1(8) pm	∢S—C—C	114.0(5)°
		∢H—C—H	109.6(14)°

The angle of rotation in the *gauche* form is 66(9)° with respect to the *syn* form. Oyanagi and Kuchitsu (1978b) pointed out the importance of steric effects as they compared the conformational properties of ethyl methyl sulphide, ethyl methyl ether (Oyanagi and Kuchitsu, 1978a) and butane (Bonham and Bartell, 1959). For these three molecules, the stability of the *gauche* form is the greatest in the case of ethyl methyl sulphide. In fact the *anti* forms in the ether analog and in butane are more stable than the corresponding *gauche* conformers by 5.1 ± 1.1 kJ mol^{-1} and 2.1 ± 0.9 kJ mol^{-1}, respectively. At the same time the H. . .H non-bond distances between the two methyl groups in these molecules increase with increasing stability of the *gauche* form. Ohno, Mitsui and Murata (1979) discussed the changes in the angle of rotation of the *gauche* form of ethyl methyl ether, ethyl methyl sulphide and ethyl methyl selenide by considering the hydrogen. . .hydrogen non-bonded repulsions between the two methyl groups.

The electron diffraction analyses of *benzene sulphenyl chloride*, C_6H_5SCl (Zaripov, Popik and Vilkov, 1980), *methyl phenyl sulphide* or thioanisole, $C_6H_5SCH_3$ (Zaripov, 1976), *benzene selenenyl bromide*, C_6H_5SeBr (Zaripov, Popik, Vilkov and Mannafov, 1980) and *methyl phenyl selenide* or selenoanisole, $C_6H_5SeCH_3$ (Zaripov, Golubinskii, Chmutova and Vilkov, 1978) all established a non-symmetrical form as the result of internal rotation. They are shown in Fig. 27 by projections representing a view along the bonds serving as axes of rotation. Of course, the angles of rotation were obtained by electron diffraction which results in certain averaging over the torsional motion, which

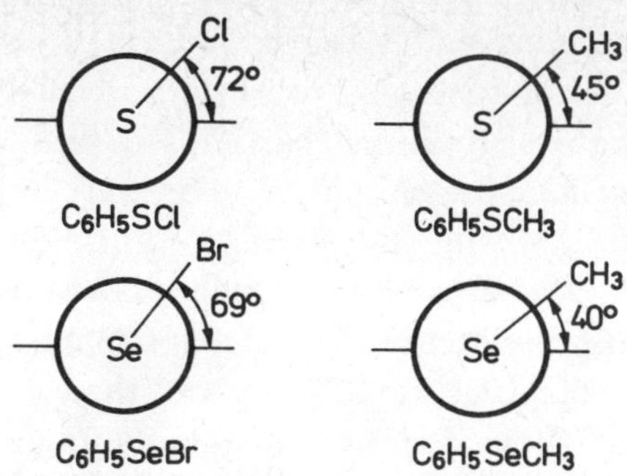

Fig. 27. The conformation of four benzene derivatives, C_6H_5SCl, $C_6H_5SCH_3$, C_6H_5SeBr, and $C_6H_5SeCH_3$, shown by projections representing views along the C_{aryl}—S(Se) bonds

TABLE 26

Geometrical parameters of benzene sulphenyl chloride, methyl phenyl sulphide, benzene selenenyl bromide, and methyl phenyl selenide from electron diffraction

Parameter	C_6H_5SCl *(a)*	$C_6H_5SCH_3$ *(b)*	C_6H_5SeBr *(c)*	$C_6H_5SeCH_3$ *(d)*	
r(C—C), mean, pm	139.5(3)	139.1(6)			
r(S—CH_3), pm		180.3(13)		195.7(12)	Se—CH_3
r(S—Cl), pm	205.1(6)		232.4(2)		Se—Br
r(S—C_6H_5), pm	176.4(12)	179.4(12)	189.9(6)	191.2(12)	Se—C_6H_5
∠C—S—C (°)	102.6(16)	105.6(20)	99.8(12)	99.6(25)	C—Se—C
Angle of rotation (°)*	71.8(34)	45(10)	68.5(30)	40(13)	

(a) Zaripov, Popik, Vilkov (1980)
(b) Zaripov (1976)
(c) Zaripov, Popik, Vilkov and Mannafov (1980)
(d) Zaripov, Golubinskii, Chmutova and Vilkov (1978)
* 0° corresponds to the eclipsed form of C_s symmetry

may have rather large amplitudes if the barrier to internal rotation is low. The vibrational spectra of benzene sulphenyl chloride and thioanisole were interpreted with a C_{2v} symmetry model (Green, 1962; 1968; Bogatyreva, Kotov and Shishkina, 1973; Remizov and Mannafov, 1976a). However, Remizov and Mannafov (1976a) noted that this assumption may be rather crude. The geometrical parameters of the four molecules studied by electron diffraction are collected in Table 26.

The microwave spectra provided evidence for the planar equilibrium conformation of thiophenol (Larsen, Pedersen and Schulz, 1974). From combined far infrared and microwave spectroscopic data Larsen and Nicolaisen (1974) calculated the barrier to internal rotation for thiophenol and phenol to be 3.2 and 14.6 kJ mol^{-1}, respectively. Much lower barriers were reported for thiophenol from molecular mechanics (Timofeeva, Kravtsov, Bokii and Struchkov, 1977) and thermodynamics (Simpson and Beynon, 1971) calculations. On the other hand according to Matsushita, Osamura,

Misawa, Nishimoto and Tsuno (1979) the very high barriers resulting from some *ab initio* molecular orbital calculations were ascribed to the fact that the small basis set used tended to overestimate the rotational barrier. This particular study was primarily aimed at determining the influence of sulphur 3d orbitals on the calculated molecular properties. It was shown that while the effect of 3d orbitals on the charge distribution was important, it had no significant influence on either the barrier to internal rotation or the shape of the energy curve. The planar structure was found to be the most stable. An assumed geometry was used throughout these calculations. On the other hand, Bock and Tabatabai (Bock, 1977) varied both the angle of rotation and the bond angle C—S—H in their CNDO calculations of the potential energy surface for thiophenol. Their interesting results are presented in Fig. 28.

Larsen and Nicolaisen (1974) discussed the origin of the low barrier in thiophenol as compared with that in phenol. They concluded that it may be a consequence of a much weaker conjugation between the sulphur lone electron pairs and the phenyl π-electrons than the corresponding conjugation between the oxygen lone pairs and the phenyl π-electrons. Matsushita, Osamura, Misawa, Nishimoto and Tsuno (1979) compared the calculated energy curves of internal rotation of C_6H_5SH and $C_6H_5SCH_3$ and also those of C_6H_5OH and $C_6H_5OCH_3$. The most interesting difference was the much flatter energy curve of $C_6H_5SCH_3$ as compared with that of C_6H_5SH.

Both planar conformers of *vinyl hydrosulphide*, $CH_2{=}CHSH$ were found to coexist in the vapour phase by microwave spectroscopy (Tanimoto, Almond, Charles,

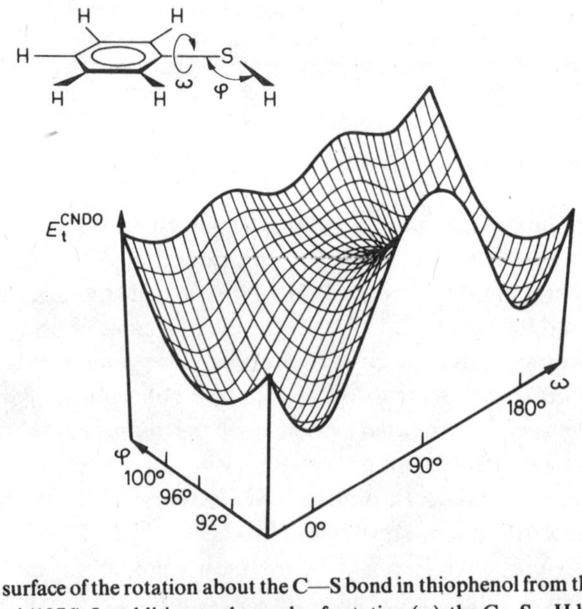

Fig. 28. Total energy surface of the rotation about the C—S bond in thiophenol from the CNDO calculations of Bock and Tabatabai (1976). In addition to the angle of rotation (ω), the C—S—H bond angle (φ) has also been varied in the calculations

Macdonald and Owen, 1979),

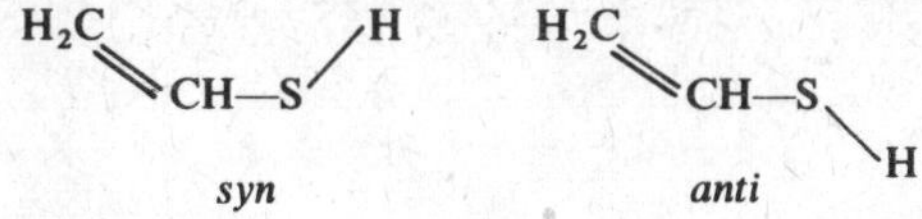

syn *anti*

with the *syn* form being more stable. Assuming the other parameters, 176.6 pm and 127° were determined for the C—S bond length and C=C—S bond angle, respectively. Although no actual values have been obtained for the *anti* form, experimental evidence suggested significant differences in the geometrical parameters of the two conformers (Tanimoto and Macdonald, 1979).

Methyl vinyl sulphide, $CH_3SCH{=}CH_2$. Its vapour-phase molecular structure has been repeatedly investigated by electron diffraction (Samdal and Seip, 1971; 1975; Derissen and Bijen, 1973; Samdal, Seip and Torgrimsen, 1979). Some of the geometrical parameters are cited in Table 27 together with those of *allenyl methyl sulphide*, $CH_3SCH{=}C{=}CH_2$, which were also determined by electron diffraction (Derissen and Bijen, 1973). It has proved satisfactory for neither molecule to use a model with only one conformer. In addition to the planar *syn* form there is a considerable amount of a *gauche* form present. The differences in results on methyl vinyl sulphide as reported from the two different laboratories have concerned primarily the relative abundance of these two forms. This difference was at least partly explained by the temperature difference in the two experiments. The angle of rotation of the non-planar conformer has been determined with large uncertainty. This, in turn, is consistent with the results of some *ab initio* calculations (Kao, 1978) according to which the non-planar conformer is associated with a very flat minimum. Another important result from these

TABLE 27
From the geometrical parameters of methyl vinyl sulphide and allenyl methyl sulphide

Parameter	$CH_2{=}CHSCH_3$		$CH_2{=}C{=}CHSCH_3$
	(a)	*(b)*	*(b)*
$r(H_2C{=}C{=})$, pm			128.2(10)
$r(C{=}C)$, pm	134.3(1)	134.2(7)	132.7(10)
$r({=}C{-}S)$, pm	175.9(8)	175.2(10)	174.5(10)
$r(C{-}S)$, pm	179.5(8)	179.4(12)	180.0(10)
∡C—S—C (°)	102.1(5)	102.5(20)	98.1(8)
∡C=C—S (°)	127.5(7)	127.0(15)	125.4(6)
More stable conformer	*syn (c)*	*syn (c)*	*syn (c)*
Angle of rotation of *gauche* form (°)	135.8(65)	94–108	45–75

(a) r_a bond lengths, r_α bond angles; Samdal, Seip and Torgrimsen (1979)
(b) r_a parameters; Derissen and Bijen (1973)
(c) 0° angle of rotation

calculations was the differences in the geometrical parameters of various rotational forms. Some of them are cited here with the *syn* conformer, $\varphi = 0°$ as reference system:

Changing parameters	Amount of changes $\varphi = 60°$	$= 120°$	$= 180°$
r(=C—S), pm	1.5	1.0	0.7
∢C=C—S (°)	−3.5	−4.9	−4.6
∢C—S—C (°)	−1.1	−2.9	−2.6
∢S—C—H (°)	3.4	4.7	4.5

It may not be possible to determine such small differences by electron diffraction but it may be very important to use them as assumptions in the structure analysis. These results from Kao (1978) have indeed been utilized in the latest electron diffraction study. Samdal, Seip and Torgrimsen (1979) used data from *ab initio* calculations, microwave spectra, and Raman spectra in addition to the electron diffraction intensities. They tested two models differing in their treatment of the molecular vibrations. A dynamic model involving large-amplitude skeletal torsion was preferred to a static model in which the skeletal torsions appeared as small harmonic oscillations about fixed equilibrium positions. The main difference between the results from the two models concerned the angle of rotation of the *gauche* form. The application of the static and dynamic models yielded 116 and 136°, respectively. Kao (1978) reported 138° for this angle from the *ab initio* calculations. Both Samdal and Seip (1975) and Kao (1978) also carried out *ab initio* calculations for vinyl hydrosulphide as well. They found minima for both the planar *syn* and a non-planar *gauche* form. As was mentioned above, microwave spectroscopic studies revealed the existence of two conformers but both of them planar (Tanimoto, Almond, Charles, Macdonald and Owen, 1979; Tanimoto and Macdonald, 1979). An earlier vibrational spectroscopic investigation identified only one planar or nearly planar *syn* form (Almond, Charles, Macdonald and Owen, 1977). Figure 29 presents the *syn* conformers of four molecules examined by Kao (1978) with some of the data of the potential energy calculations in Fig. 30. Kao (1978) discussed in detail the correlation between the conformational properties and the electronic structure of these molecules.

Phenyl vinyl sulphide, CH_2=$CHSC_6H_5$. Its electron diffraction analysis by Zaripov and Mannafov (1975) resulted in the determination of a mean S—C bond length of 174.9 pm and the following two bond angles: ∢C—S—C = 109.2°, ∢C=C—S = 128°. The conformation of this molecule is characterized by two angles of rotation φ_1 and φ_2.

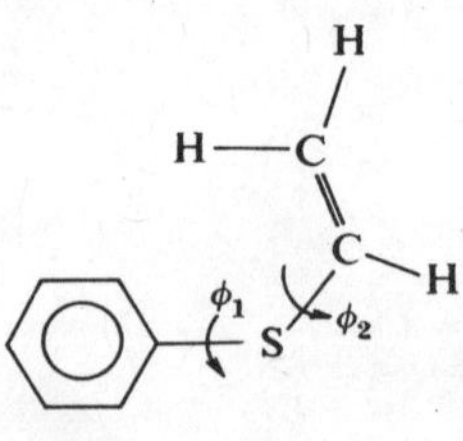

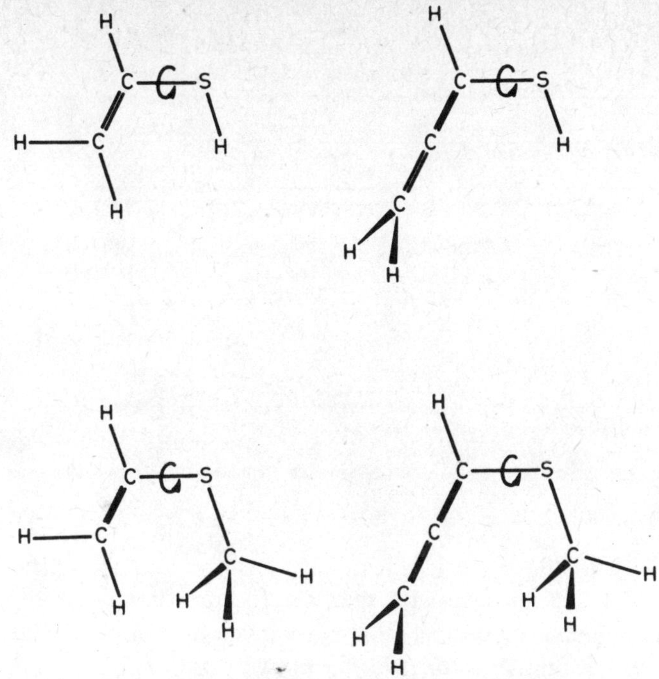

Fig. 29. The *syn* conformation of vinyl hydrosulphide, allenyl hydrosulphide, methyl vinyl sulphide, and allenyl methyl sulphide

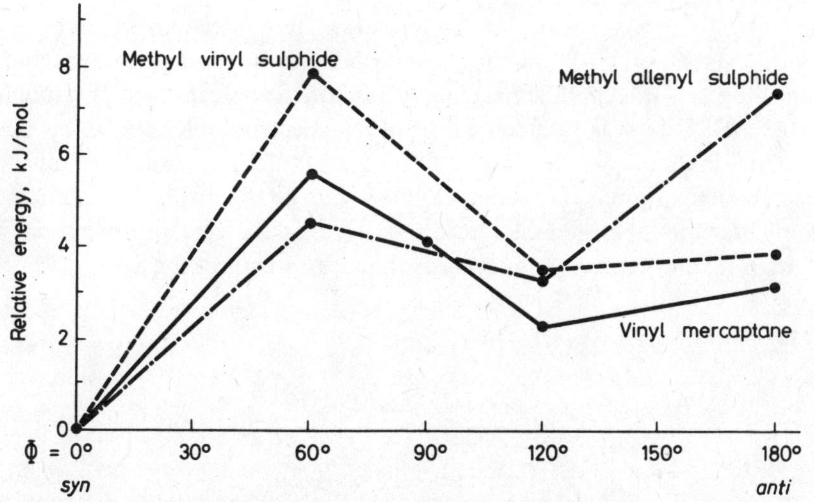

Fig. 30. Selected values of the potential energy oi internal rotation about the C—S bond in vinyl hydrosulphide, methyl vinyl sulphide, and allenyl methyl sulphide from the *ab initio* calculations of Kao (1978)

TABLE 28
Geometrical parameters of methylthioethyne and 2,5-dithiahex-3-yne

Parameter	$H_3CSC{\equiv}CH$ (a)	$H_3CSC{\equiv}CH$ (b)	$H_3CSC{\equiv}CSCH_3$ (c)
r(C≡C), pm	121.5(3)	120.5(7)	
r(≡C—S), pm	168.3(3)	168.5(5)	167.1(2)
r(C—S), pm	181.4(3)	181.3(2)	180.6(2)
∠C—S—C (°)	99.9(8)	99.9(2)	102.7(2)
∠S—C≡C (°)	184(3)	182(0.3)	
Tilt of the methyl group (°)	3 *(d)*	2.5(5)	

(a) Electron diffraction, r_a parameters, van Eijck (1971)
(b) Microwave spectroscopy, r_s parameters, den Engelsen (1969)
(c) Electron diffraction, r_a parameters, Beagley, Ulbrecht, Katsumata, Lloyd, Connor and Hudson (1977)
(d) Assumed

Although one single conformer was selected to approximately reproduce the experimental data, it seems not unlikely that other forms may also be present. Remizov and Mannafov (1977) established the presence of two rotational forms of phenyl vinyl sulphide on the basis of liquid-phase vibrational spectra. They estimated the energy difference between the two conformers to be 1.5 ± 0.4 kJ mol^{-1}. Such an energy difference would correspond to appreciable amounts of both forms under the conditions of the electron diffraction experiment.

Both electron diffraction (van Eijck, 1971) and microwave spectroscopy (den Engelsen, 1967; 1969) were employed to determine the molecular structure of *methylthioethyne*, $CH_3SC{\equiv}CH$. The two sets of results are in good agreement and both of them are cited in Table 28. The electron diffraction investigation of *2,5-dithiahex-3-yne*, $CH_3SC{\equiv}CSCH_3$ was augmented by other spectroscopic studies by Beagley, Ulbrecht, Katsumata, Lloyd, Connor and Hudson (1977). The results are reproduced in Table 28 with the original uncertainties, which may be somewhat underestimated. It is interesting that the experimental electron scattering data were approximated with a model containing one conformer as regards the rotation about the S—C≡C—S axis,

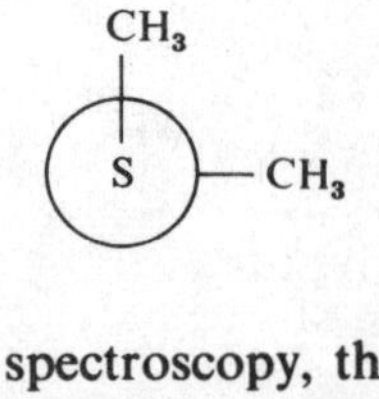

As determined by microwave spectroscopy, the sulphur bond configurations of *methyl thiocyanate*, CH_3SCN (Dreizler, Rudolph and Schleser, 1970) and *ethyl thiocyanate*, C_2H_5SCN (Bjørseth and Marstokk, 1972) are very similar

	CH_3SCN (r_s)	CH_3CH_2SCN (r_0)
$r(H_3C—S)$, pm	182.4(2)	182.0 ($H_2C—S$)
$r(S—C\equiv)$, pm	168.4(3)	169.0
∢C—S—C	99° 2′(5′)	101.0°

For ethyl thiocyanate only the *gauche* form could be identified. Two parallel independent microwave spectroscopic studies reported structural data on *methyl selenocyanate*, CH_3SeCN (*a:* Sakaizumi, Kohri, Ohashi and Yamaguchi, 1978; *b:* Landsberg, 1978). Some of the parameters determined are as follows

	a	*b*
$r(H_3C—Se)$, pm	195.0	195.6
$r(Se—C\equiv)$, pm	185.1	183.7
∢C—Se—C (°)	95.57	96.1

The C≡N bonds are 116–117 pm long in all thiocyanates and selenocyanates. The C≡N bond stretching force constant of C_2H_5SCN was given as 1673 N m^{-1} (Bjørseth, 1974).

Two kinds of rotational spectra were observed in the microwave spectroscopic investigation of *monothioformic acid* (Hocking and Winnewisser, 1975; 1976a; 1976b; 1976c; 1977). The rather small inertial defect, small centrifugal distortion effect and other indications pointed to the presence of rather rigid planar structures. Varying the geometrical parameters and fitting them to the experimental data it was established that the spectra originated from two HC(O)SH isomers,

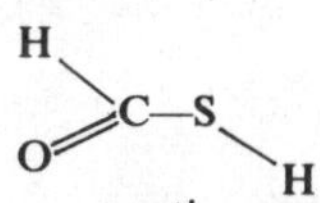

anti

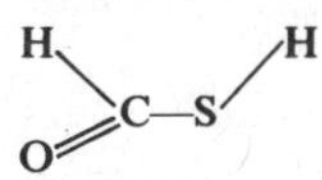

syn

Later on more and more isotopic species were studied and eventually a complete substitution structure (r_s) was determined as given in Table 29. There are some important differences between some of the geometrical parameters of the two conformers. The H—S—C and H—C—S bond angles are larger by 2°, the O═C—S bond angle is smaller by 3° and the S—H bond is some 2 pm shorter in the *syn* form as compared with the *anti* conformer. Hocking and Winnewisser (1976a; 1976b) discussed these differences in terms of intramolecular atom–atom interactions.

Dithioformic acid was first identified in its microwave spectrum (Bak, Nielsen and Svanholt, 1978). Both planar conformers

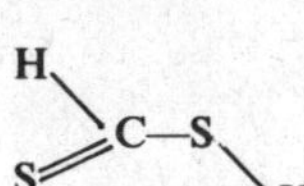

anti

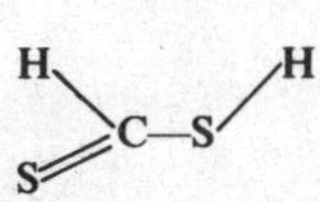

syn

TABLE 29
The molecular geometries of monothioformic acid (*syn* and *anti*) and dithioformic acid (*anti*) as determined by microwave spectroscopy

Parameter	HCOSH (*a*)		HCSSH (*b*)
	syn	*anti*	*anti*
r(C—H), pm	110.4(3)	110.4(3)	110(2) [assumed]
r(C=O), pm	120.3(3)	120.5(3)	162.5(4) C=S
r(C—S), pm	177.1(3)	176.8(3)	173.3(5)
r(S—H), pm	133.5(2)	135.4(2)	135.7(2)
∢O=C—S (°)	122.5(3)	125.9(3)	127.8(3) S=C—S
∢H—S—C (°)	94.9(2)	92.5(2)	94.3(2)
∢H—C—S (°)	114.4(2)	111.0(2)	111(2) [assumed]

(*a*) r_s structure, Hocking and Winnewisser (1977)
(*b*) "Iterative" structure, Bak, Nielsen, Svanholt and Christiansen (1979)

were observed, the *anti* being the more stable by about 4.2 kJ mol^{-1}. The geometrical parameters of the *anti* form were determined experimentally (Bak, Nielsen, Svanholt and Christiansen, 1979). They are listed in Table 29. These parameters are neither r_s nor r_0, but are the results of an iterative procedure. The microwave spectroscopic study was augmented by *ab initio* calculations of both the *syn* and *anti* geometries, energy difference, barrier to conversion, etc.

The microwave spectra of only two isotopic species were analysed for *thioacetic acid* (Naito, Ohashi and Yamaguchi, 1977). It was not even possible to distinguish between the two conformers with planar skeleton.

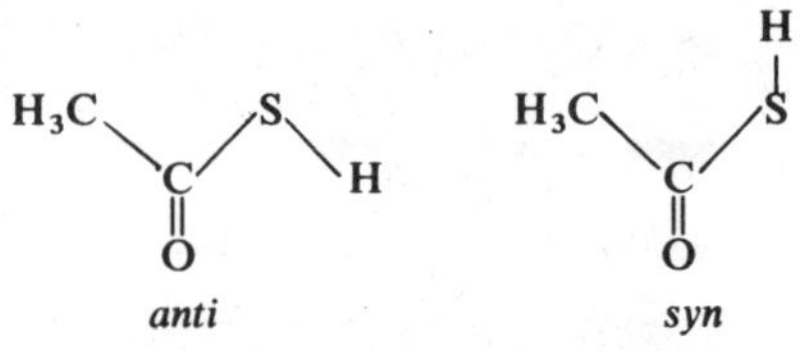

Only the r_s coordinates of the thiol hydrogen could be determined.

A series of low-resolution microwave spectroscopic studies have provided valuable information on the conformational properties of alkyl thioesters (Silvia, True and Bohn, 1978; 1979; True and Bohn, 1979). They have been shown to display similarities as well as contrasts with the conformational isomerism of the corresponding oxyesters (*cf.*, e.g., True and Bohn, 1976).

The geometrical parameters of three somewhat related molecules have been collected in Table 5:

Dimethyl trithiocarbonate,	$S{=}C(SCH_3)_2$
Dimethyl dithiocarbonate,	$O{=}C(SCH_3)_2$
Bis(methylthio)ethylene,	$H_2C{=}C(SCH_3)_2$

The conformational behaviour of the first two has been commented upon in the discussion on molecules with C=S bond of one-coordinated sulphur. Their prevailing form of internal rotation about the =C—S bonds is *syn–syn* (*cf.* Fig. 3). On the other hand, an electron diffraction investigation (Jandal, Seip and Torgrimsen, 1976) established a *syn–gauche* form to be prevailing for 1,1-bis(methylthio)ethylene. Further forms also exist. Theoretical calculations including those *ab initio* (De Alti, Decleva and Sgamellotti, 1979) confirmed the possibility of the coexistence of several conformers for 1,1-bis(methylthio)ethylene. They predicted, however, the *syn–syn* form to be the most stable in contrast to the electron diffraction results for this molecule.

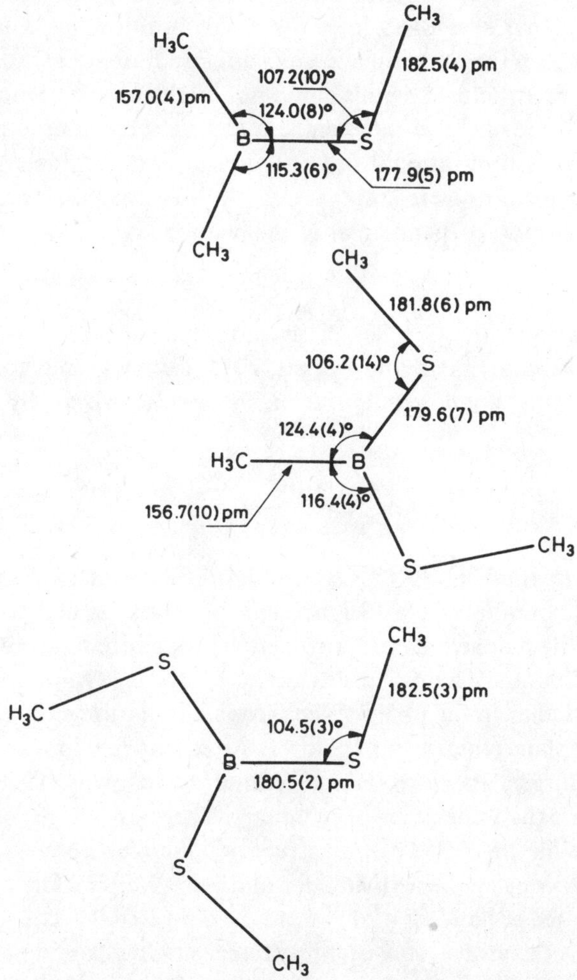

Fig. 31. The geometrical parameters of methylthio dimethylborane (Brendhaugen, Nilssen and Seip, 1973), bis(methylthio)methylborane (Lindøy, Seip and Seip, 1976), and tris(methylthio)borane (Johansen, Nilssen, Seip and Siebert, 1973) from electron diffraction

A series of molecular structures with boron–sulphur bonds has been determined by electron diffraction:

Methylthio dimethylborane,	$CH_3SB(CH_3)_2$	(Brendhaugen, Nilssen and Seip, 1973)
Bis(methylthio)methylborane,	$(CH_3S)_2BCH_3$	(Lindøy, Seip and Seip, 1976)
Tris(methylthio)borane,	$(CH_3S)_3B$	(Johansen, Nilssen, Seip and Siebert, 1973)

The structures which have an essentially planar skeleton are shown in Fig. 31 together with the geometrical parameters. The experimental data could also be approximated well by models with slightly non-planar carbon–boron–sulphur skeleton. The non-planar structures from the electron diffraction analysis, however, may be a consequence of the out-of-plane vibrations and torsional motion around the B—S bond. Some fine structural details could not be elucidated because of the strong correlation among some of the parameters, e.g. among those characterizing bonds of similar length. Although vibrational data (Goubeau and Wittmeier, 1952; Vahrenkamp, 1971) and corrections were employed in all three analyses, the information on the intramolecular motion perpendicular to the plane of the heavy-atom skeleton was rather uncertain.

Tris(methylseleno)borane, $(CH_3Se)_3B$. Its molecular structure as determined by electron diffraction (Lindøy, Seip and Siebert, 1975) is very similar to that of the thio analog. The selenium bond configuration is characterized by the following parameters:

$$r_a(\text{Se—B})\ \ 193.6(2)\ \text{pm} \quad r_a(\text{Se—C})\ \ 195.4(4)\ \text{pm}$$
$$\angle\text{B—Se—C} \quad 102.5(5)^\circ$$

Tris(trifluoromethylthio)amine, $(F_3CS)_3N$. Figure 32 summarizes the main results of the electron diffraction analysis of this molecule by Marsden and Bartell (1977). The NS_3 skeleton is nearly planar. The trifluoromethyl groups are staggered with respect to the adjacent S—N bonds. The molecule does not have C_3 symmetry. The N—S—C planes are perpendicular to the plane of the three sulphur atoms. Two CF_3 groups are situated above this plane and the third below it. An extensive vibrational spectroscopic investigation by Bürger, Pawelke, Haas, Willner and Downs (1978) also provided evidence for no more than one plane of symmetry. The results were consistent with the electron diffraction findings. The overall structure described above was supported by ^{19}F N.M.R. spectroscopic evidence (Marsden and Bartell, 1977). The barrier to internal rotation was estimated to be 25 kJ mol^{-1} from the N.M.R. data. The planarity or near-planarity of the NS_3 skeleton is consistent with the very low basicity of $(F_3CS)_3N$. The results on the conformational behaviour of tris(trifluoromethylthio)amine were in agreement with conclusions from semi-empirical molecular orbital calculations on the hypothetical $N(SF)_3$ molecule (Marsden and Bartell, 1977).

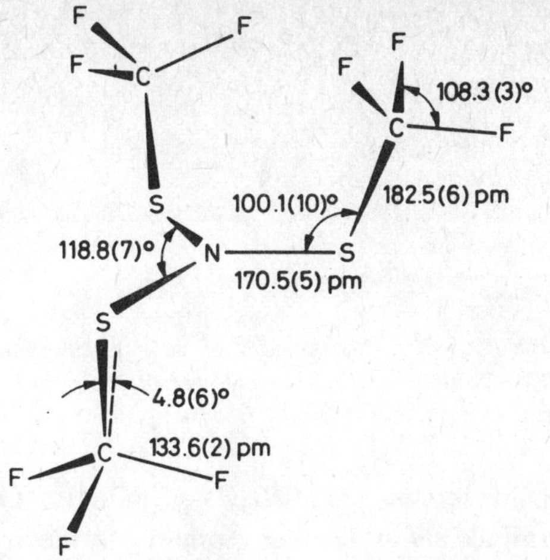

Fig. 32. Molecular model and the geometrical parameters of tris(trifluoromethylthio)amine from the electron diffraction analysis by Marsden and Bartell (1977)

The *anti*/*gauche* ratio slightly above room temperature in the vapour of *S-methyldichloro thiophosphate*, $CH_3SP(O)Cl_2$ was determined to be 7/3

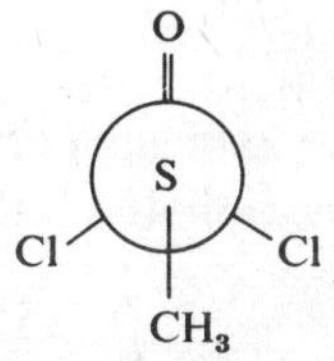

anti

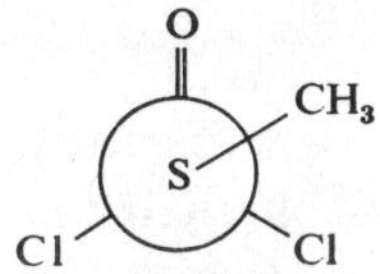

gauche

by electron diffraction (Naumov and Bezzubov, 1976). On the other hand, the *gauche* conformer was found to be the more stable one for $CH_3OP(O)Cl_2$ (Bezzubov and Naumov, 1976). The sulphur bond configuration is characterized by the following parameters in $CH_3SP(O)Cl_2$:

r(P—S) 209.7(15) pm r(C—S) 182.9(30) pm
∢C—S—P 117.4(30)°

Trimethyl trithiophosphite, $P(SCH_3)_3$. Its prevailing rotational form has C_s symmetry with *syn–syn–anti* orientations and a smaller amount of the molecules may have a *syn–*C_{3v} configuration according to a recent electron diffraction study (Tuzova, Naumov, Galiakberov, Ofitserov and Pudovik, 1981). The sulphur bond configuration is characterized by

r_a(P—S) 211.5(4) pm r_a(C—S) 183.4(7) pm
and ∢P—S—C 115.3(7)°

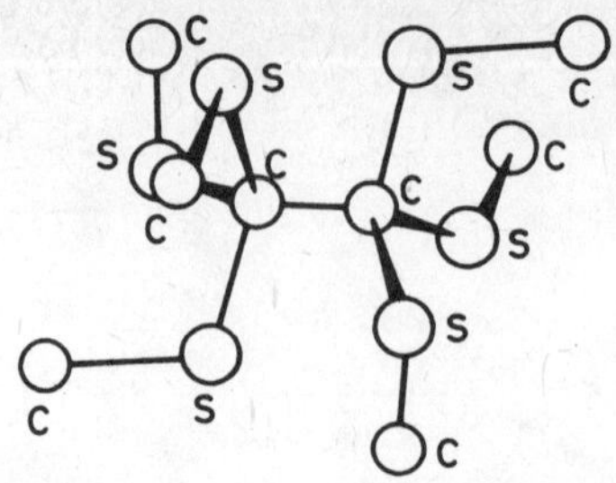

Fig. 33. Molecular model of the $(CS)_3CC(SC)_3$ skeleton of hexakis(trifluoromethylthio)ethane after Oberhammer, Haas and Schlosser (1979)

Hexakis(trifluoromethylthio)ethane, $(F_3CS)_3C$—$C(SCF_3)_3$. Oberhammer, Haas and Schlosser (1979) determined its molecular geometry by electron diffraction. The molecular model of the $(CS)_3C$—$C(SC)_3$ skeleton with overall C_i symmetry is shown in Fig. 33. The CF_3 groups were assumed to be staggered with respect to the adjacent S—C bonds. A very long C—C bond was found, but unfortunately with large uncertainty: r_g(C—C) 162.4(37) pm. Some other parameters:

r_g(C—F)	134.0(3) pm	∢F—C—F	107.3(2)°
$r_g(S—C)_{mean}$	183.2(4) pm	∢S—C—S	113.9(3)°
∢C—C—S	104.5(3)°		
∢C—S—C	110.6(12)°		

The structure analysis was partially hindered by strong correlation among the parameters and the lack of reliable calculated vibrational spectroscopic quantities.

Trifluoromethyl selenium cyanide, F_3CSeCN. The electron diffraction data (Marsden and Sheldrick, 1971b) were consistent with a staggered conformation but also with another one obtained from the former by a 45° rotation. The geometrical parameters (r_a) were the same in the two forms, except the C—Se—C bond angle which was determined to be 92.2(20)° and 103.0(25)°, respectively. The other parameters: C—F 133.2(7), C—Se 198.4(20), Se—C≡ 185.4(16), C≡N 115.2(20) pm, Se—C—F 109.6(12)°. The staggered form with large-amplitude torsional motion about the C—Se bond seemed to be more probable.

2.3. STRUCTURAL VARIATIONS IN THE SULPHIDE SERIES

2.3.1. SULPHUR–CARBON BOND LENGTHS

The bond lengths and bond angles characterizing the sulphur bond configuration in symmetrically substituted sulphide molecules are collected in Table 30. The bond lengths calculated from the covalent radii and electronegativity (χ) corrections are also

TABLE 30
Comparison of the molecular geometry of sulphides and ethers

Compound		Bond length (pm) Experimental	Bond length (pm) Calculated (a)	Bond length (pm) Calculated (b)	Δr (pm) (a)	Δr (pm) (b)	Bond angle (°) Experimental	$r(X\ldots X)$, pm Experimental	$r(X\ldots X)$, pm Calculated (c)	$\Delta r(X\ldots X)$, pm
SF_2	(d)	159.2	156	156.5	3	2.7	98.2	240.7	216	25
$S(CF_3)_2$	(d)	181.9	181	178.7	1	3	97.3	273.1	250	23
SCl_2	(d)	200.6	200	199.6	1	1.0	103.0	314.0	288	26
$S(CN)_2$	(d)	170.1	181	178.7	−11	−8.6	95.6	252.0	250	2
$S(CH_3)_2$	(d)	180.7	181	178.7		2.0	99.2	274.9	250	25
$S(C_6H_5)_2$	(d)	177.1	181	178.7	−4	−1.6	103.7	278.5	250	28
$S[N(CH_3)_2]_2$	(d)	168.8	170	172.8	−1	−4.0	116.5	287.1	228	59
$S(SiH_3)_2$	(d)	213.6	218	215.3	−4	−1.7	97.4	320.9	310	11
$S(GeH_3)_2$	(d)	220.9	223	218.6	−2	2.3	98.8	335.8	316	20
OF_2	(e)	140.5	126	143.6	14	−3.1	103.1	220.1	216	4
OCl_2	(f)	169.3	161	172.1	8	−2.8	111.2	279.5	288	−8
$O(CH_3)_2$	(g)	141.0	135	143.9	6	−2.9	111.7	233.4	250	−17
$O(SiH_3)_2$	(h)	163.4	169	170.3	−6	−6.9	144.1	310.7	310	1
$O(GeH_3)_2$	(i)	176.6	174	173.6	3	3	126.5	315.4	316	−1

(a) $r(A\text{—}B) = r_A + r_B - c|\chi_A - \chi_B|$, Pauling (1960)
(b) $r(A\text{—}B) = r_A + r_B - 7.3|\chi_A - \chi_B|^2$, Robiette (1973), Pauling (1960), Glidewell (1976)
(c) The sum of the 1,3 non-bonded radii (*cf.* Table 2)
(d) The sources are given earlier
(e) Morino and Saito (1966)
(f) Beagley, Clark and Hewitt (1968)
(g) Blukis, Kasai and Myers (1963)
(h) Almenningen, Bastiansen, Ewing, Hedberg and Traetteberg (1963)
(i) Glidewell, Rankin, Robiette, Sheldrick, Beagley and Cradock (1970)

given. Both the original Schomaker and Stevenson (1941) expression

$$r(\mathrm{A-B}) = r_\mathrm{A} + r_\mathrm{B} - c\,|\chi_\mathrm{A} - \chi_\mathrm{B}| \qquad [*]$$

with the data from the book of Pauling (1960) and the expression suggested by Robiette (1973)

$$r(\mathrm{A-B}) = r_\mathrm{A} + r_\mathrm{B} - 7.3\,|\chi_\mathrm{A} - \chi_\mathrm{B}|^2 \qquad [**]$$

with the covalent radii from Glidewell (1976) were applied. The experimental and calculated bond lengths differ very little, so the sulphur bond in the sulphides can be considered to be essentially single bond. The S—C bond of sulphur dicyanide with a considerable shortening is the only real exception. It has been noted that there is probably a partial bond formed between the two carbon atoms in this molecule.

While the C—S bonds of sulphides may be considered to be single bonds, the C═S bonds of one-coordinated sulphur are usually considered to be double bonds. The bond order in itself, however, is not a well-defined parameter and it is more useful to follow the bond order variations in series of molecules. Some time ago Colapietro, Domenicano and Vaciago (1976) suggested a second-order relationship between the CS bond length and bond order *(n)*:

$$r(\mathrm{CS}) = an^2(\mathrm{CS}) + bn(\mathrm{CS}) + c$$

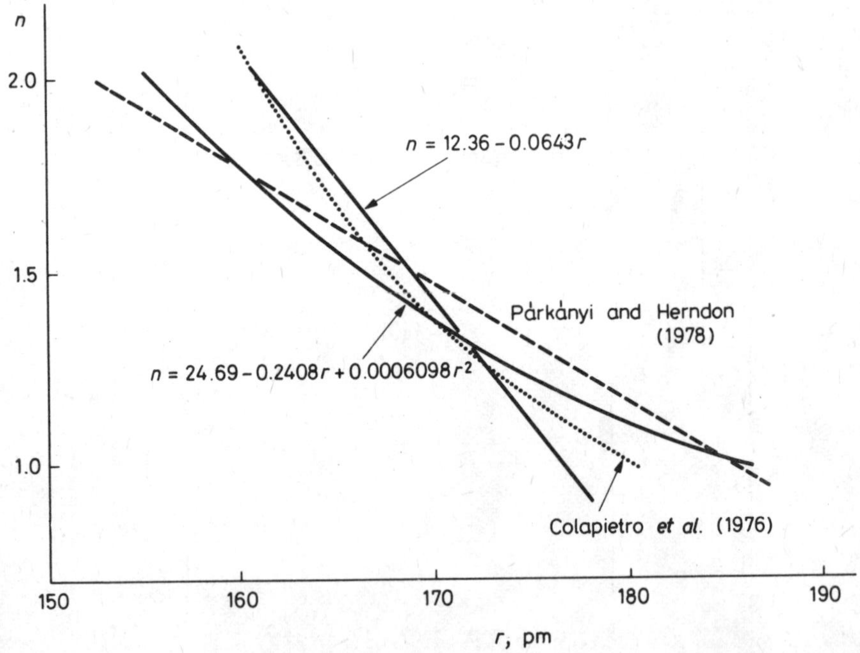

Fig. 34. Various bond length/bond order relationships for the carbon–sulphur bonds

postulating the following points:

r(CS)	180 pm	n(CS) = 1.00
r(CS)	171 pm	n(CS) = 1.33
r(CS)	161 pm	n(CS) = 2.00

The first point was associated with the single bond $C(sp^2)$—S and the third with the double bond $C(sp^2)$=S on the basis of a large body of experimental (mainly crystal-phase x-ray diffraction) data. The middle point was estimated from the structural data of the trithiocarbonate anion. The relationship suggested by Colapietro, Domenicano and Vaciago (1976) is shown in Fig. 34 (with $a = 11.1$, $b = -52.3$, and $c = 221.2$).

Let us consider now some vapour-phase data for $C(sp^2)S$ bonds again. The following three points may be, for example, postulated

$H_2C{=}C(SCH_3)_2$	r(CS)	176.7 pm	n(CS) = 1.00
$S{=}C(SCH_3)_2$	$r(CS)_{mean}$	171.3 pm	n(CS) = 1.33
$S{=}CH_2$	r(CS)	161.1 pm	n(CS) = 2.00

These three points lie on a good straight line, $r = -15.5\,n + 192.1$. Or, carbon

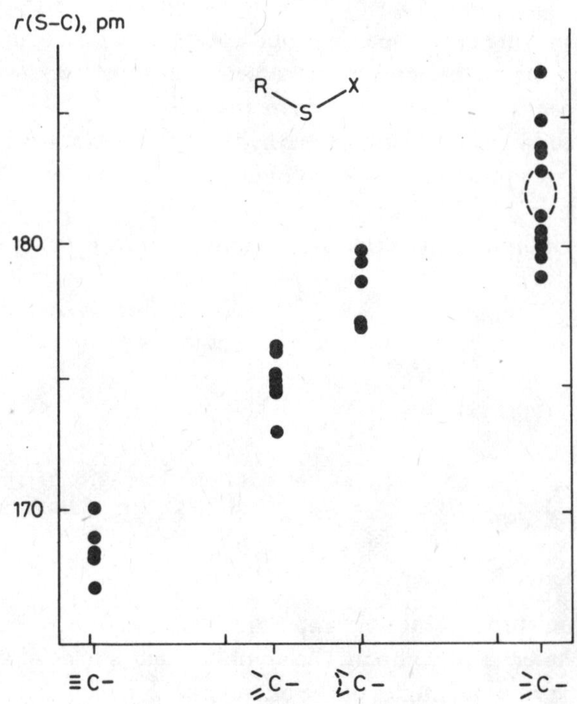

Fig. 35. Variation of S—C bond lengths in sulphides in different environments (Rozsondai and Hargittai, 1980)

hybridization disregarded, take some extreme CS bond lengths for the first and third points

CF_3SO_2Cl*	r(CS)	185.7 pm	n(CS) = 1.00
S=C(SCH_3)$_2$	$r(CS)_{mean}$	171.3 pm	n(CS) = 1.33
CS_2	r(CS)	155.3 pm	n(CS) = 2.00

These points yield the following relationship: $r = 19.3\,n^2 - 88.3\,n + 254.7$. The above two relationships are also shown in Fig. 34, as well as another one suggested by Párkányi and Herndon (1978) for relating bond length and π-bond-order: $r = -33.4\,n_\pi + 185.5$.

The sulphur–carbon bond seems to vary considerably with the changes of its environment. Let us consider now some vapour phase data on R—S—X sulphides in Fig. 35 (Rozsondai and Hargittai, 1980). Not all the data used here have been necessarily discussed in the previous sections. The spread of the C—S bond lengths is remarkable indeed, remembering that all these bonds are formally considered to be single bonds.

2.3.2. SULPHUR BOND ANGLES

Most of the bond angles are smaller than 100° in the sulphide molecules. The bond angle of SF_2 is smaller than that of SCl_2 in agreement with the prediction of the VSEPR model. Applying the same model and considering the group electronegativity difference between the methyl and trifluoromethyl groups, the C—S—C bond angle of $S(CF_3)_2$ is expected to be smaller than that of $S(CH_3)_2$. A rather large bond angle was observed in hexakis(trifluoromethylthio)ethane whose origin may be steric effects. Another, exceptionally large sulphur bond angle was determined in the dimethylamino derivative (*cf.* Fig. 16). A little smaller N—S—N bond angles were determined in crystalline *N*, *N'*-thiobis(dicyclohexylamine)

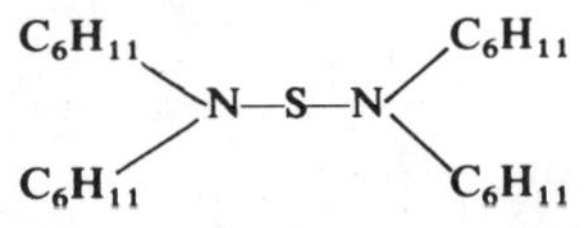

r(S—N) 165.7(4) pm
∢ N—S—N 110.7(2)°

and a heptasulphurimide derivative

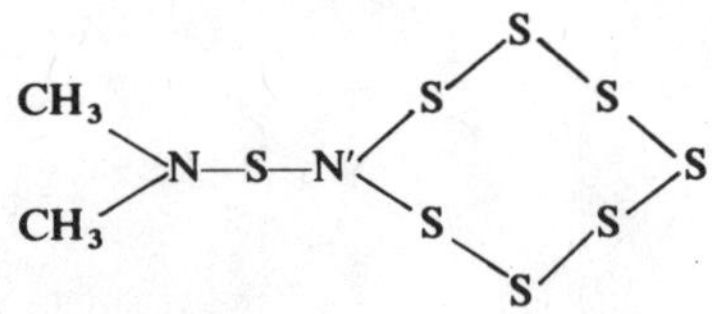

r(S—N) 160.7(5) pm
r(S—N') 173.4(5) pm (exocyclic)
∢ N—S—N 110.6(3)°

by x-ray diffraction (Bruce, Gillespie and Slim, 1978).

It is of special interest to compare the sulphur bond angles of sulphides with the corresponding oxygen bond angles of the analogous oxygen derivatives. The relevant

* See later in the discussion of sulphone molecules.

geometrical data of some ether molecules are also given in Table 30. The oxygen bond angle is always considerably larger than the corresponding sulphur bond angle. The very large oxygen angles of disiloxane or digermoxane are usually interpreted as a consequence of the considerable double-bond character of the Si—O and Ge—O bonds originating from $d\pi$—$p\pi$ interaction. An alternative explanation may also be given, however.

Consider, for example, the relative bond shortening of 6 pm in OF_2 as compared with that in SF_2 relative to the calculated single bonds (using expression [**]). Similarly, this relative bond shortening is 4 pm in the molecule-pair OCl_2, SCl_2, or 5 pm in $O(CH_3)_2$, $S(CH_3)_2$, or 5 pm as well in $O(SiH_3)_2$, $S(SiH_3)_2$, and there is no relative shortening in the molecule pair of the germyl derivatives. Thus disiloxane is not really an exceptional case as compared with, e.g., dimethyl ether. This is, of course, not a proof for the absence of $d\pi$–$p\pi$ interactions in disiloxane but it indicates that the bond shortening may not necessarily be so directly linked with the opening of the bond angle.

The bond angle variations observed in analogous sulpides and ethers may be interpreted in a straightforward manner utilizing the so-called atom–atom non-bonded interactions sometimes also called 1,3 interactions following Bartell (1960) and Glidewell (1976).

In addition to the S—X bond lengths and X—S—X bond angles, the corresponding X. . . X non-bond distances are also given in Table 30 together with the values twice the non-bonded radii from Table 2 (these will be the calculated X. . . X distances). The respective interatomic distances of the oxygen-containing molecules are also listed. The differences between the experimental and calculated X. . . X distances show that even the very large oxygen bond angles barely separate the two X atoms by a distance corresponding to twice the non-bonded radii. Thus the 1,3 interactions may play decisive role in establishing the geometry of the ether molecules. On the other hand, a much more modest role for the 1,3 interactions is predicted by the large experimental X. . . X distances in the sulphides. Note at the same time that the distance between the two carbon atoms is roughly twice the carbon non-bonded radius in sulphur dicyanide.

As the selenium bonds are even longer than the sulphur bonds, the 1,3 interactions are expected to be even less important in forming the selenium bond configurations in the selenium derivatives. This expectation is in agreement with the data shown below:

	r(X. . .X), pm		Δr(X. . .X), pm
	Observed	Calculated	
$Se(CH_3)_2$	289.8	250	40
$Se(SiH_3)_2$	339.6	310	30
$Se(GeH_3)_2$	344.5	316	29

The striking differences between the analogous sulphur and oxygen bond angles may also be observed in the mixed substituted sulphides and ethers as shown by three examples here:

A=S or O	∢C—S—C (°)	∢C—O—C (°)
$CH_2{=}CHACH_3$	104.6(8)	118.3(11) *(a)*
$C_6H_5ACH_3$	105.6(7)	120.0(20) *(b)*
$CH_3ACH_2CH_3$	97.1(11)	111.9(5) *(c)*

(a) Samdal and Seip (1975)
(b) Seip and Seip (1973)
(c) Oyanagi and Kuchitsu (1978a)

Further examples refer to boron derivatives:

A=S or O	∢B—S—C (°)	∢B—O—C (°)
$(CH_3)_2BACH_3$	107	126.0 *(a)*
$B(ACH_3)_3$	104.5	121.4 *(b)*

(a) Gundersen (1978)
(b) Gundersen (1976)

Considering the 1,3 non-bonded interactions, it is of interest to discuss not only the sulphur bond configuration, but also the geometry of the NS_3 skeleton in tris(trifluoromethylthio)amine following Marsden and Bartell (1977). The nitrogen bond configuration is nearly planar similarly to trisilyl amine (Si—N—Si $119.7 \pm 0.3°$, Beagley and Conrad, 1970; Hedberg, 1955) but in contrast to trimethyl amine (C—N—C $110.9 \pm 0.6°$, Beagley and Medwid, 1977) or nitrogen trichloride (Cl—N—Cl 107.8 $\pm 0.4°$, Cazzoli, Favero and Dal Borgo, 1974). The X. . .X non-bond distances of the NX_3 skeletons indicate that the 1,3 interactions may be important in accounting for the geometrical variations in these systems

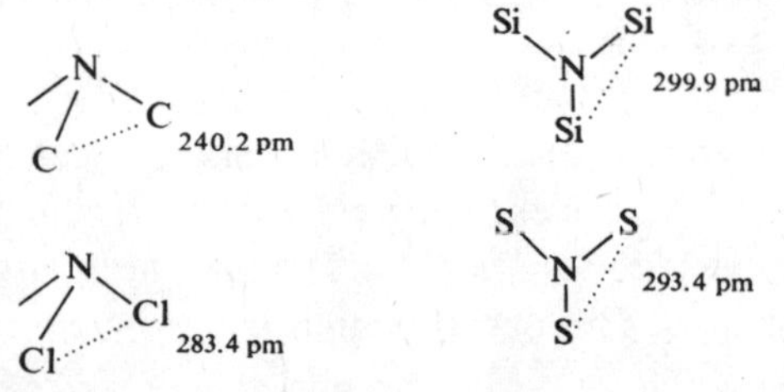

Of course, it would not be correct to overemphasize the importance of 1,3 interactions, since the physical reality is certainly more complex than can be described by a single and so much oversimplified effect. At the same time, the geometrical variations discussed above may be considered as clearly reflecting the consequences of the atom–atom interactions. Also, they should not be looked upon as evidence for the absence of $d\pi$–$p\pi$ interactions. Obviously, different interactions may dominate in different structures. It is also possible that the 1,3 interactions create favourable geometrical conditions for $d\pi$–$p\pi$ interactions in some structures.

2.3.3. THE TILT OF THE METHYL GROUP

A tilt of the methyl group has been observed in the methyl sulphide derivatives. The methyl tilt is also characteristic for molecules in which the methyl group is linked to oxygen or nitrogen, while it is of much lesser importance when the methyl group is linked to carbon or silicon. The tilt means that the methyl symmetry axis does not coincide with the axis of the $H_3C—X$ bond, $X=S$, O or N, and the methyl group is tilted towards the lone electron pairs of the X atom.

It is not surprising, of course, that the methyl group loses its C_{3v} symmetry in an asymmetrical environment. It may have not been possible earlier to detect this effect but with the application of the modern experimental techniques, especially microwave spectroscopy, there is a growing body of experimental evidence for methyl tilt. This tilt usually amounts to about 3°. The C—H bond of the methyl group that is *anti* with respect to the lone electron pairs of the X atom is usually some 1 pm longer than the other C—H bond(s). Some experimental data are listed here on the methyl tilt:

CH_3OH	3.3°	(Lees and Baker, 1968)
CH_3SH	2.2°	(Kojima and Nishikawa, 1957; Kojima, 1960)
CH_3SeH	1.5°	(Thomas, 1973)
CH_3SCl	3.7°	(Alekseev and Velichko, 1967)
$(CH_3)_2O$	2.6°	(Blukis, Kasai and Myers, 1963)
$(CH_3)_2S$	2.75°	(Pierce and Hayashi, 1961)
$(CH_3)_2Se$	2.6°	(Pandey and Dreizler, 1977)
$CH_3SC{\equiv}CH$	2.5°	(Den Engelsen, 1969)
CH_3NH_2	2.9°	(Takagi and Kojima, 1971)
CH_3NHCl	2.9°	(Caminati, Cervellati and Mirri, 1974)

The absence of the methyl tilt (at least within experimental error) is shown by the microwave spectroscopic results (r_s) on the dimethyl silane structure (Pierce, 1961) according to which ∢C—Si—C 110°59′ ±10′ and the angle made by the two symmetry axes of the methyl groups is 110°59′ ±1°. Some slight methyl tilt, 0.35°, has been determined in dimethyl diazirine, 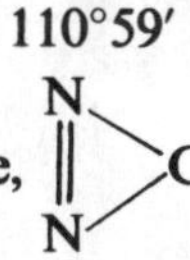 $C(CH_3)_2$ by microwave spectroscopy (r_0 parameter, Wollrab, Scharpen, Ames and Merritt, 1968) and a larger one, 2.0(5)°, in acetone by a combined analysis of electron diffraction and microwave spectroscopic data (r_z parameter, Iijima, 1972).

Ab initio calculations using the force method (Flood, Pulay and Boggs, 1977) produced surprisingly accurate methyl tilts in a series of molecules. Since structural *differences* were determined in this work most of the assumptions used in the calculations cancel out. The methyl tilt was satisfactorily accounted for by the proposed localized molecular orbital model. Interactions between the methyl C—H bonds and the bonds of the adjacent atom on one hand, and between the methyl C—H bonds and the lone electron pairs of the adjacent atom on the other hand, have been suggested to be of importance, with apparently the former prevailing.

Even more considerable tilts of trihalomethyl groups have been observed in molecules like CF_3SF and CF_3SCl (*cf.* Figs. 20b, 20c, 20d).

2.3.4. HINDERED ROTATION OF THE METHYL GROUP

The methyl groups take a staggered conformation in all the sulphide derivatives investigated and internal rotation of the group is hindered. Pierce and Hayashi (1961) have already examined the general trends in the variations of the barrier heights for molecules in which the methyl group is linked to group IV, V or VI elements. Such studies may greatly facilitate our understanding the origin of potential barriers. The observations of Pierce and Hayashi (1961) have subsequently received further support from more recent experimental data. They noted that with the C—X bond lengthening (X group IV element) the methyl rotational barrier decreases, as e.g. in the series CH_3—CH_3, CH_3—SiH_3, CH_3—GeH_3, the barriers are 11.5, 7.5 and 5.2 kJ mol^{-1}, respectively. Similar observation was made for molecules involving group VI elements like dimethyl sulphide and dimethyl ether. On the other hand for CH_3SH and CH_3OH the opposite trend occurs. Table 31 illustrates these observations with more recent data after Sakaizumi, Kohri, Ohashi and Yamaguchi (1978). It is also seen that taking the molecules with the same group VI element (A), the rotational barrier decreases with increasing C—A bond length. Another observation by Pierce and Hayashi (1961) was that the methyl rotational barrier is not sensitive to chemical substitution in the adjacent group when group IV element is linked to the methyl group. Thus, for

TABLE 31

C—A bond lengths and barrier heights of methyl rotation about the C—A bonds in some simple molecules after Sakaizumi, Kohri, Ohashi and Yamaguchi (1978)*

A=	O	S	Se
C—A bond lengths (pm)			
CH_3AH	142.7(7)	181.9(5)	195.9
CH_3ACN		181.8	195.0
CH_3ACH_3	141.0(3)	180.2(2)	194.3(1)
Barrier heights (kJ mol^{-1})			
CH_3AH	4.5	5.3(1)	4.2(2)
CH_3ACN		6.7(3)	5.2(2)
CH_3ACH_3	11.4	8.92(1)	6.3(1)

* For original references see cited paper

instance, the barrier is 13.5 ± 1.2 kJ mol^{-1} in CH_3CH_2X molecules and 5.0 ± 0.6 kJ mol^{-1} in CH_3COX molecules where X may be H, F, Cl, Br or CN. There may be exceptions but the deviations are much smaller than the difference between the rotational barriers of, e.g., CH_3OH and CH_3OCH_3 or CH_3SH and CH_3SCH_3 (*cf.* Table 31).

Thomas (1973) observed that the methyl rotational barrier in methane selenol is smaller than in methane thiol and is essentially the same as in methanol. It was argued that for understanding the relationship between geometry and rotational barrier it is not enough to consider only the length of the bond about which the rotation takes place. In the series CH_3AH (A = O, S, Se) the bond angles are of great importance

$\measuredangle$C—O—H	108°32′ ±29′ (Lees and Baker, 1968)
$\measuredangle$C—S—H	$96.5 \pm 1°$
$\measuredangle$C—Se—H	$95.5 \pm 1°$

Using a simple model with tetrahedral angles and no tilt for the methyl group, Thomas (1973) calculated the proton–proton repulsions in the staggered (V_{st}) and eclipsed (V_{ecl}) form

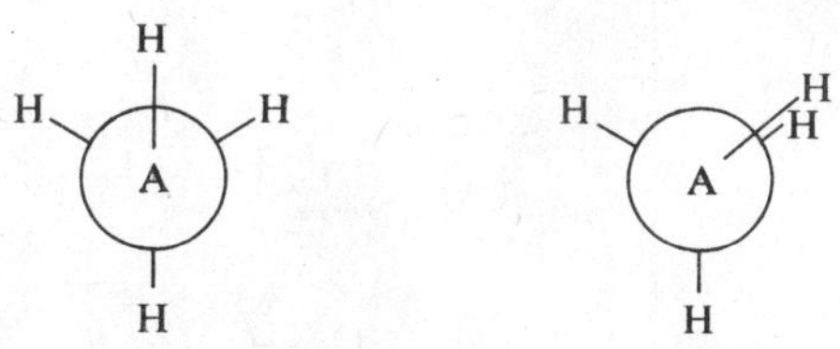

staggered *eclipsed*

Comparison of these repulsions (by their difference $\Delta V = V_{ecl} - V_{st}$) and the rotational barriers, or rather their respective ratios shows

	ΔV ratios	Barrier ratios
CH_3OH	0.89	0.84
CH_3SH	1	1
CH_3SeH	0.88	0.80

that the interactions between bonds or atoms at the two ends of the bond about which the rotation takes place contribute predominantly to the rotational barrier and it is not likely that the barrier is an intrinsic feature of this bond itself. In this connection it seems to be of interest to refer to a study on the variation of the rotational barriers in the series Cl_3CCCl_3, Cl_3CSiCl_3, Cl_3CGeCl_3, $Cl_3SiSiCl_3$ (Hargittai and Brunvoll, 1978). Figure 36a shows that the rotational barrier gradually decreases with the lengthening of the central bond. It is also shown, however, that a considerable portion of the total barrier originates from the van der Waals potential. The variation of the van der Waals potential with the angle of rotation is shown in Fig. 36b. The above observations are in

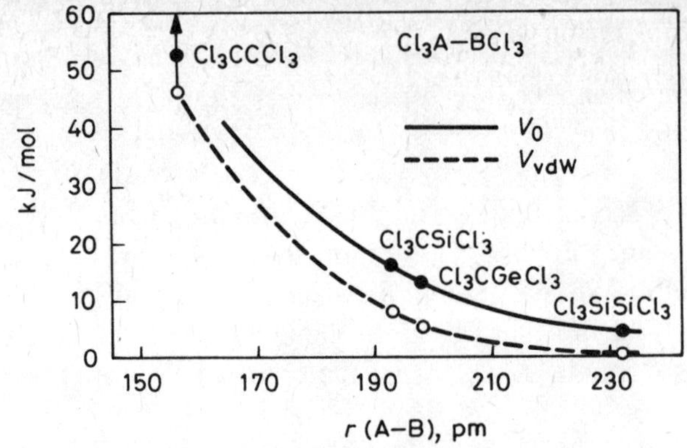

Fig. 36a. The barriers to internal rotation in $Cl_3A—BCl_3$ molecules and the length of the A—B bonds; V_0 is the total rotational barrier and V_{vdw} is the estimated van der Waals interaction energy (Hargittai and Brunvoll, 1978)

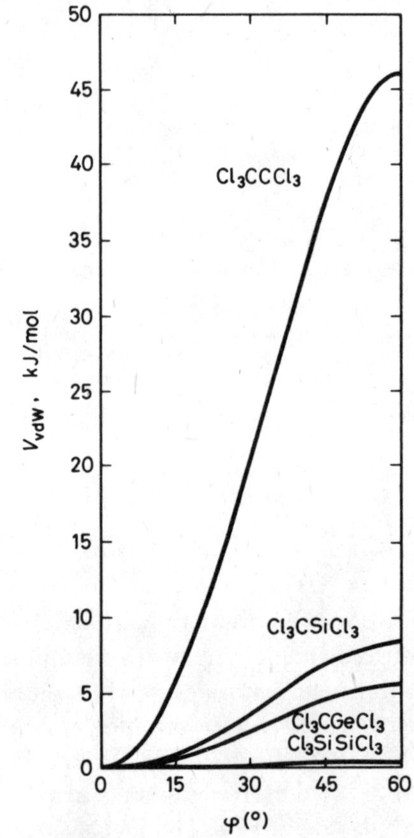

Fig. 36b. Variation of the chlorine–chlorine van der Waals energy in $Cl_3A—BCl_3$ molecules as a function of the angle of rotation ($\varphi = 0°$ eclipsed form, $\varphi = 60°$ staggered form), after Hargittai and Brunvoll (1978)

agreement with the results of Oyanagi and Kuchitsu (1978b) concerning the changes in the conformational behaviour in the series of butane, ethyl methyl ether, ethyl methyl sulphide, as discussed earlier.

2.4. DISULPHANES

Dihydrogen disulphide (disulphane), S_2H_2. Its molecular geometry was determined both by electron diffraction and by microwave spectroscopy (Winnewisser, Winnewisser and Gordy, 1968; Winnewisser and Haase, 1968; Winnewisser, 1972). The results from different techniques are consistent. Table 32 contains the r_s parameters. The conformation of disulphane and derivatives is illustrated by Fig. 37.

Dimethyl disulphide, $S_2(CH_3)_2$. Two independent electron diffraction studies (Beagley and McAloon, 1971 and Yokozeki and Bauer, 1976) and a microwave spectroscopic investigation (Sutter, Dreizler and Rudolph, 1965) elucidated its molecular structure. The electron diffraction analysis by Yokozeki and Bauer (1976) involved a detailed examination of the intramolecular motion and its influence on the average molecular geometry. The geometrical parameters are given in Table 32. An early electron diffraction study by Donohue and Schomaker (1948) determined the

TABLE 32
Geometrical parameters of symmetrically substituted disulphide and diselenide molecules

XSSX		r(S—S), pm	r(S—X), pm	∢S—S—X (°)	∢XS—SX (°)
HSSH	r_s *(a)*	205.5(1)	132.7(3)	91.3(5)	90.6(5)
CH_3SSCH_3	r_a *(b)*	202.2(4)	180.6(2)	104.1(4)	83.9(14)
	r_g *(c)*	202.9(3)	181.6(3)	103.2(2)(∢$_\alpha$)	85.3(37)(∢$_\alpha$)
	r_0 *(d)*	203.8(20)	181.0(20)	102.8(15)	84.7(30)
FSSF	r_0 *(e)*	188.8(15)	163.5(15)	108.3(10)	87.9(20)
ClSSCl	r_a *(f)*	193.1(5)	205.7(2)	108.2(3)	84.8(13)
	r_0 *(g)*	193.7(6)	[205.7]	107.92(1)	84.88(15)
BrSSBr	*(h)*	198(4)	224(2)	105(3)	84(11)
$(CH_3)_2BSSB(CH_3)_2$	r_a *(i)*	207.8(4)	180.5(5)	105.3(4)	120(6)

XSeSeX		r(Se—Se), pm	r(Se—X), pm	∢Se—Se—X (°)	∢XSe—SeX (°)
$CH_3SeSeCH_3$	r_g *(j)*	232.6(4)	195.4(5)	98.9(2)	87.5(40)
$CF_3SeSeCF_3$	r_a *(k)*	292.2(10)	201.8(20)	98.0(5)	84.5(30)

(a) Winnewisser (1972)
(b) Beagley and McAloon (1971)
(c) Yokozeki and Bauer (1976)
(d) Sutter, Dreizler and Rudolph (1965)
(e) Kuczkowski (1964)
(f) Beagley, Eckersley, Brown and Tomlinson (1969)
(g) Yamazaki, Mogi, Koyama and Yamaguchi (1979)
(h) Hirota (1958)
(i) Johansen, Seip and Siebert (1975)
(j) D'Antonio, George, Lowrey and Karle (1971)
(k) Marsden and Sheldrick (1971c)

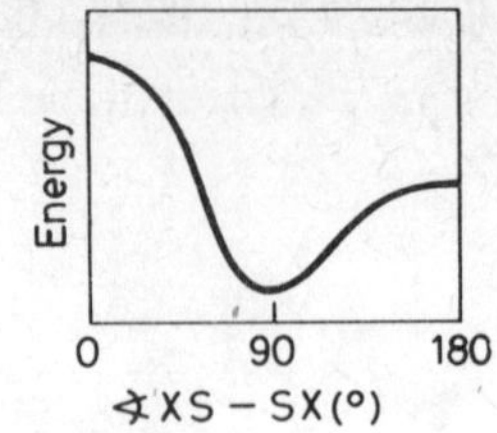

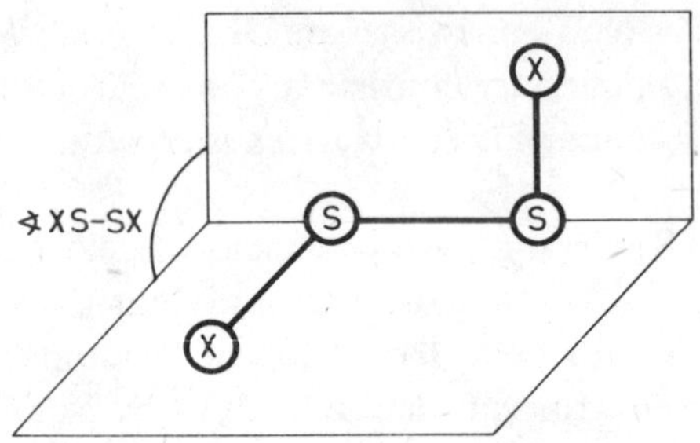

Fig. 37. Conformation of disulphanes and the total potential energy as a function of the angle of rotation after Steudel (1975) with further references therein

geometry of *dimethyl trisulphide*, $S_3(CH_3)_2$. The molecule has a chain C—S—S—S—C skeleton. The following parameters characterize the sulphur bond configuration:

r(C—S)	178(4) pm	∢S—S—S	104 (5)°
r(S—S)	204(2) pm	∢C—S—S	104 (5)°

The C—S bonds were found to perform large-amplitude oscillation about a mean position, ca. 106° out of the plane of the three sulphur atoms.

The molecular structure of *1,2-difluorodisulphide*, S_2F_2, is known from a microwave spectroscopic study by Kuczkowski (1964). *1,2-Dichlorodisulphide*, S_2Cl_2, has been repeatedly investigated by electron diffraction (Beagley, Eckersley, Brown and Tomlinson, 1969; and references therein) and recently also by microwave spectroscopy (Yamazaki, Mogi, Koyama and Yamaguchi, 1979). The latter was performed for the $S_2{}^{35}Cl_2$ species only and the electron diffraction S—Cl bond length was assumed in the analysis. The geometrical parameters of S_2F_2 and S_2Cl_2 are given in Table 32 as well as those of *1,2-dibromodisulphide*, S_2Br_2 from an early electron diffraction study (Hirota, 1958). The uncertainties of the microwave spectroscopic parameters of S_2Cl_2 include estimates from the errors of the rotational constants and the assumed r(S—Cl). Thus they do not include estimates on the effects of intramolecular motion.

Bis(dimethylboryl)disulphane, $(CH_3)_2BSSB(CH_3)_2$. The skeleton around each B—S bond is planar according to the electron diffraction structure analysis (Johansen, Seip and Siebert, 1975). The parameters characterizing the geometry of the boron–sulphur skeleton are given in Table 32. The angle of rotation about the S—S bond is larger than in the other disulphanes.

The conformational behaviour of disulphanes was discussed by Steudel (1975) and by Johansen, Seip and Siebert (1975) involving also the then available results of quantum chemical calculations. A schematic representation of the total energy as a function of the angle of rotation about the XS—SX bond is given in Fig. 37. Accordingly, the *syn* orientation is the least stable. The *anti* forms of disulphane and dimethyl disulphane are more stable than the *syn* form by 12–16 and 30 kJ mol^{-1}, respectively.

The length of the central S—S bond is very sensitive to the substituent changes according to the data of Table 32. The C—C bond length in a similar series of 1,2-disubstituted ethane derivatives appears to be rather constant, *cf.* Table 33, with the notable exception of 1,2-dibromoethane. There is obviously a correlation between the S—S bond lengths and the X ligand electronegativities (χ_X) in the XSSX disulphane series as is seen in Fig. 38.

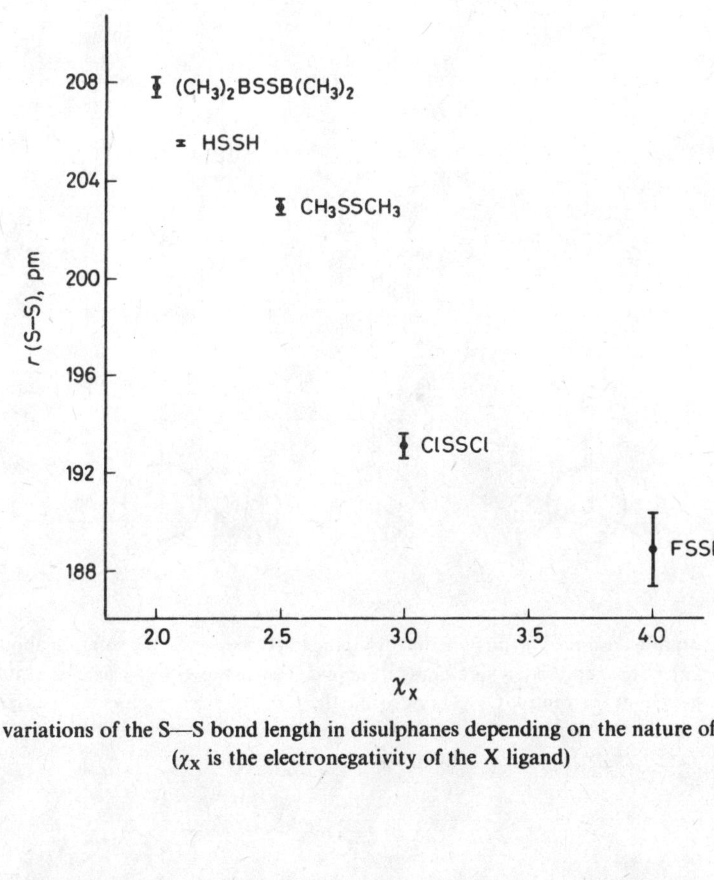

Fig. 38. The variations of the S—S bond length in disulphanes depending on the nature of the substituent (χ_X is the electronegativity of the X ligand)

Ethyl methyl disulphide, $CH_3SSC_2H_5$, is so far the only unsymmetrically substituted disulphane derivative whose vapour-phase molecular geometry has been determined. Yokozeki and Bauer (1976) examined the presence of a series of rotational isomers (see Fig. 39) in their electron diffraction analysis. Their work was paralleled by a low-

TABLE 33
The length of the central C—C bonds in X—CH_2—CH_2—X molecules

X—CH_2—CH_2—X	r_g(C—C), pm	
H—CH_2—CH_2—H	153.3(2)	*(a)*
CH_3—CH_2—CH_2—CH_3	153.3(2)	*(b)*
F—CH_2—CH_2—F	153.7(5)	*(c)*
Cl—CH_2—CH_2—Cl	153.5(3)	*(d)*
Br—CH_2—CH_2—Br	150.8	*(e)*

(a) Iijima (1973)
(b) Mean value of the C—C bond lengths; Bradford, Fitzwater and Bartell (1977)
(c) From r_a; van Schaick, Geise, Mijlhoff and Renes (1973)
(d) From r_a; Kveseth (1975)
(e) From r_a; Fernholt and Kveseth (1978)

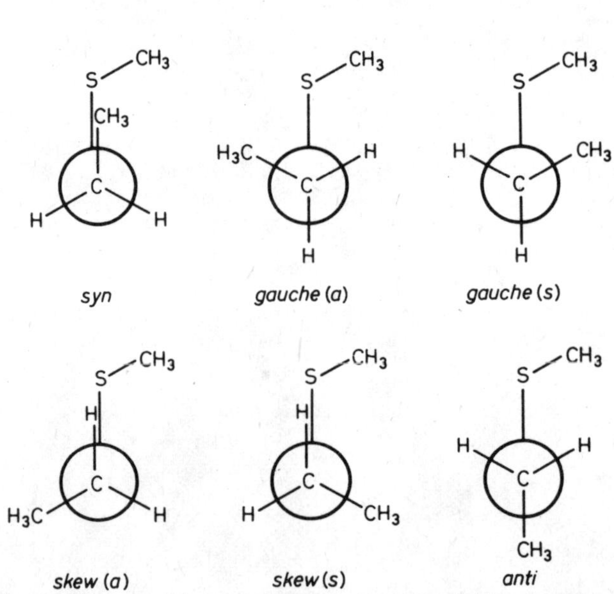

Fig. 39. Possible rotational isomers of ethyl methyl disulphide with respect to the rotation about the central C—S bond. The projections represent a view along this bond. The names *syn*, *gauche*, *skew*, and *anti* denote the rotational forms about the central C—S bond while the parenthesized letters characterize the mutual orientation of the two methyl groups. This figure was prepared after Yokozeki and Bauer (1976) but with differing notation

frequency Raman spectroscopic investigation (Van Wart, Cardinaux and Scheraga, 1976) finding three coexisting isomers with respect to the rotation about the C—S bond. Earlier spectroscopic studies established fewer conformers but examined the presence of selected forms only (Sugeta, Go and Miyazawa, 1973; Scott and El-Sabban, 1969). According to spectroscopic and theoretical results (Van Wart and Scheraga, 1977; and references therein) the most stable form is the *syn* or one that is slightly (0–30°) twisted towards the *gauche(a)* form. There is another energy minimum corresponding to the *anti* form. The electron diffraction data unambiguously showed the presence of the *anti* conformer and it was proven that at least one, possibly more rotational isomers are also present. The experimental data, however, could be equally well approximated by various mixtures of conformers. The geometrical parameters of ethyl methyl disulphide are very much the same as those of the dimethyl derivative:

r_g(C—S)	181.7(4) pm	$\measuredangle_\alpha$S—S—C	103.2(3)°
r_g(S—S)	203.1(3) pm	$\measuredangle_\alpha$C—S—S—C	84(7)°

Dimethyl diselenide, $Se_2(CH_3)_2$ and *bis(trifluoromethyl)diselenide*, $Se_2(CF_3)_2$ have similar conformational properties according to the electron diffraction analyses by D'Antonio, George, Lowrey and Karle (1971) and Marsden and Sheldrick (1971c), respectively. In both molecules the $CH_3(CF_3)$ groups are unsymmetrically placed with respect to the C—Se—Se planes. The geometrical parameters are cited in Table 32. The observed bond length differences are in agreement with the electronegativity differences of the methyl and trifluoromethyl groups.

2.5. NON-AROMATIC RING MOLECULES

Non-aromatic ring molecules containing two-coordinated sulphur atom in the ring are discussed in this section. So far the vapour-phase molecular geometry of the following substances has been determined:

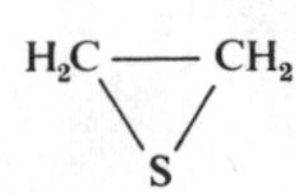

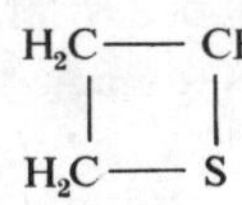

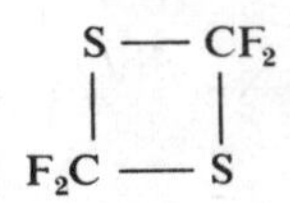

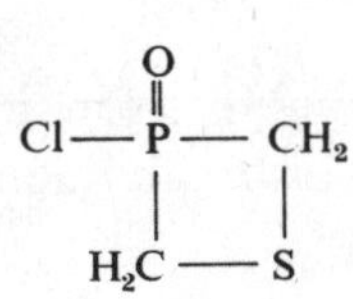

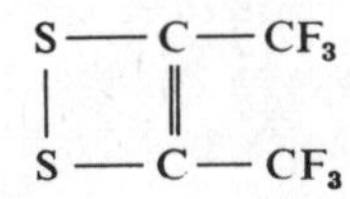

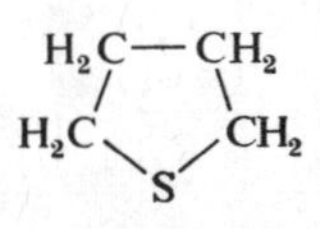

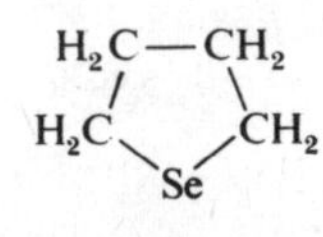

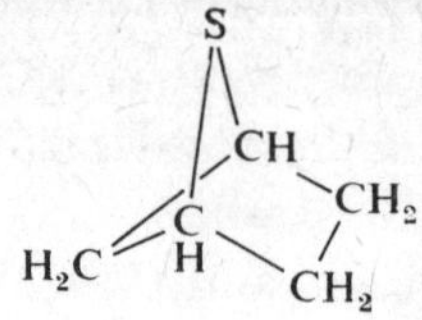
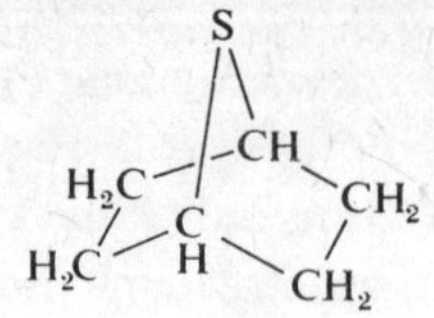

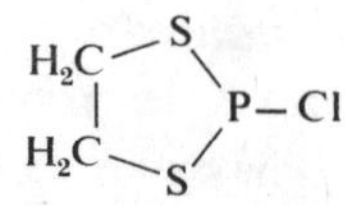

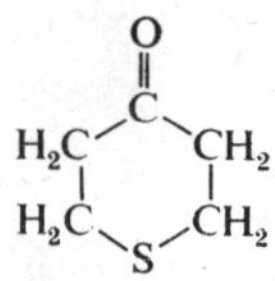
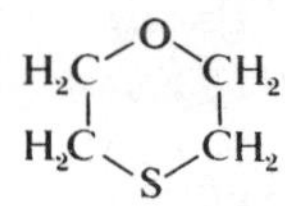
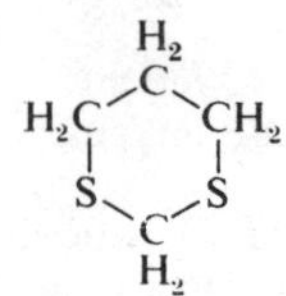
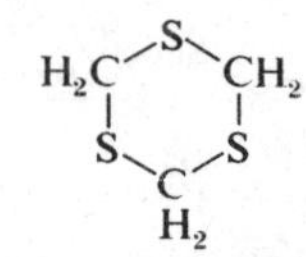

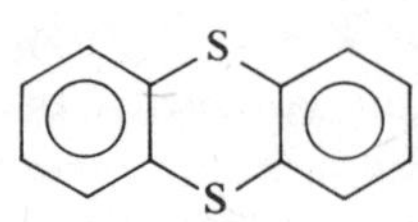

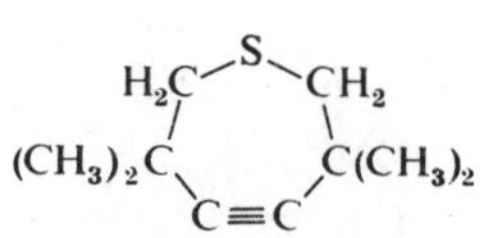

The majority of the available data originate from electron diffraction studies. Because of the compactness of the ring structures, their determination is often hindered by the strong correlation among the parameters. Also, the complicated intramolecular vibrations present problems.

2.5.1. THREE-MEMBERED RINGS

Ethylene sulphide, $(CH_2)_2S$. Its substitution structure (r_s) was determined by microwave spectroscopy (Okiye, Hirose, Lister and Sheridan, 1974). The bond lengths and the bond angles are the following:

r(C—C)	148.4(3) pm	∢C—S—C	48°16′
r(C—S)	181.5(3) pm	∢C—C—S	65°52′
r(C—H)	108.3(2) pm	∢H—C—H	115°50′

The angle between the plane of the CH_2 group and the C—C bond is 151°47′. The data of four isotopic species have been utilized in the determination of this structure. According to Okiye, Hirose, Lister and Sheridan (1974), the above parameters may deviate only by a few tenths of a picometer or a few minutes, respectively, from the parameters characterizing the equilibrium structure (r_e).

Cyclohexene sulphide,

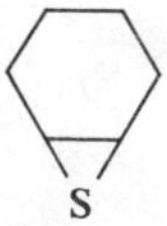

or 7-thiabicyclo[4.1.0]heptane. Only a single isotopic species was examined in its microwave spectroscopic investigation by Kewley (1973). The three available rotational constants provided some basis for a limited amount of information on the conformation. Assuming some reasonable geometrical parameters, utilizing among them the data on ethylene sulphide (from an earlier microwave spectroscopic study by Ikeda, Kewley and Curl, 1972), a twisted half-chair conformation well reproduced the experimental rotational constants. The heavy-atom skeleton of this model is shown in Fig. 40. Cyclohexene sulphide and cyclohexene oxide (Cunningham, Boyd, Myers, Gwinn and Le Van, 1951) were found to have similar conformational properties.

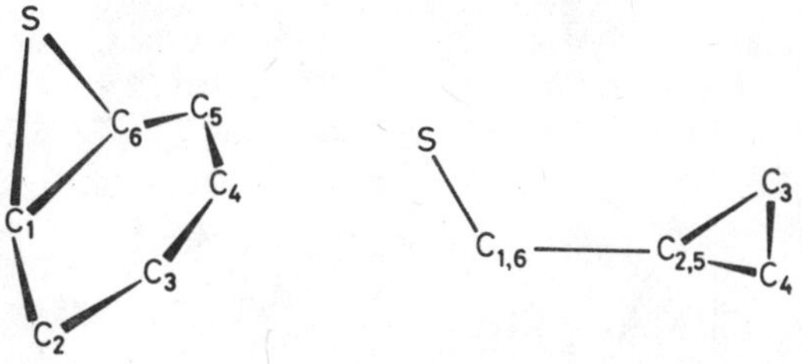

Fig. 40. The arrangement of the heavy-atom skeleton of cyclohexene sulphide after Kewley (1973)

Interesting structural variations can be observed by comparing the molecular geometry of ethylene sulphide with that of other heterocyclopropanes. According to the data of Table 34, some of the heteroatom–carbon atom bonds have some double bond character. In the heteroatoms series C, N, O and Si, P, S, this double bond character is the largest in the case of the least electronegative heteroatom. Another observation concerns the rapid decrease of the carbon–carbon bond length with increasing electronegativity of the heteroatom in the same two series as is also seen in Fig. 41.

TABLE 34
Bond lengths in heterocyclopropanes

X /\\ H_2C—CH_2		r(C—C), pm	r(C—X), pm	r(C—X), pm calculated *(a)*	Δr(C—X), pm
H_2 C /\\ H_2C—CH_2	*(b)* r_a	151.0(2)	151.0(2)	153.4	−2.4
H N /\\ H_2C—CH_2	*(c)* r_s	148.1(3)	147.5(3)	147.5	0.0
O /\\ H_2C—CH_2	*(d)* r_s	146.6(2)	143.1(2)	143.9	−0.8
H_2 Si /\\ H_2C—CH_2	*(e)*	152.0(3)	182.6(3)	190.0	−7.4
H P /\\ H_2C—CH_2	*(f)* r_s	150.2(5)	186.7(5)	185.8	0.9
S /\\ H_2C—CH_2	*(g)* r_s	148.4(3)	181.5(3)	178.7	3.2

(a) $r(A—B) = r_A + r_B - 7.3\,|\chi_A - \chi_B|^2$, Pauling (1960) and Glidewell (1976)
(b) Bastiansen, Fritsch and Hedberg (1964)
(c) Bak and Skaarup (1971)
(d) Hirose (1974)
(e) Delker, Wang, Stucky, Lambert, Haas and Seyfert (1976)
(f) Bowers, Beaudet, Goldwhite and Tang (1969)
(g) Okiye, Hirose, Lister and Sheridan (1974)

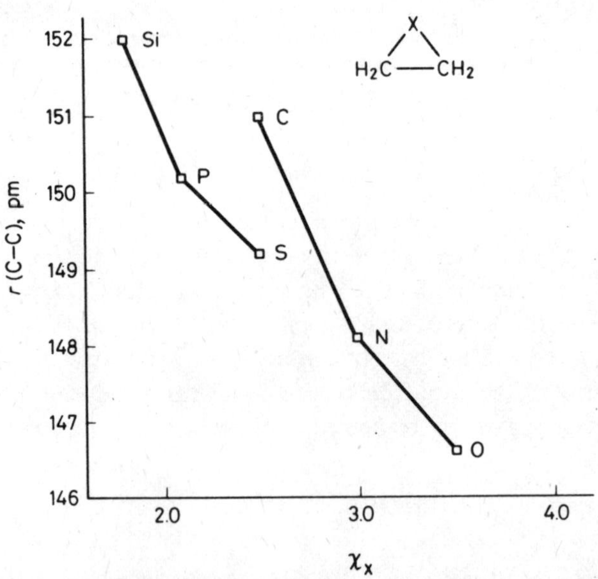

Fig. 41. The C—C bond lengths in heterocyclopropanes and the electronegativities of the heteroatoms

2.5.2. FOUR-MEMBERED RINGS

Trimethylene sulphide, $(CH_2)_3S$. In the electron diffraction structure analysis Karakida and Kuchitsu (1975) utilized the experimental data of an earlier microwave spectroscopic study by Harris, Harrington, Luntz and Gwinn (1966). Already this latter work established that the ring-puckering motion of trimethylene sulphide is described by a double-minimum potential function with 274 cm^{-1} barrier at the planar conformation. The average ring puckering (*cf.* Fig. 42) was found by the electron diffraction analysis to amount to $26 \pm 2°$. The other geometrical parameters are:

r_g(C—S)	184.7(2) pm	$\sphericalangle_{av}$C—S—C	76.8(3)°
r_g(C—C)	154.9(3) pm	$\sphericalangle_{av}$H—C_α—H	112(4)°
r_g(C_α—H)	109(2) pm	$\sphericalangle_{av}$H—C_β—H	114(5)°
r_g(C_β—H)	112(4) pm		

As part of the electron diffraction investigation, the molecular force field has been determined and the mean amplitudes of vibration have been calculated. The latter were in excellent agreement with the electron diffraction results.

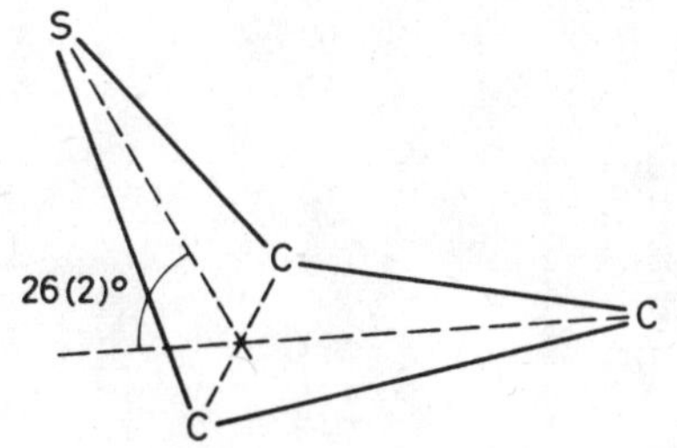

Fig. 42. The puckering of the four-membered ring of trimethylene sulphide

Tetrafluoro-1,3-dithietane, $\underline{SCF_2SCF_2}$. Within a short time two papers reported the results of two independent electron diffraction analyses. The reports by Smith and Seip (1976) and by Chiang and Lu (1977) are in agreement concerning the shape of the molecule. There is, however, serious discrepancy as regards the size of the molecule. The structural results are cited in Table 35. Both data sets were found to be consistent with a model of D_{2h} symmetry, thus the four-membered ring is planar. There is not much difference between the angles determined in the two studies and they are also consistent with the bond angles of crystal-phase tetrachloro-1,3-dithietane

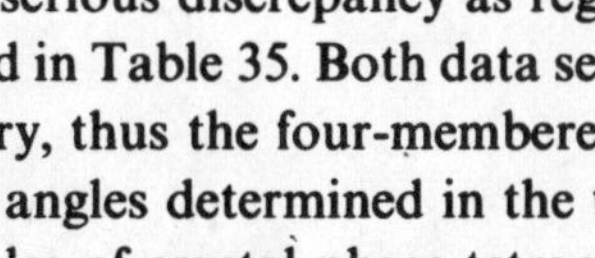

as determined by x-ray diffraction (Krebs and Beyer, 1969), and with the bond angles of 1,3-dithietane

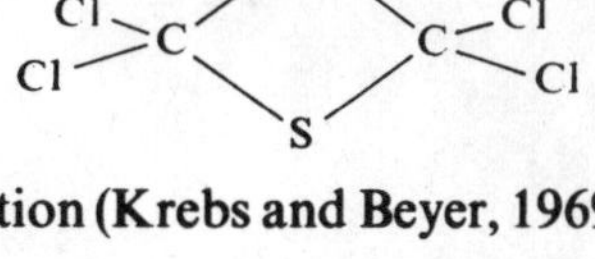

TABLE 35
Structural parameters of 1,3-dithietane and derivatives

	Tetrafluoro-1,3-dithietane gas-phase electron diffraction		Tetrachloro-1,3-dithietane crystal-phase x-ray diffraction	1,3-Dithietane *ab initio* MO calculations; Kao (1977)		Tetrafluoro-1,3-diselenetane electron diffraction	
	Smith and Seip (1976)	Chiang and Lu (1977)	Krebs and Beyer (1969)	STO–3G basis	44–31G basis	Wehrung, Oberhammer, Haas, Koch and Welcman (1976)	
Bonds (pm)							
C—S	182.0(2)	178.5(2)	180.4	181.3	189.8	C—Se	196.8(4)
C—F	134.4(2)	131.4(2)	176.9 (C—Cl)				135.3(3)
Other distances (pm)							
F5...F6	215.4(4)	209.0					215(1)
C...C	240.2(5)	237.1	241.3				261(4)
S...F	265.6(3)	260.8				Se...F	277.9(2)
S...S	272.8(4)	266.9	268.4			Se...Se	298.1(3)
C...F	337.6(5)	333.6					355(1)
F5...F7	399.1(6)	396.5					419(3)
F5...F8	453.8(6)	448.2					466(1)
Angles (°)							
C—S—C	82.7(2)	83.2(3)	83.9	83.8	84.3	C—Se—C	[83]*
F—C—F	106.5(2)	105.3	107.1 (Cl—C—Cl)				106.3(8)
S—C—S	[97.1]*	96.8	96.1	96.2	95.7	Se—C—Se	98.5(4)
S—C—F	[113.3]*	113.7(2)	113.4 (S—C—Cl)			Se—C—F	[112.3]*

* Calculated from the other parameters

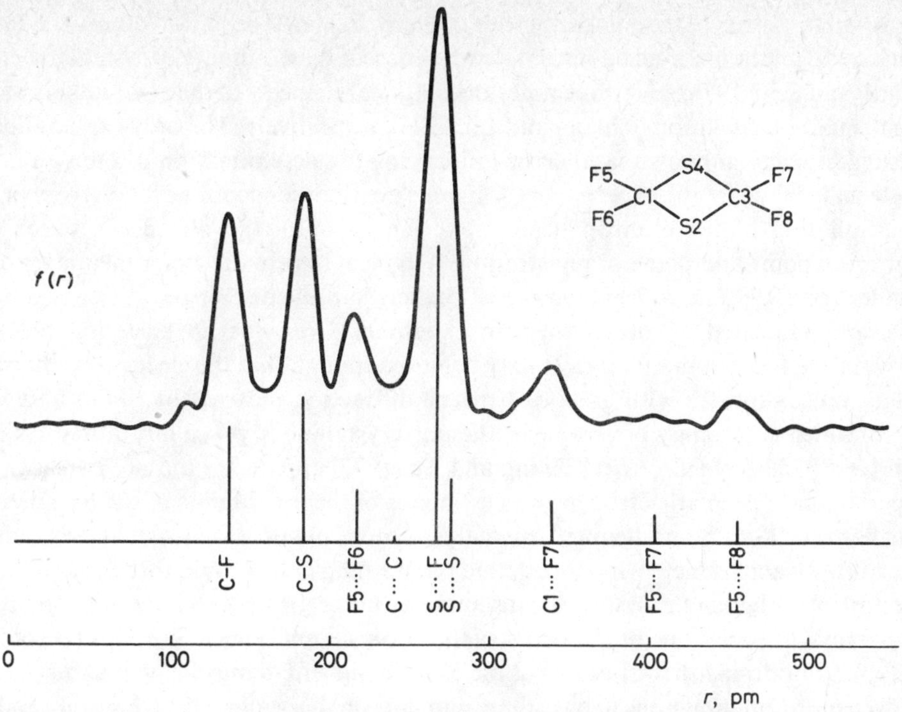

Fig. 43. Radial distribution of tetrafluoro-1,3-dithietane after Smith and Seip (1976) with the assignment of the maxima and the numbering of atoms

as were obtained from *ab initio* molecular orbital calculations by Kao (1977). The discrepancy between the results of the two electron diffraction studies appears in the interatomic distances. Therefore, not only the bond lengths but also the other distances are given in Table 35. The numbering of atoms is shown in Fig. 43 together with a radial distribution and the assignment of its peaks. It is possible to choose five individual maxima on the radial distribution curve each of which originate from the contributions of a single interatomic interaction. Accordingly, the corresponding interatomic distances can be considered to be especially well determined. The ratio of the five interatomic distances, as determined in the two studies cited (Smith and Seip/Chiang and Lu) is as follows:

C—F	1.023
C—S	1.020
F5...F6	1.031
C1...F7	1.012
F5...F8	1.012
Mean	1.020
σ	0.008

Thus there is a 2% systematic deviation on the average between the two sets of distances. This deviation is much larger than the uncertainties cited in Table 35 from the original papers. These uncertainties contain standard deviations from the least-squares calculation corrected for data correlation and an estimate of systematic error (Smith and Seip, 1976), and three times the diagonal elements of the error matrix of the least-squares calculation (Chiang and Lu, 1977), respectively. The only explanation is to suggest some unknown fatal error influencing the determination of the scattering angle scale in one of the experiments. Obvious possibilities could be fatal errors in the determination of the electron beam wave length or of the distance between the scattering point and plane of registration. A typical systematic experimental error is smaller than 0.2‰, i.e. at least ten times smaller than the discrepancy discussed here. The communicated details of the two experiments provide no basis for deciding between the two parameter sets. It may be noted though that the electron beam wave length was calibrated with gaseous benzene diffraction patterns by Smith and Seip (1976) which is certainly preferable to the polycrystalline MgO calibration of Chiang and Lu (1977). Note also that Chiang and Lu (1977) apparently did not consider the experimental systematic error in their estimates of the error limits of the parameters. Furthermore, Smith and Seip (1976) analysed more plates and apparently examined the vibrational parameters in more detail, but nothing in this could throw any light on the possible origin of the systematic discrepancy in the distances. However, it seems to be possible to prefer one of the two structures on stereochemical grounds (Hargittai, 1979b). Considering the variations of the C—S bond length may be helpful but it may be even more unambiguous to base the arguments on the geminal F. . . F distances. For the analogous *tetrafluoro-1,3-diselenetane* (*cf.* Table 35)

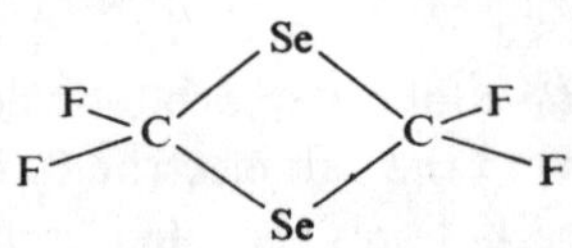

the F. . . F distance was reported to be 215(1) pm from electron diffraction by Wehrung, Oberhammer, Haas, Koch and Welcman (1976). Further, the mean value of F. . . F distances calculated from C—F bond lengths and F—C—F bond angles of 35 molecules containing CF_3 groups as compiled by Yokozeki and Bauer (1975) (see their Table 12) and five more recently studied molecules* is 216.2 pm with a standard deviation of 0.8 pm. The F. . . F distance is never smaller than 214 pm in these 40 molecules. The remarkable stability of the F. . . F distances points to the importance of non-bonded interactions of fluorine atoms. The mean *r*(F. . . F) of 216 pm is in excellent agreement with the fluorine 1,3 non-bond radius of 108 pm postulated by Bartell (1960) many years ago (*cf.* Table 2). The 1,3 F. . . F distance in tetrafluoro-1,3-dithietane as reported by Smith and Seip (1976) is consistent with this value. On the other hand, the value of 209 pm (Chiang and Lu, 1977) seems to be far too low considering the other

* CF_3OOF, CF_3OOH, CF_3OOCl (Marsden, DesMarteau and Bartell, 1977), CF_3OOCF_3 (Marsden, Bartell and Diodati, 1977), CF_3SO_2Cl (Brunvoll, Hargittai and Kolonits, 1978).

data on F...F distances including that in tetrafluoro-1,3-diselenetane. Concluding, the considerations of the F...F non-bonded interactions render the size of the tetrafluoro-1,3-dithietane molecule, as determined by Smith and Seip (1976), to be preferred. The electron diffraction analysis by Smith and Seip (1976) was paralleled by a vibrational spectroscopic investigation of tetrafluoro-1,3-dithietane by Klaboe and Smith (1978).

The controversy concerning the size of the tetrafluoro-1,3-dithietane molecule cannot diminish the structural interest of the planarity of its four-membered ring. The four-membered rings of cyclobutane and its substituted derivatives are non-planar unless they are attached to groups influencing the ring conformation otherwise (*cf.*, e.g., Bastiansen, Seip and Boggs, 1971 and references therein). The conformational properties of four-membered rings containing heteroatoms are indeed very diverse.

There are two opposing effects influencing the conformation of the four-membered ring. One is the so-called Bayer-strain which is smallest when the bond angles are nearest to their "natural" values, for example the ideal tetrahedral angle being the "natural" value for the C—C—C bond angle. From this point of view planar arrangement of the ring would be the most favourable for cyclobutane. This planarity, however, would result in great repulsions between the hydrogen atoms of neighbouring methylene groups (Pitzer-strain). These repulsions thus lead to a non-planar conformation. Kao (1977) determined a planar four-membered ring of 1,3-dithietane (1,3-dithiacyclobutane) from *ab initio* calculations and interpreted the difference between the conformational behaviour of cyclobutane and the 1,3-dithia derivative by means of the above effects. It was argued that the changes in the bond angles, accompanying the ring puckering, lead to a larger amount of strain in the systems containing a sulphur atom than in the hydrocarbons (for relevant references see Kao, 1977). This effect, i.e. the Bayer-strain would then suggest a less puckered ring for the sulphur derivative as compared with cyclobutane. As regards the Pitzer-strain, i.e. the strain caused by the eclipsing position of the hydrogen atoms, this is expected to be smaller in the —CH_2—S— than in the —CH_2—CH_2— bonding system. Of course, from these qualitative considerations it is expected that the ring-puckering is smaller in 1,3-dithietane than in cyclobutane itself. It is not possible, however, to predict the exact amount of this difference. Recent infrared and Raman matrix-isolation spectroscopic studies of Kalasinsky, Block, Powers and Harris (1979) concluded C_{2v} molecular symmetry for 1,3-dithietane with a puckered four-membered ring. The spectra of the annealed solid, on the other hand, were consistent with D_{2h} molecular symmetry. Earlier the lack of absorption lines in the microwave spectra was interpreted as a result of a planar or nearly planar ring structure (Block, Corey, Penn, Renken and Sherwin, 1976). It may well be that 1,3-dithietane is another quasi-planar molecule with low-frequency large-amplitude out-of-plane motion and double-minimum potential whose conformation and intramolecular motion certainly calls for further theoretical as well as experimental studies.

3-Oxo-3-chloro-1,3-thiaphosphetane, $CH_2SCH_2P(O)Cl$. In its electron diffraction structure analysis (Naumov and Semashko, 1971) it was difficult to distinguish between the two conformations which differ in the orientation of the exocyclic ligands of the phosphorus atom.

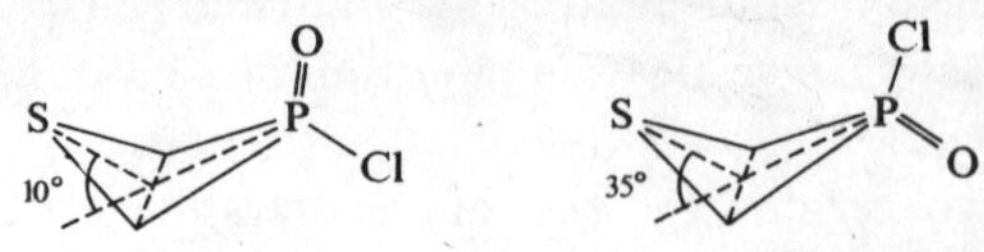

(*a*) (*b*)

Finally, the experimental data could be best approximated by a four-to-one mixture of models (a) and (b). Two different puckering angles were established in the two models. Unfortunately, the cited paper did not discuss the vibrational parameters (except those for the bonds) which are so important in the conformational analysis. The above statements on the conformational properties of 3-oxo-3-chloro-1,3-thiaphosphetane should probably be considered with caution. As for the geometrical parameters, the following bond lengths and bond angles are given here after Naumov and Semashko (1971):

r(P—C)	183.6(15) pm	∢C—P—C	86(2)°
r(C—S)	179.6(15) pm	∢O=P—Cl*	118(3)°
r(P—Cl)	203.2(15) pm	∢C—P—Cl*	103(3)°
r(P=O)	144.1(10) pm		

It is noteworthy that detailed spectroscopic studies were performed by Arbuzov, Remizov and Nuretdinova (1973) on 3-chlorothietane, $CH_2CHClCH_2S$, extending to all three phases. It was established that (*i*) there is only one conformer and it is the same in all phases, (*ii*) this conformer cannot be a form with non-planar ring and axial chlorine, and (*iii*) the stable

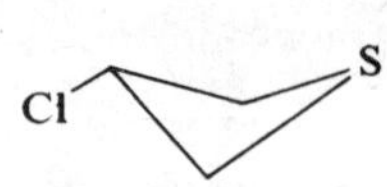

conformer probably has a non-planar ring and an equatorial chlorine, but it cannot be distinguished from the form with a planar ring on the basis of the available experimental data.

1,2-Bis(trifluoromethyl)dithiete. Its electron diffraction structure analysis by Hencher, Shen and Tuck (1976) unambiguously established the four-membered dithiete ring structure (**I**) *vs.* the acyclic dithione arrangement (**II**)

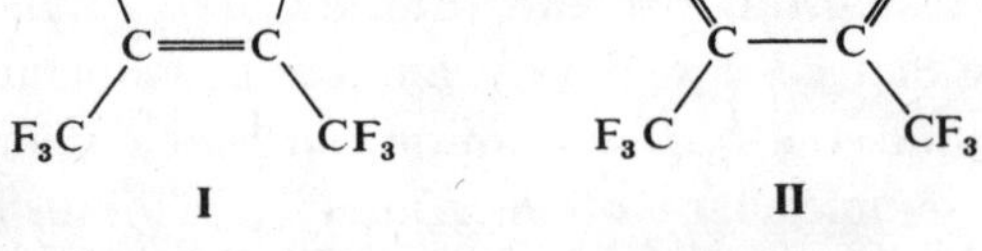

I **II**

* In the original paper (Naumov and Semashko, 1971) there are probably some misprints in the designation of the bond angles. They are listed here as thought to be correct.

The molecule as a whole was assumed to have C_2 symmetry with local C_{3v} symmetry for the CF_3 groups. The four-membered ring was found to be planar within the experimental error. Some of the more important geometrical parameters:

r(S—S)	205(1) pm	∢C=C—S	100.8(6)°
r(C—S)	173(1) pm	∢C=C—C	122.9(7)°
r(C=C)	140(3) pm	∢C—C—F	110.8(6)°
r(C—C)	150(1) pm		
r(C—F)	132.6(3) pm		

The rather long S—S bond indicates no appreciable participation in the electron delocalization. Hencher, Shen and Tuck (1976) compared the structure of the free 1,2-bis(trifluoromethyl)dithiete with that of its transition metal complex derivatives. They concluded that the sulphur–sulphur linkage is broken in the process of complexation and the molecule acts as a bidentate ligand. The rest of the complex structures show no appreciable differences from the structure of the free molecule.

2.5.3. FIVE-MEMBERED RINGS

Tetrahydrothiophene and *tetrahydroselenophene*. Their geometrical parameters are presented in Table 36 from electron diffraction (Náhlovská, Náhlovský and Seip, 1969 and 1970, respectively) and in case of the seleno derivative from microwave

TABLE 36

Geometrical parameters of tetrahydrothiophene and tetrahydroselenophene

Parameter A = S or Se	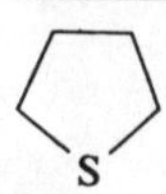 (b)	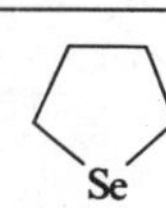 (b)	(c)
r(C—A), pm	183.9(2)	197.5(3)	196.3(2)
r(C2—C3), pm *(a)*	} 153.6(2)	} 153.8(4)	154.9(3)
r(C3—C4), pm *(a)*			152.7(2)
∢C—A—C	93.4(5)°	89.1(5)°	90°44′ ± 7′
∢A—C—C	106.1(4)°	105.8(3)°	104°58′ ± 10′
∢C—C—C	105.0(5)°	106.0(7)°	106°52′ ± 12′
∢CC—AC	14.8(5)°	15.4(5)°	
∢CC—CA	−40.4(12)°	−42.7(14)°	
∢CC—CC	52.5(16)°	56.9(17)°	

(a) The numbering of atoms starts with atom A

(b) Electron diffraction, r_a parameters, Náhlovská, Náhlovský and Seip (1969 and 1970)

(c) Microwave spectroscopy, r_s parameters, 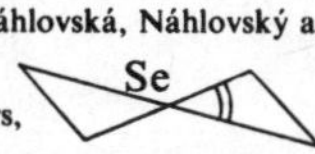

the angle marked was given as 29°44′ ± 14′, Mamleev, Pozdeev and Magdesieva (1976)

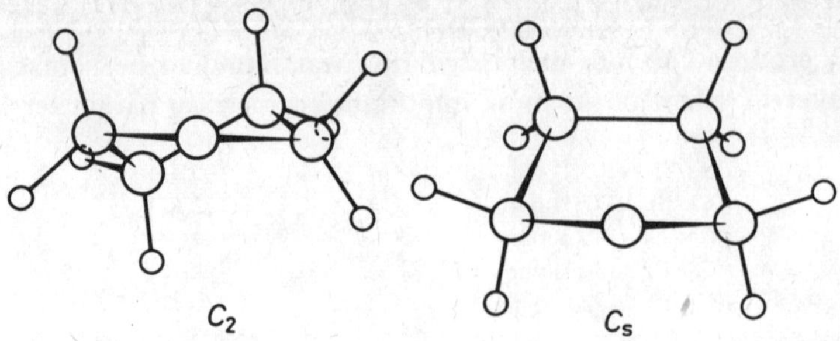

Fig. 44. C_2 symmetry "half-chair" and C_s symmetry "envelope" conformations of tetrahydrothiophene and tetrahydroselenophene

spectroscopy as well (Mamleev, Pozdeev and Magdesieva, 1976). An important merit of the microwave spectroscopic results is that they distinguish between the two different carbon–carbon lengths while the electron diffraction analysis could yield only mean parameters. Both tetrahydrothiophene and tetrahydroselenophene have well-defined rotational forms, thus they differ from the structures of cyclopentane (Adams, Geise and Bartell, 1970) or tetrahydrofuran (Almenningen, Seip and Willadsen, 1969; Geise, Adams and Bartell, 1969) characterized by pseudorotation. All the experimental

TABLE 37

Geometrical parameters of 5-thiabicyclo[2.1.1]hexane *(a)* and 7-thiabicyclo[2.2.1]heptane *(b)* after Fukuyama, Oyanagi and Kuchitsu (1976)

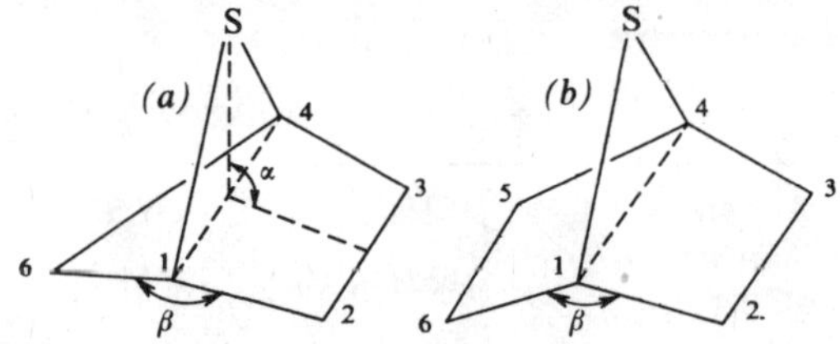

Parameter	(a)	(b)
r(C1—C2), pm	} 153.9(16)	153.8(6)
r(C2—C3), pm		156.1(15)
r(C1—C6), pm	156.4(24)	
r(C—S), pm	186.5(4)	183.7(4)
r(C—H), pm	111.0(8)	111.5(8)
∡C—S—C (°)	69.5(6)	80.2(8)
α (°) *(c)*	115.2(17)	
β (°) *(c)*	116.1(24)	115.0(12)

(a) r_g bond distances, r_α bond angles
(b) r_g bond distances, r_{av} bond angles
(c) Dihedral angles

studies as well as molecular mechanics calculations (Náhlovská, Náhlovský and Seip, 1969; 1970) established the C_2 symmetry conformer to be the most stable. The conformer with C_s symmetry could not be brought into agreement with the experimental data. The two conformers are shown in Fig. 44. The C_2 form is more stable than the C_s form by 8 to 12 kJ mol^{-1} for tetrahydrothiophene according to molecular mechanics calculations. This difference is 14 to 16 kJ mol^{-1} for tetrahydroselenophene.

5-Thiabicyclo[2.1.1]hexane and *7-thiabicyclo[2.2.1]heptane* (or 7-thianorbornane). Their structural formulae are illustrated in Table 37 along with their geometrical parameters as obtained from the electron diffraction work of Fukuyama, Oyanagi and Kuchitsu (1976). The C—S bond of 5-thiabicyclo[2.1.1]hexane is particularly long, it is longer by about 3 pm than that in other five-membered rings (*cf.* Fig. 45) which themselves have longer C—S bonds than, e.g., dimethyl sulphide by some 3 pm. A similar observation can be made for the analogous oxygen derivatives, *viz.* their C—O

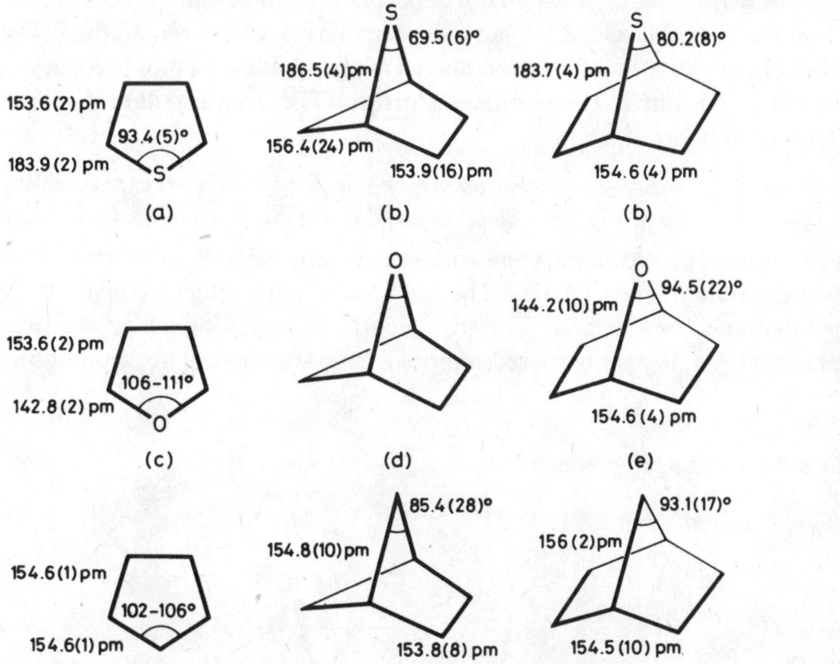

Fig. 45. Comparison of the geometry of five-membered ring molecules

(a) Náhlovská, Náhlovský and Seip (1969)
(b) Fukuyama, Oyanagi and Kuchitsu (1976)
(c) Almenningen, Seip and Willadsen (1969); Geise, Adams and Bartell (1969)
(d) Not yet studied
(e) Oyanagi, Fukuyama, Kuchitsu, Bohn and Li (1975)
(f) Adams, Geise and Bartell (1970)
(g) Chiang (1971)
(h) Yokozeki and Kuchitsu (1971a)

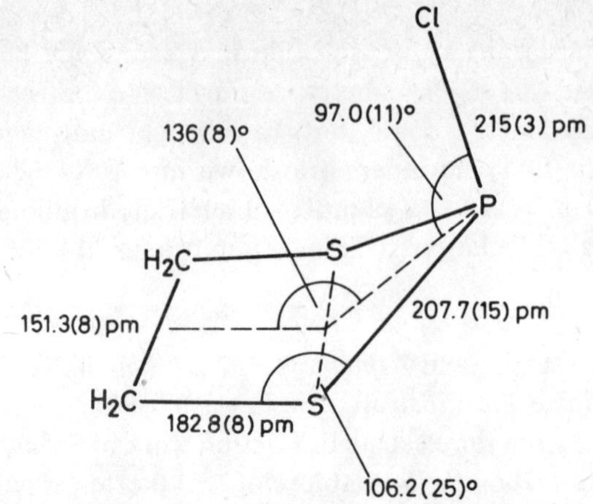

Fig. 46. The envelope conformation of 2-chloro-1,3-dithia-2-phospholane with axial P—Cl bond and the most important geometrical parameters (Schultz, Hargittai, Martin and Robert, 1974)

bonds (*cf.* Fig. 45) are longer than the C—O bond of dimethyl ether (141 pm, see Table 30). Considering all this, it would not be surprising to determine a particularly long C—O bond in 5-oxabicyclo[2.1.1]hexane which has not yet been studied. The carbon–carbon bond lengths of five-membered rings containing no heteroatom are not different to within the experimental errors. The relevant data are collected for comparison in Fig. 45.

2-Chloro-1,3-dithia-2-phospholane, $\underline{CH_2CH_2SP(Cl)S}$. The experimental electron diffraction data (Schultz, Hargittai, Martin and Robert, 1974) were consistent with a phosphorus-edge envelope conformation having axial P—Cl bond. Other forms, however, could not be excluded. The geometrical parameters are given in Fig. 46. The uncertainties are quite large. The strong correlation among the geometrical and also vibrational parameters hindered a more accurate structure determination.

2.5.4. SIX-MEMBERED RINGS

4-Thiacyclohexanone,

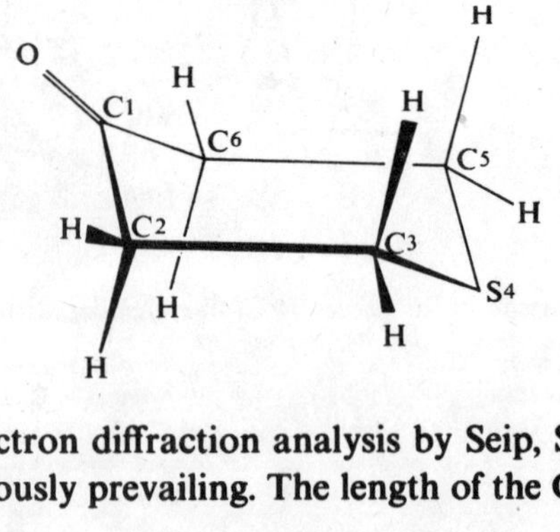

According to the electron diffraction analysis by Seip, Seip and Smith (1976) the chair form is unambiguously prevailing. The length of the C—S bonds is 180.4(3) pm and the C—S—C bond angle is 97(2)°. The angles of rotation are the following:

C6—C1—C2—C3	50.0(15)°
C1—C2—C3—S4	57.1(16)°
C2—C3—S4—C5	58.0(29)°

Seip, Seip and Smith (1976) have performed molecular mechanics calculations which also pointed to the stability of the chair form. In fact, this model with C_s symmetry was found to be more stable than the model with C_2 symmetry by some 14 kJ mol^{-1}. There is an abundance of different results and opinions concerning the conformation of 1,4-cyclohexanedione (see Dowd, Dyke and Klemperer, 1970 and references therein). The most obvious possibilities are the chair form (a) and a twist boat form (b)

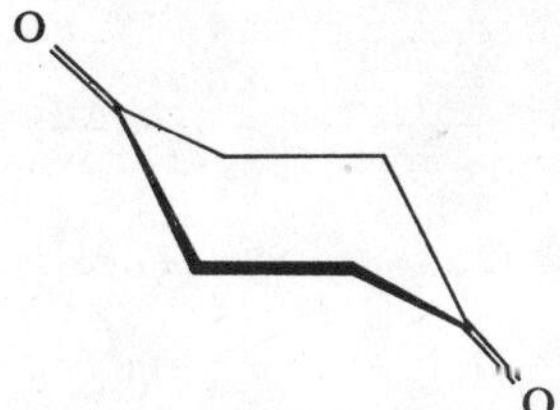

(a)

(b)

Dowd, Dyke and Klemperer (1970) established that the molecule was non-polar by the molecular beam electric deflection method. This observation is consistent with both models (a) and (b). Discussing stereochemical analogs, Dowd, Dyke and Klemperer (1970) gave preference to the chair model. This would suggest close similarity between the conformational behaviour of 4-thiacyclohexanone and 1,4-cyclohexanedione.

1,4-Oxathiane,

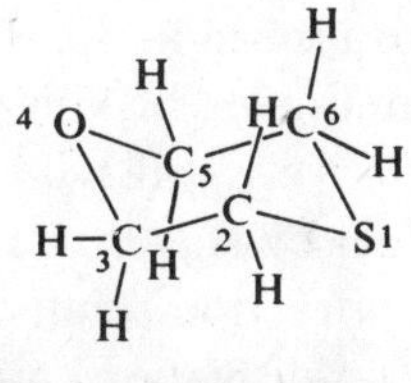

The results of an electron diffraction analysis by Schultz, Hargittai and Hermann (1972) are given in Table 38. Subsequently these results received confirmation from the microwave spectroscopic investigation by Kitchin, Avirah, Malloy and Cook (1975). Schultz, Hargittai and Hermann (1972) unambiguously determined the chair form to be prevailing in the vapour phase. The molecular conformation was characterized by the following angles of rotation:

C6—S1—C2—C3	51.2(28)°
S1—C2—C3—O4	60.1(12)°
C2—C3—O4—C5	63.5(20)°

According to the electron diffraction data, no appreciable amount of other conformers

TABLE 38
Geometrical parameters of 1,4-oxathiane

Parameter	Electron diffraction *(a)*		Microwave spectroscopy (fitted parameters) *(b)*
	Parameter	Uncertainty	
r(C—S), pm	182.6	0.4	182.9
r(C—O), pm	141.8	0.4	142.0
r(C—C), pm	152.1	0.6	152.6
r(C—H), pm	109.3	0.4	109.5
∢C—S—C (°)	97.1	2.0	95.7
∢C—O—C (°)	115.1	2.2	115.6
∢H—C—H (°)	124.2	3.5	120.9
τ (°) *(c)*	−55.9	3.5	−56.4
β (°) *(c)*	+46.9	2.0	+48.6

(a) Schultz, Hargittai and Hermann (1972)
(b) Kitchin, Avirah, Malloy and Cook (1975)
(c) The angles β (next to atom S) and τ after Kitchin, Avirah, Malloy and Cook (1975):

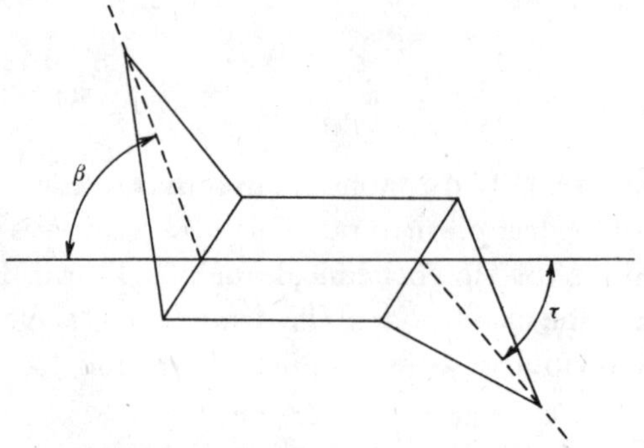

was present (the nozzle temperature applied in the electron scattering experiment was about 60 to 80 °C). The mean amplitudes of vibration determined in the electron diffraction analysis suggested a relatively stiff six-membered ring for 1,4-oxathiane. The calculated vibrational amplitudes by Hagen, Hargittai and Schultz (1975) were in good agreement with the electron diffraction results. Spectroscopic calculations were also carried out for 1,4-dioxane, 1,4-dithianc, 1,4-diselenane, 1,4-oxaselenane, 1,4-oxatellurane, and 1,4-thiaselenane in the same investigation. The relative stiffness of the rings of all these substances is indicated by the small values of the mean vibrational amplitudes associated with interactions between the two heteroatoms in the ring (6–8 pm) and their practically zero perpendicular vibrational correction terms.

The excellent agreement between the geometry from electron diffraction and the rotational constants from microwave spectroscopy on 1,4-oxathiane is well demonstrated by the very small deviations between the calculated and observed rotational constants

ΔA	11 MHz	(or 0.25%)
ΔB	22 MHz	(or 0.72%)
ΔC	4 MHz	(or 0.20%)

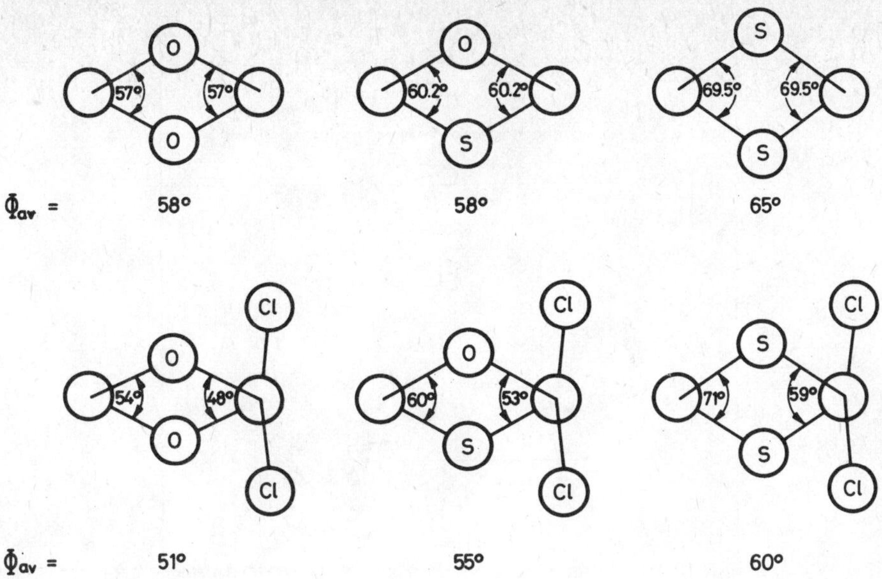

Fig. 47. Dihedral angles and average puckering angles of vapour-phase 1,4-dioxane, 1,4-oxathiane, and 1,4-dithiane and their crystal-phase *trans*-2,3-dichloro derivatives (*cf.* Romers, Altona, Buys and Havinga, 1969)

This agreement refers to the two independent sets of data from the two techniques without applying any fitting procedure. As a matter of fact it is not meaningful to expect any agreement better than, say, 1% since, as a consequence of the difference in the averaging procedure over the intramolecular vibrations the electron diffraction and microwave spectroscopic results correspond to somewhat different physical meanings. In spite of this, presumably because the already excellent agreement arose their interest, Kitchin, Avirah, Malloy and Cook (1975) even fitted the geometrical parameters to the experimental rotational constants. The parameters exactly reproducing these rotational constants are cited in Table 38. Here it must be emphasized, however, that this set of parameters may be only one of slightly different parameter sets, exactly reproducing the rotational constants.

The results on the 1,4-oxathiane structure provided further evidence for the observation by Romers, Altona, Buys and Havinga (1969) concerning the puckering of some non-aromatic heterocycles. According to this observation, as illustrated by Fig. 47, the presence of substituents on the ring carbon atoms decrease the puckering of six-membered rings containing heteroatoms in 1,4-positions. The structural data on *trans*-2,3-dichloro derivatives refer to the crystalline phase, thus the question remains whether and how much these molecular geometries are influenced by intermolecular interactions.

1,3-Dithiane, $\underline{SCH_2SCH_2CH_2CH_2}$. Its chair conformation and geometrical parameters were determined by electron diffraction (Adams and Bartell, 1977). Some

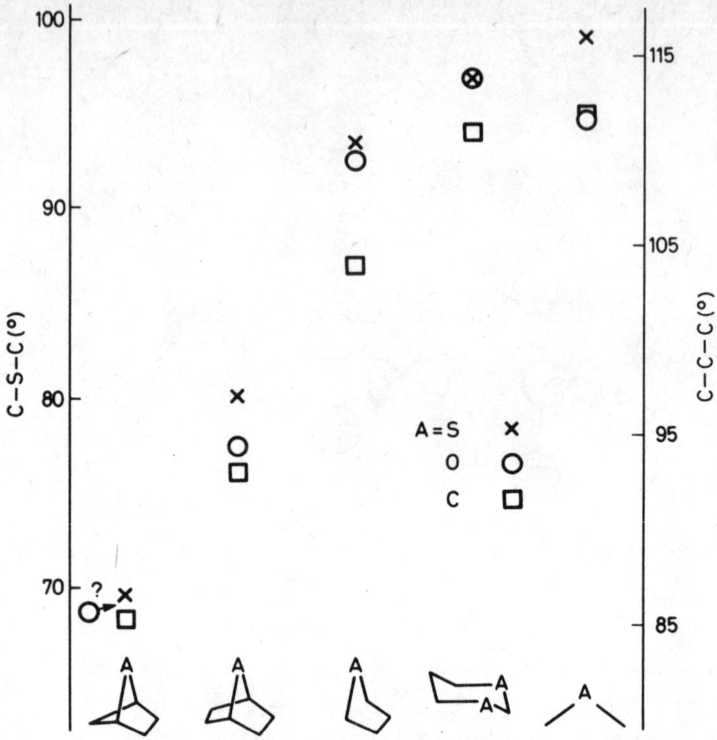

Fig. 48. Comparison of C—S—C, C—O—C, and C—C—C bond angles in three series of analogous molecules, following, in part, Kuchitsu (1972; 1976)

bond lengths (r_g) and bond angles:

r(C—S)	181.2(3) pm	∢C—S—C	98.1(7)°
r(C—C)	153.3(5) pm	∢S—C—S	115.0(3)°
	∢S—C—C	114.9(4)°	

The angles of rotation were reported as follows:

S—C—S—C	61.4°
C—S—C—C	58.3°
S—C—C—C	64.1°

Augmenting the observations of Adams and Bartell (1977) it is noted that the C—S—C bond angle of 1,3-dithiane is smaller than that of the strain-free dimethyl sulphide, and is larger than that of the even more strained five-membered ring systems. This is illustrated in Fig. 48 following in part Kuchitsu (1972) and extending similar observation for the oxygen-derivatives. The C—S as well as the C—O bond lengths are compared in Fig. 49. It is seen that the opening of the bond angles is accompanied by a shortening of the bonds as going towards the strain-free acyclic molecules. On the basis

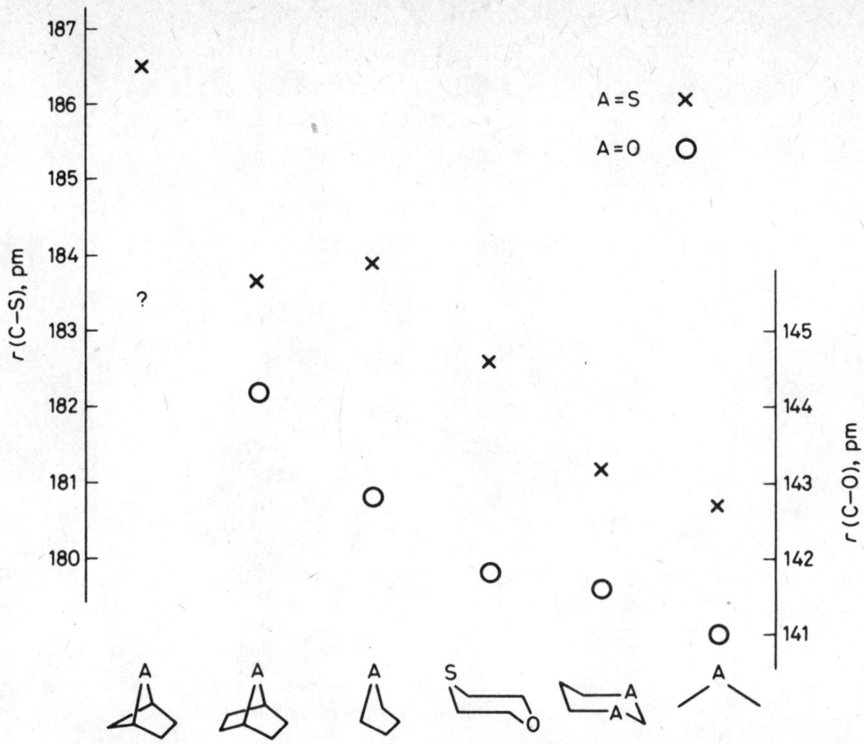

Fig. 49. Comparison of C—S and C—O bond lengths in two series of analogous molecules

of the available structural data on other molecules and considering merely the geometrical variations, it is tempting to predict the C—O bond length and the C—O—C bond angle of 5-oxabicyclo[2.1.1]hexane whose structure has not been investigated to date:

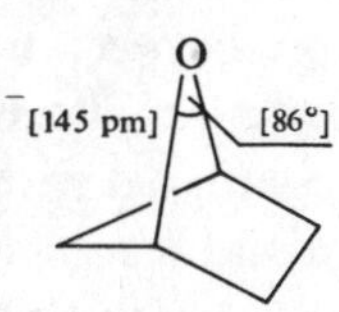

Pauling (1949) gave the following exact relationship between the average angle of rotation (Φ_{ave}) and the average bond angle (Θ_{ave}) of cyclohexane:

$$\cos \Phi_{ave} = -\cos \Theta_{ave}/(1+\cos \Theta_{ave})$$

It was then shown by Geise, Altona and Romers (1967) that this relationship held approximately for other six-membered chair-form rings as well. The average angles of rotation are collected in Fig. 50 for several six-membered rings containing sulphur or oxygen as heteroatoms. In obtaining these average angles, the above approximate relationship or the reported angles of rotation were utilized. As discussed by Adams

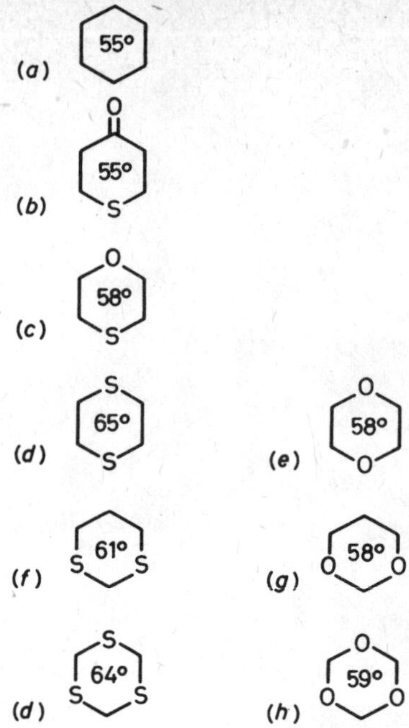

Fig. 50. Characterization of the conformation of six-membered rings of non-aromatic molecules by the average angle of rotation. Sources of data used:

(a) Bastiansen, Fernholt, Seip, Kambara and Kuchitsu (1973)
(b) Seip, Seip, and Smith (1976)
(c) Schultz, Hargittai and Hermann (1972)
(d) Hassel and Viervoll (1947)
(e) Bastiansen, Seip and Boggs (1971)
(f) Adams and Bartell (1977)
(g) Schultz and Hargittai (1974)
(h) Clark and Hewitt (1971)

and Bartell (1977) the average angle of rotation characterizes the ring puckering and is an important parameter determined by the relative magnitude of the interactions between bonded atoms as well as those between non-bonded ones. According to the average angles of rotation, the introduction of a heteroatom into the six-membered ring increases the ring puckering as compared with the conformation of cyclohexane (Bastiansen, Fernholt, Seip, Kambara and Kuchitsu, 1973). This effect is seen already as oxygen is introduced and is enhanced with the participation of sulphur.

Diphenylene disulphide (thianthrene),

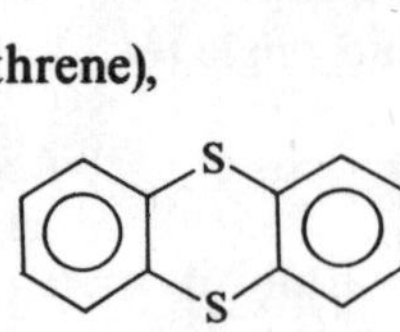

Gallaher and Bauer (1975) determined its molecular structure by electron diffraction. The experimental data were consistent with a model of C_{2v} symmetry with undistorted benzene rings and planar $C_6H_4S_2$ moieties. The latter feature is not a consequence of

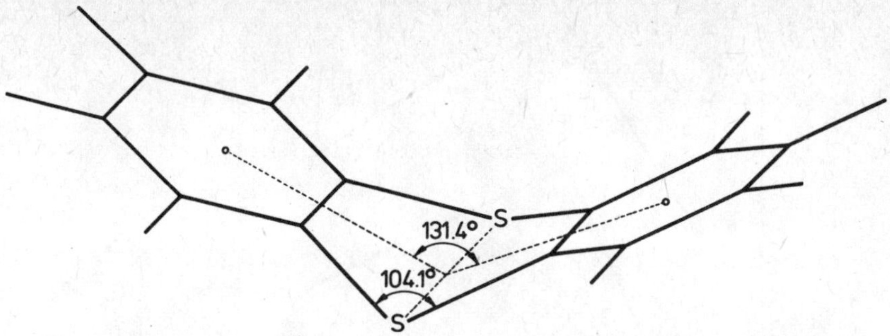

Fig. 51. The molecular model and some geometrical parameters of diphenylene disulphide after Gallaher and Bauer (1975)

the C_{2v} symmetry. The two $C_6H_4S_2$ halves are folded along the S...S axis. The molecular model is shown in Fig. 51.

The results of Gallaher and Bauer (1975) may indicate that the thianthrene molecules perform large-amplitude "butterfly" motion around the S...S axis. The apparently complicated intramolecular motion hindered the accurate and unambiguous determination of the C—S—C bond angle. A value of 104° was obtained for this angle as averaged over the intramolecular motion, while probably a smaller (ca. 102°) angle would correspond to the minimum of the potential energy. Even this means a larger C—S—C angle than that in the strain-free dimethyl sulphide. From this Gallaher and Bauer (1975) drew conclusion on the effects from a possible repulsion between the two benzene rings. At the same time the C—S bond of thianthrene (177.0 ±0.3 pm) is much shorter than that of dimethyl sulphide or the C—S bonds in six-membered ring systems. Similarly, the C—S and C—O bonds of the crystal-phase phenoxathiin (Hosoya, 1966a)

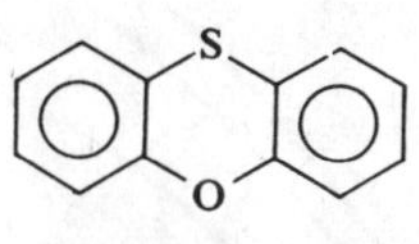

viz. 175 pm and 139–140 pm, respectively, are considerably shorter than the analogous bonds in 1,4-oxathiane.

3,3,6,6-Tetramethyl-1-thiacycloheptyne,

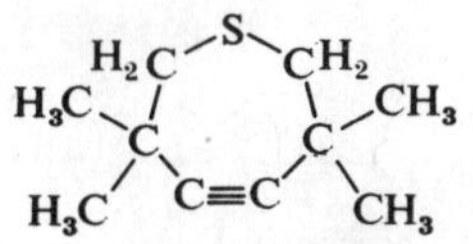

is the only seven-membered ring system containing sulphur whose vapour-phase molecular geometry has been elucidated (Haase and Krebs, 1972). The electron

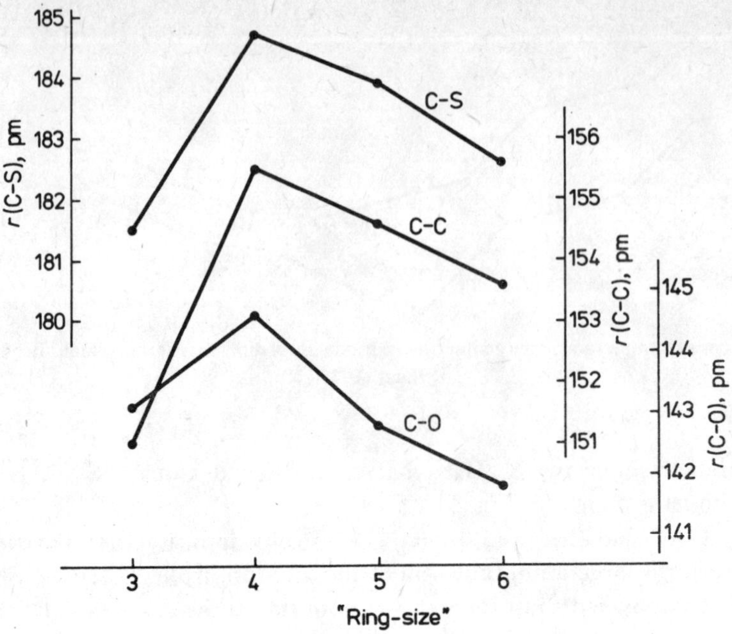

Fig. 52. Variations of the C—S, C—C and C—O bond lengths with increasing ring-size (from three-membered rings to six-membered rings)

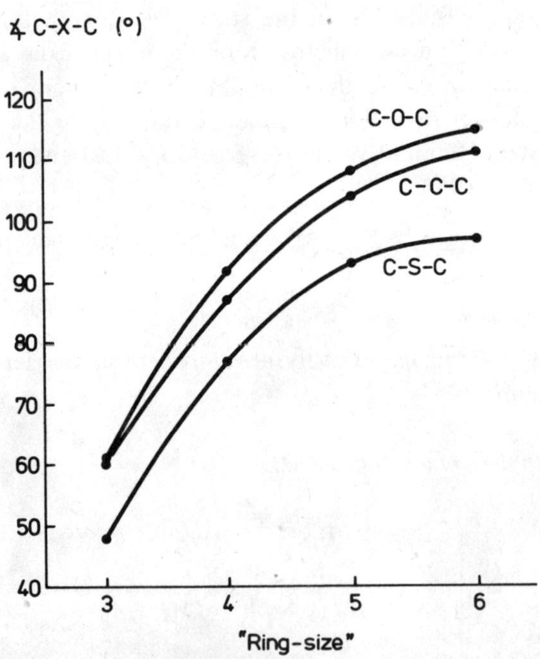

Fig. 53. Variations of the heteroatom bond angles with increasing ring-size (from three-membered rings to six-membered rings) following, in part, Mastryukov, Osina and Vilkov (1975)

diffraction results characterized the sulphur bond configuration with the following data: r(C—S) 181.6(9) pm and ∢C—S—C 104(2)°.

Concluding the discussion of the structures of non-aromatic heterocycles containing sulphur, the C—S bond lengths and C—S—C bond angles are compared with increasing ring size in Figs. 52 and 53, respectively. In complete agreement with the observations by Kuchitsu (1972; 1976) and Mastryukov, Osina and Vilkov (1975), the longest C—S bond occurs in the four-membered ring. As for the bond angle, it gradually opens as going from the three-membered ring towards the six-membered ring. These features appear in other molecule series as well, and the data on carbocyclic and oxygen-containing heterocyclic molecules are also demonstrated in Figs. 52 and 53.

2.6. AROMATIC RING MOLECULES

The molecules discussed in this section may be looked upon having an aromatic ring in which a —CH=CH— group is substituted by a sulphur atom. As was stated by Párkányi and Herndon (1978), the C—S bonds are 170–178 pm long in these aromatic rings and are rather insensitive to the other structural changes of the heterocycle. This means that the sulphur may not participate to any considerable extent in the delocalized π-electron system and, accordingly, these heterocycles may be less aromatic than generally supposed. Although the aromatic character may be formal, all molecules discussed in this section certainly have one common peculiarity, *viz.* their five-membered rings are unambiguously planar according to the experimental data.

TABLE 39
From the geometrical parameters of thiophene, selenophene and tellurophene

Parameter A = S, Se or Te	 S		 Se	 Te
	(a)	*(b)*	*(c)*	*(d)*
r(C—A), pm	171.7(4)	171.4(2)	185.5(2)	205.5
∢C—A—C (°)	92.0(5)	92.2(2)	87.8(2)	82.5
r(C—C), pm	142.9(20)	142.3(3)	143(5)	
r(C=C), pm	136.8(6)	137.0(3)	136.9(3)	

(a) Electron diffraction, r_a parameters, Bonham and Momany (1963); Harshbarger and Bauer (1970)
(b) Microwave spectroscopy, r_s parameters, Bak, Christensen, Hansen-Nygaard and Rastrup-Andersen (1961)
(c) Microwave spectroscopy, r_s parameters, Pozdeev, Akulinin, Shapkin and Magdesieva (1970)
(d) Microwave spectroscopy, r_0 parameters, Brown and Crofts (1973)

Thiophene. Its molecular geometry was repeatedly determined by electron diffraction (Bonham and Momany, 1963; Harshbarger and Bauer, 1970) and by microwave spectroscopy (Bak, Christensen, Hansen-Nygaard and Rastrup-Andersen, 1961). The structural parameters (given in Table 39) originating from the two techniques are in excellent agreement. Since these parameters are subject to different averaging over the intramolecular vibrations, this agreement is an indirect indication that the thiophene molecule is relatively stiff. A complete substitution structure is available for *selenophene* as well (Pozdeev, Akulinin, Shapkin and Magdesieva, 1970). On the other hand, the bond configuration of tellurium in *tellurophene* could only be determined by assumptions for other parameters of this molecule (Brown and Crofts, 1973). The parameters of selenophene and tellurophene are given also in Table 39.

2-Chlorothiophene(**1**) and *2-bromothiophene*(**2**),

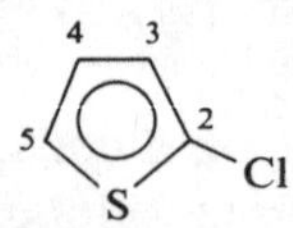

1

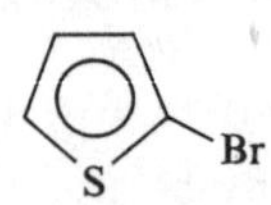

2

According to the electron diffraction analysis by Harshbarger and Bauer (1970), the C_{2v} symmetry of the thiophene ring is distorted in these molecules. It is true there has been an attempt to interpret the data* by a C_{2v} symmetry ring; this could be achieved by utilizing vibrational amplitudes that seem to be rather unrealistic (Derissen, Kocken and van Weelden, 1971). Further refinement of the original experimental data by Karl and Bauer (1972) yielded, among others, the following parameters:

	(1)	**(2)**
r(C2—S), pm	172.2(4)	174.9(8)
r(C5—S), pm	171.7(5)	166.8(9)
∢C—S—C (°)	91.0(2)	91.6(5)
∢S—C2—C3 (°)	112.9(3)	110.7(10)
∢S—C5—C4 (°)	111.0(4)	115.2(11)
r(C—Cl, Br), pm	171.3(4)	184.5(8)

2-Nitrothiophene and *3-nitrothiophene*,

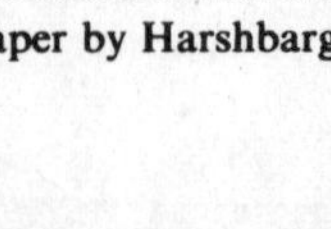

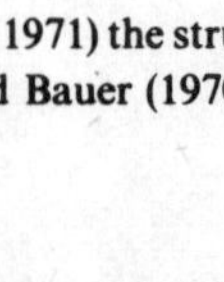

* In this attempt (Derissen, Kocken and van Weelden, 1971) the structure refinement was carried out on data taken from figures of the paper by Harshbarger and Bauer (1970).

The microwave spectra of both molecules (Ralowski and Ljunggren, 1977 and Isvén, Ralowski and Ljunggren, 1977, respectively) provided evidence for planar molecular structures. The barriers hindering the rotation of the nitro groups were tentatively estimated to be 20 and 16 kJ mol^{-1}, respectively.

3-Cyanothiophene,

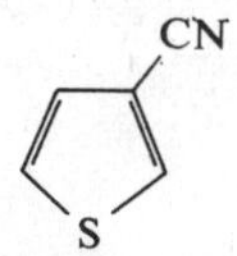

The available microwave specroscopic data were used by Wiese and Sutter (1977) to examine possible structural changes in the part of the thiophene ring adjacent to the CN substituent. This was done by assuming that the other parts of the ring and the C≡N bond remained unchanged as compared with thiophene and benzonitrile, respectively. A considerable shortening of the C2C3 bond (to 134.6 pm) was thus established paralleled by a decrease of the C2C3C(N) angle (to 121.2°). These structural changes led to a good agreement with the experimental rotational constants and the authors concluded that the 3-cyano substitution caused major change only in the adjacent double bond indeed.

2-Thiophene sulphonyl chloride,

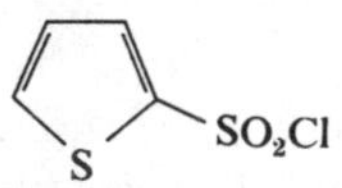

The electron diffraction investigation (Brunvoll, Hargittai, Székely and Pappalardo, 1980) was primarily aimed at determining the molecular conformation and the sulphur bond configuration of the sulphonyl chloride part. The respective results will be discussed with the other sulphone molecules. As regards the size of the thiophene ring, the results are rather uncertain but consistent with the structure of the thiophene molecule itself (*cf.* Table 39). The asymmetry of the ring due to substitution could not be determined. The following geometrical parameters (some of them mean values) have been found for the ring:

r(C=C)mean	137(3) pm	∢(C=C—C)mean	113(1)°
r(C—C)	142(3) pm	∢(C=C—S)mean	111(1)°
r(C—S)mean	172(2) pm	∢C—S—C	92(1)°
r(C—H)mean	107(2) pm		

2,2′-Dithienyl (Fig. 54). The electron diffraction investigation by Almenningen, Bastiansen and Svendås (1958) was greatly facilitated by the then available microwave spectroscopic data on the thiophene structure (Bak, Christensen, Rastrup-Andersen and Tannenbaum, 1956). Some of the geometrical parameters determined are shown in

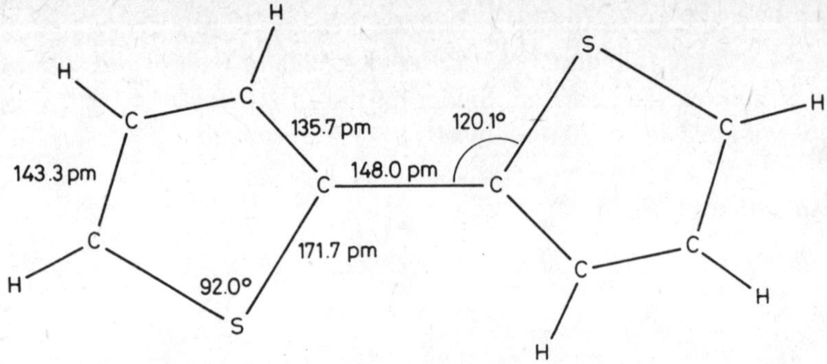

Fig. 54. Of the geometrical parameters of 2,2′-dithienyl after Almenningen, Bastiansen and Svendås (1958)

Fig. 54. The least-squares structure refinement yielded $\varphi = 34°$ for the torsional angle. A good agreement with the experimental radial distribution could also be achieved by models with larger φ values or mixtures of models with different torsional angles. It was supposed that the potential energy function describing the rotation about the central bond may have two minima ($\varphi = 34°$ and $\varphi = 85°$) with a small maximum in between.

Thiazole,

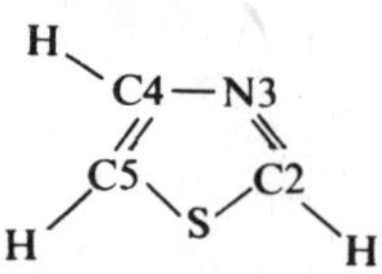

The complete substitution structure of this molecule was determined by microwave spectroscopy (Nygaard, Asmussen, Høg, Maheshwari, Nielsen, Petersen, Rastrup-Andersen and Sørensen, 1971). Some characteristic geometrical parameters are cited here:

r(S—C)	172.4(2) pm	∢C—S—C	89.3(3)°
r(C2—N)	130.4(3) pm	∢S—C—N	115.2(2)°
r(C4—N)	137.2(2) pm	∢S—C—C	109.6(2)°

1,2,5-Thiadiazole and *1,3,4-thiadiazole,*

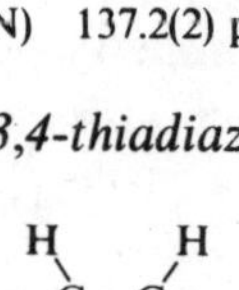

Their molecular geometries have been accurately determined by electron diffraction and microwave spectroscopy. The geometrical parameters and references are collected in Table 40. The relative rigidity of these molecules is indicated by the small vibrational corrections. Markov and Stølevik (1970) found the r_a and r_α parameters of 1,3,4-

TABLE 40
Geometrical parameters of 1,2,5-thiadiazole and 1,3,4-thiadiazole and their selenium analogs

Parameter	1,2,5-Thiadiazole, C_{2v} (a)	1,2,5-Thiadiazole, C_{2v} (b)	1,2,5-Selenadiazole, C_{2v} (c)	
r(N—S), pm	163.2(5)	163.1(3)	180	Se—N
r(C—N), pm	132.9(8)	132.8(3)		
r(C—C), pm	141.3(10)	142.0(3)		
r(C—H), pm	108.0(10)	107.9(5)		
∢N—S—N (°)	99.4(2)	99.55(20)	94.5(20)	N—Se—N
∢S—N—C (°)	106.5(4)			
∢C—C—N (°)		113.8(2)		
∢C—C—H (°)		126.2(5)		

Parameter	1,3,4-Thiadiazole, C_{2v} (d)	1,3,4-Thiadiazole, C_{2v} (e)	1,3,4-Selenadiazole, C_{2v} (f)	
r(C—S), pm	172.2(6)	172.1(3)	186.8(20)	C—Se
r(C—N), pm	130.4(10)	130.2(3)		
r(N—N), pm	138.1(16)	137.1(3)		
r(C—H), pm	108.1(28)	107.9(5)		
∢C—S—C (°)	86.4(4)	86.4(2)	81.8(15)	C—Se—C
∢S—C—N (°)	114.8(5)	114.6(2)	114.8(15)	Se—C—N
∢C—N—N (°)	112.0(4)	112.2(2)		
∢S—C—H (°)	124.1(30)	121.9(7)		
∢N—C—H (°)	121.1(30)	123.5(7)		

(a) Electron diffraction, r_g; Momany and Bonham (1964)
(b) Microwave spectroscopy, r_s; Dobyns and Pierce (1963)
(c) Microwave spectroscopy, r_0; Blackman, Brown, Burden and Kent (1967)
(d) Electron diffraction, r ; Markov and Stølevik (1970)
(e) Microwave spectroscopy, r_s; Nygaard, Hansen and Sørensen (1971)
(f) Microwave spectroscopy, r_0; Levine, Krugh and Gold (1969)

thiadiazole to be practically equal. The analogous selenium compounds have been investigated by microwave spectroscopy. Only some of the parameters could be determined (*cf.* Table 40) while many others had to be assumed.

1,3,4-Oxathiazol-2-one,

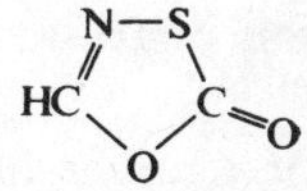

A combined structural investigation of this molecule was carried out by Bak, Nielsen, Svanholt, Almenningen, Bastiansen, Fernholt, Gundersen, Nielsen, Cyvin and Cyvin (1978) involving microwave, infrared, Raman spectroscopic and electron diffraction data as well as a normal coordinate analysis. The coplanarity of the molecule has been unambiguously established. The following bond lengths and bond angles were consistent with both the electron diffraction intensities and the rotational constants C—H 110.2(19), C=O 119.2(2), N=C 128.6(2), HC—O 135.6(3), OC—O 140.2(3), S—N 169.0(2), S—C 176.7(2) pm, and O—C—H 114.5(31), O=C—O 122.6(3), C—O—C 110.8(2), C—S—N 93.8(1)°.

Trithiapentalenes and its analogs, where X, Y, Z are O, S, Se, etc., are

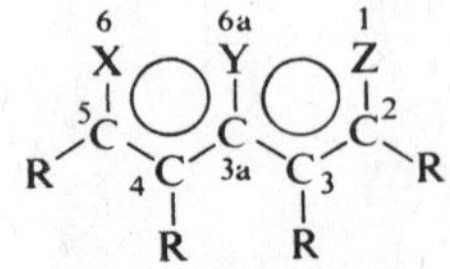

aromatic and the interesting bonding of the XYZ fragment was discussed by Gleiter and Hoffmann (1968). Recently two of these structures have been subjected to gas-phase investigation: *1,6,6a-trithia(6a-S(IV))pentalene*

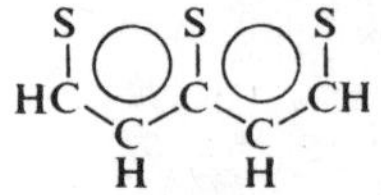

by electron diffraction (Shen and Hedberg, 1974) and *1,6-dioxa-6a(S(IV))-thiapentalene*

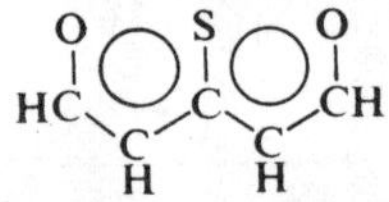

by microwave spectroscopy (Pedersen, Skaarup and Pedersen, 1977). Both studies concluded that these molecules are planar and have C_{2v} symmetry. The structural parameters determined by electron diffraction suggest relatively loose S—S "bonds" and relatively stiff S1 . . . S6 interactions. This observation was interpreted as evidence for the "bell-clapper" effect suggested by Gleiter and Hoffmann (1968). The gas-phase findings were in contrast to the x-ray diffraction crystallographic results according to which the S—S bond lengths are different even in the case of symmetrical substitution (for references see the cited papers on the gas-phase studies). On the other hand, the microwave spectroscopic study unambiguously determined that the 1,6-dioxa-6a(S(IV))-thiapentalene molecule has a single-minimum structure in the vapour phase. The electron diffraction investigation of trithiapentalene yielded a series of bond lengths and bond angles.

In conclusion it is noted that the amount of reliable experimental data on the molecular geometry of aromatic ring molecules is steadily growing. Eventually this will facilitate the observation and understanding of the structural variations in this class of compounds. Kao and Radom (1979) have recently communicated the results of *ab initio* molecular orbital calculations on a series of 2- and 3-monosubstituted thiophene derivatives. They give an interesting discussion of geometry, conformational behaviour, electronic structure and their interrelationship.

2.7. INORGANIC RING MOLECULES

Structural data on free inorganic ring molecules with sulphur in the ring are extremely scarce. There has been an early visual electron diffraction study of the vapour of elemental sulphur (Lu and Donohue, 1944). A reinvestigation with modern techniques is certainly warranted. In addition, it is desirable to parallel such an electron diffraction analysis with a mass spectrometric control of the vapour composition. Obviously strong theoretical interest (*cf.*, e.g., Kao and Allinger, 1977; Kao, 1977; Herman and Weiss, 1975; Salahub, Foti and Smith, 1978) should stimulate experimental work in this area. The molecular structure of *cyclo*-hexasulphur, *cyclo*-heptasulphur and *cyclo*-decasulphur has been recently determined in the crystal phase (Steidel, Pickardt and Steudel, 1978, Steudel, Reinhardt and Schuster, 1977 and Reinhardt, Steudel and Schuster, 1978, respectively, see also Donohue, 1978).

The vapour of selenium at 450 °C was found to contain 75% hexamer and 25% dimer (Barzdain and Alekseev, 1968). The Se—Se bond distance (234 pm) and Se—Se—Se bond angle (102°) were determined by electron diffraction for the chair-form hexamer

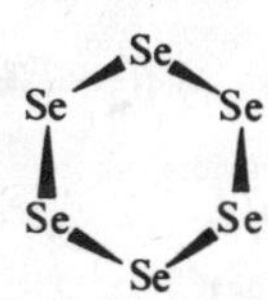

The electron diffraction data were consistent with a C_{3v} model for the P_4S_3 molecule (Akishin, Rambidi and Ezhov, 1960)

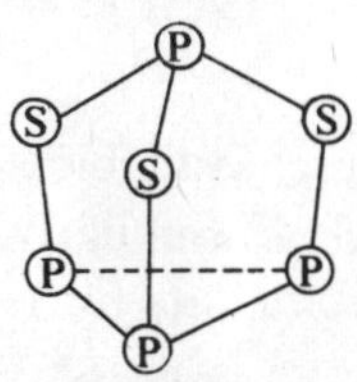

The following geometrical parameters have been communicated:

$$r(\text{P—S}) \quad 217(2)\ \text{pm} \qquad r(\text{P—P}) \quad 221(2)\ \text{pm}$$
$$\angle\,\text{P—S—P} \quad 105(3)°$$

Dichloro-1,2,4-trithia-3,5-diborolane and *dimethyl-1,2,4-trithia-3,5-diborolane.* The five-membered rings in both molecules are at least approximately planar according to electron diffraction (Almenningen, Seip and Vassbotn, 1973 and Seip, Seip and Siebert,

TABLE 41
Geometrical parameters (r_a) of dichloro-1,2,4-trithia-3,5-diborolane and dimethyl-1,2,4-trithia-3,5-diborolane as determined by electron diffraction

Parameter	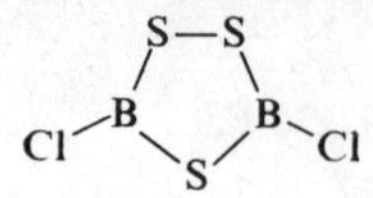 (a)	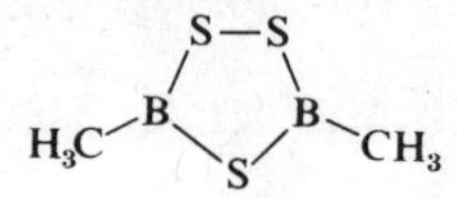 (b)
r(S—S), pm	206.9(3)	207.6(3)
r(B—S) mean, pm	179.4(5)	180.3(3)
r(B—Cl), r(B—C), pm	175.6(9)	156.9(5)
∢B—S—B (°)	96.9(6)	101.6(4)
∢S—B—S (°)	121.7(5)	117.7(2)
∢B—S—S (°)	99.9(3)	101.5(4)

(a) Almenningen, Seip and Vassbotn (1973)
(b) Seip, Seip and Siebert (1973)

1973, respectively). The geometrical parameters are collected in Table 41. The B—S—S—B chains of these ring molecules are characterized by similar bond lengths as that of the $(CH_3)_2BSSB(CH_3)_2$ molecule (*cf.* Table 32).

2.8. MOLECULES WITH =S= OR ≡S— BOND SYSTEMS

Molecules in which the sulphur atom is two-coordinated and has formal double or triple bonds are discussed here:

SO_2	S_2O	CH_2SO
HNSO	CH_3NSO	C_6H_5NSO
ClNSO	$S(NCH_3)_2$	
NSF	NSCl	

Very accurate geometrical data are available for *sulphur dioxide* and *disulphur monoxide*. They are collected in Table 42. Similarly, accurate data for selenium dioxide and the only available rather uncertain data for tellurium dioxide are also shown for comparison.

The microwave spectra of eight isotopic species of *sulphine*

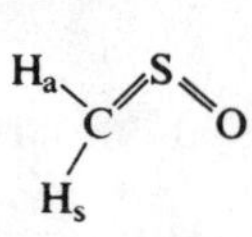

or methanethial oxide were analysed by Penn and Olsen (1976). The following substitution parameters were determined:

TABLE 42
The geometrical parameters of sulphur dioxide and its analogues

Parameter	SO_2 (a)	SO_2 (b)	S_2O (c)	SeO_2 (d)	TeO_2 (e)
r(S=O), pm	143.076(13)	143.1(2)	146.4(3)		
r(Se=O), pm				160.76(6)	
r(Te=O), pm					183(2)
∢O=S=O (°)	119.33(1)	118.5(10)			
∢S=S=O (°)			118.3(2)		
∢O=Se=O (°)				113.83(8)	

(a) Microwave spectroscopy, r_e, Saito (1969a)
(b) Electron diffraction, r_a; Clark and Beagley (1971)
(c) Microwave spectroscopy, r_s; Tiemann, Hoeft, Lovas and Johnson (1974)
(d) Microwave spectroscopy, r_e; Takeo, Hirota and Morino (1970)
(e) Electron diffraction; Zasorin, Zharskii, Pinaev, Kupreev and Spiridinov (1974)

$r(C—H_s)$	108.5 pm	∢H—C—H	121.86°
$r(C—H_a)$	107.7 pm	$∢S—C—H_s$	122.51°
r(C=S)	161.0 pm	$∢S—C—H_a$	115.63°
r(S=O)	146.9 pm	∢C=S=O	114.72°

The most interesting feature of this structure is the asymmetry of the carbon bond configuration. Compared with thioformaldehyde (or methanethial, see section on $C(sp^2)$=S bonds with one-coordinated sulphur), the hydrogen atom *anti* to the oxygen atom is shifted in sulphine about 5° towards the sulphur atom, while the position of the *syn* hydrogen is essentially unchanged. Another interesting feature is the identical C=S bond lengths in sulphine and thioformaldehyde. *Ab initio* molecular orbital calculations (Flood and Boggs, 1976) essentially reproduced both these structural peculiarities. The sulphur–oxygen π-bond was found close to being a lone pair of the oxygen atom as shown in Fig. 55. Accordingly, the sulphur–oxygen bond may be considered as a coordinate bond in agreement with the observation of Penn and Olsen (1976) that with decreasing X electronegativities in the X=S=O series, SO_2, HNSO, S_2O, CH_2SO, the SO bond length increases and the bonding changes from the type $X^-—S^+=O$ to $X=S^+—O^-$. The latter bonding arrangement also explains the constancy of the C=S bond lengths as going from CH_2=S to CH_2=S=O. The attachment of the oxygen to CH_2=S changes very little the bonding in the rest of the molecule. The *ab initio* calculations provided a feasible interpretation of the changes in the hydrogen positions. As illustrated by Figs. 56a, 56b* the sulphur lone electron pair is somewhat shifted towards the oxygen atom as the weak SO bond excercises less repulsion in sulphine than the second sulphur lone pair in thioformaldehyde does. This

* The captions of the respective figures (Figs. 2 and 3) in the original paper of Flood and Boggs (1976) were apparently interchanged.

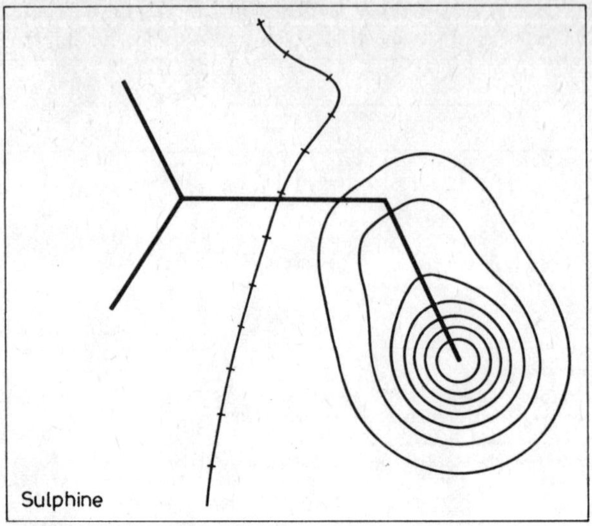

Fig. 55. Plot of the sulphur–oxygen π-bond in sulphine, H_2CSO, plotted 0.9 a.u. above the molecular plane. Zero contour is hatched, difference between contours is 0.05 electrons after Flood and Boggs (1976)

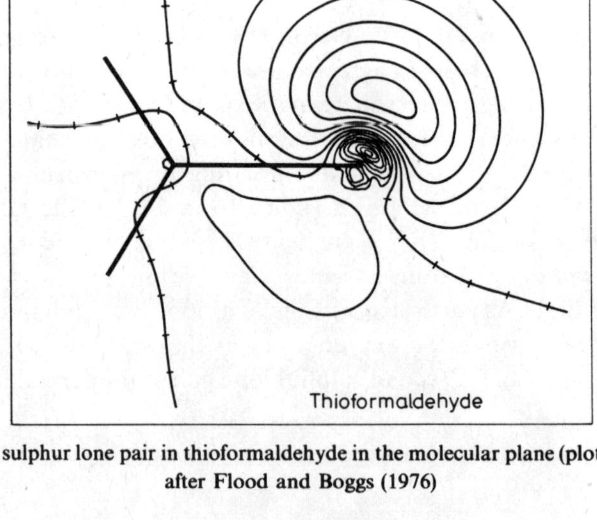

Fig. 56a. Plot of the sulphur lone pair in thioformaldehyde in the molecular plane (plot values as in Fig. 55) after Flood and Boggs (1976)

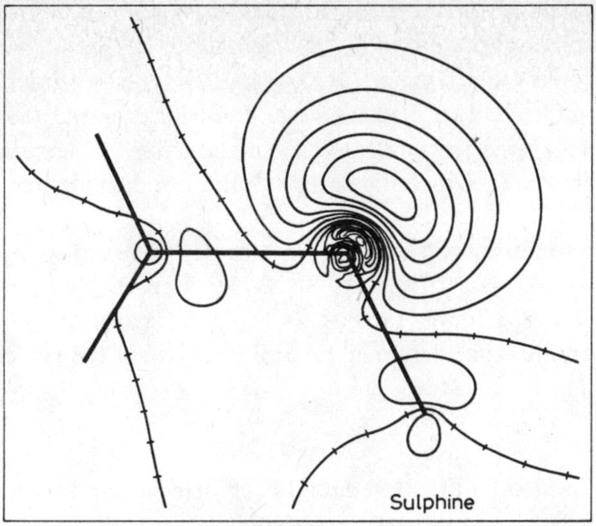

Fig. 56b. Plot of the sulphur lone pair in sulphine in the molecular plane (plot values as in Fig. 55). Flood and Boggs (1976)

TABLE 43
Geometrical parameters of XNSO (X = Cl, CH_3 or H) molecules

Compound	r(X—N) pm	r(N=S) pm	r(S=O) pm	∢N=S=O (°)
ClNSO *(a)*	169.6(3)	155.9(4)	144.5(4)	116.0(6)
CH_3NSO *(b)*	142.1(5)	152.5(4)	146.6(4)	117(2)
HNSO *(c)*	102.9(10)	151.2(5)	145.1(5)	120.4(3)

(a) Electron diffraction, r_a; Oberhammer (1970)
(b) Electron diffraction r_a; Beagley, Chantrell, Kirby and Schmidling (1975)
(c) Microwave spectroscopy, r_s; Kirchhoff (1969)

shift of the sulphur lone pair allows the opening of the HCH angle by letting the *anti* hydrogen change its position. Note that the geometrical changes followed in a very sensitive way the changes in the electronic structure.

Thionylimide, HNSO. Its substitution molecular geometry (r_s) was determined by a microwave spectroscopic study of Kirchhoff (1969). The geometrical parameters are given in Table 43. The uncertainties cited according to Kirchhoff (1969) reflect the indeterminacy of the physical meaning of the r_s parameters and are about an order of magnitude larger than those which would characterize the reproducibility of these parameters.

N-methyl-thionylimide, CH_3NSO and *N-chloro-thionylimide*, ClNSO. The electron diffraction determination of their molecular geometry (Beagley, Chantrell, Kirby and Schmidling, 1975 and Oberhammer, 1970, respectively) was hindered by the strong correlation among the structural parameters. Thus, for example, the contributions of all bonds (except C—H) appear under the same maximum of the radial distribution of both molecules. The data characterizing the sulphur bond configurations are collected in Table 43.

The molecular conformation of XNSO molecules was studied also by vibrational spectroscopy (X = F, Cl, Br, I: Eysel, 1970; X = H: Richert, 1961) and by microwave spectroscopy (X = H: Kirchhoff, 1969; X = CH_3: Rao, Yardley and Curl, 1965). The electron diffraction and spectroscopic data all established the *syn* conformation

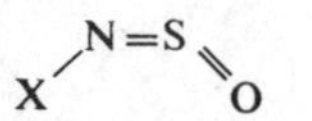

According to the electron diffraction data, the chlorine atom deviates about 35° from the plane of the NSO chain (Oberhammer, 1970)

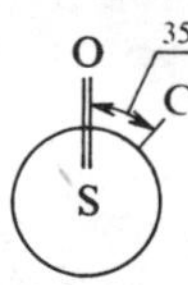

The *syn* conformation was found to be present exclusively for HNSO and CH_3NSO by *ab initio* molecular orbital calculations (Bak, Svanholt and Larsen, 1977).

There was some controversy over the conformation of *N-sulphinylaniline*, C_6H_5NSO ("*anti*" Grunwell and Danison, 1971 *vs.* "*syn*" van Woerden and Bijl-Vlieger, 1973). In fact, there may be four conformers, all of them having C_s symmetry as shown in Fig. 57. A recent microwave spectroscopic investigation by Caminati, Mirri and Maccagnani (1977) has unambiguously established the *syn*-1 form. Further evidence for this conformer was obtained from vibrational and low-resolution microwave spectra (Bak, Svanholt and Larsen, 1977). A series of *N*-sulphinylanilines was also studied in solution by ^1H and ^{13}C N.M.R. spectroscopy (Meij, Oskam and Stufkens, 1979). All these compounds were found to be in a *syn* form, and all except one (the mesityl derivative) in the *syn*-1 form.

Although only one isotopic species of *N*-sulphinylaniline has been studied by microwave spectroscopy (Caminati, Mirri and Maccagnani, 1977), a limited amount of geometrical information could be derived in addition to establishing the conformation. A series of test calculations showed the bond angles C—N═S and N═S═O to be somewhat larger than the analogous angles in HNSO and CH_3NSO.

There seems to be some contradiction in the variations of the S═O stretching frequencies and S═O bond lengths of the XNSO molecules. While the S═O bond of the methyl derivative was found to be 2 pm longer than that of ClNSO, the S═O stretching frequency of the latter was reported to be some 30 cm^{-1} smaller than that of

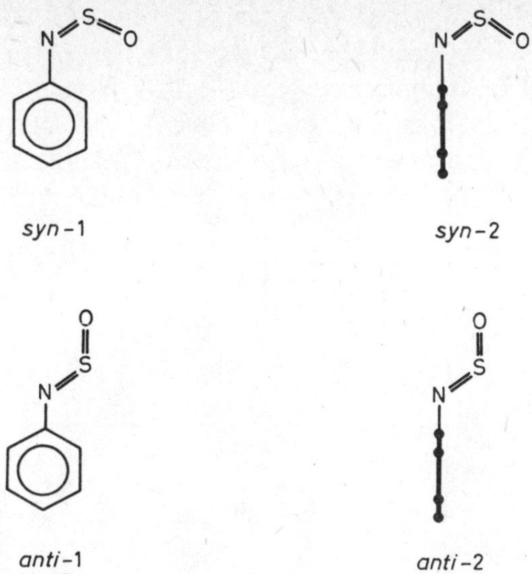

Fig. 57. Four conformations of C_s symmetry for *N*-sulphinylaniline

the former. A difference of the opposite sign (and of larger magnitude) would be expected. Here we refer to the empirical correlation relationships established between the S=O bond lengths and stretching frequencies to be discussed later. Note that the vibrational frequencies used for the above comparison originated from liquid-phase spectra. The stretching frequencies determined in the vapour phase and the liquid phase may be quite different and this difference may vary considerably from compound to compound. Thus, for example, the difference $\Delta\nu = \nu(\text{vapour}) - \nu(\text{liquid})$ amounts to 25, 29, 42 and 37 cm^{-1} in the compound series FNSO, ClNSO, BrNSO, and INSO. Returning to the 2 pm difference in the S=O bond lengths of CH_3NSO and ClNSO, it is the more surprising since it is not the ligand adjacent to the sulphur–oxygen linkage that is changed but the one attached to nitrogen. The variations of these bond stretching force constants are shown in Fig. 58 (data from Richert, 1961, are also used). It is mentioned, however, that the calculation procedure used by Eysel (1970) has been subjected to much criticism (*cf.* Averbukh, Mayants and Shaltuper, 1969; Tchir and Spratley, 1975). In the case of the S=O bond stretching frequency and bond lengths variations, if the empirical relationship

$$\lg \nu(\text{S}{=}\text{O}) = 4.301 \lg r(\text{S}{=}\text{O}) + 3.770$$

established for the sulphoxides (see later) is used, the $\nu(\text{S}{=}\text{O}) = 1221\ cm^{-1}$ frequency reported for ClNSO (Eysel, 1970) corresponds to $r(\text{S}{=}\text{O}) = 144.4$ pm in excellent agreement with the experimental result (*cf.* Table 43). The same relationship yields $r(\text{S}{=}\text{O}) = 143.3$ pm for CH_3NSO from its stretching frequency (Eysel, 1970) *vs.* the reported experimental result of 146.6 pm (Beagley, Chantrell, Kirby and Schmidling, 1975, *cf.* Table 43).

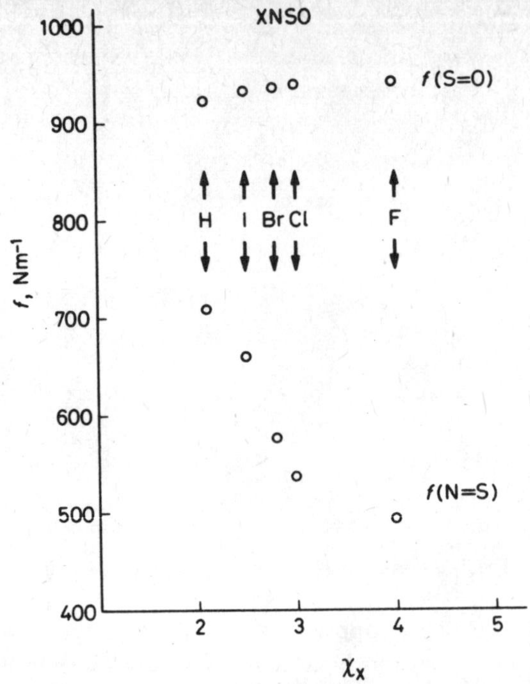

Fig. 58. Variations in the S=O and N=S bond stretching force constants depending on the X ligand electronegativity (χ_X) in the XNSO series with X = H, I, Br, Cl, or F

Dimethylsulphurdiimine, $S(NCH_3)_2$. Various physical techniques were employed to establish its vapour-phase molecular conformation (*cf.* Kuyper, Isselmann, Mijlhoff, Spelbos and Renes, 1975). The pre-eminence of two forms, *viz. cis–trans* and *trans–trans* has been established:

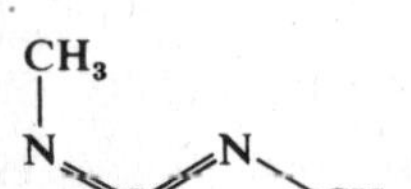

cis–trans

trans–trans

An electron diffraction study by Kuyper, Isselmann, Mijlhoff, Spelbos and Renes (1975) showed this molecule to be present predominantly in the (planar) *cis–trans* conformer in the vapour phase. The two C—N=S bond angles in the two halves of the molecule were found to be considerably different. The following geometrical parameters were determined (r_a):

r(S=N)	153.2(10) pm	∢N=S=N	113.6(16)°
r(C—N)	146.4(18) pm	∢C—N=S(*cis*)	116.5(12)°
r(C—H)	111.0(16) pm	∢C—N=S(*trans*)	124.3(20)°

Thiazyl chloride, NSCl and *thiazyl fluoride*, NSF. These are relatively new compounds for which even the atom connectivity order has not been established up to some time ago. The atomic arrangement of the chlorine derivative, NSCl, was unambiguously established by independent electron diffraction (Emken and Hedberg, 1973) and microwave spectroscopic analyses (Beppu, Hirota and Morino, 1970). Thus, for example, the electron diffraction radial distribution (reproduced in Fig. 59)

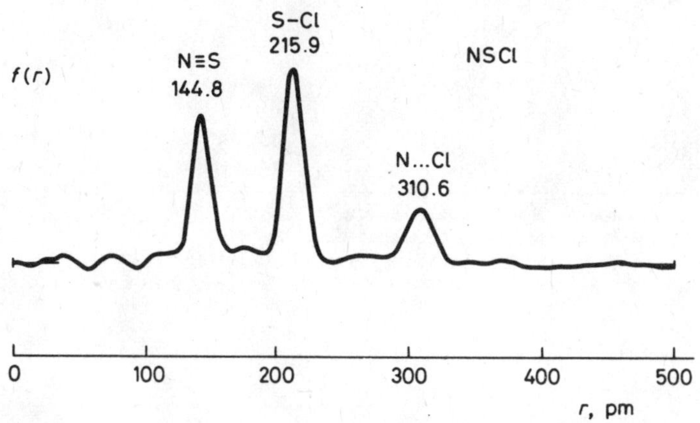

Fig. 59. Radial distribution of thiazyl chloride from the electron diffraction investigation by Emken and Hedberg (1973)

provided direct evidence for establishing the atomic arrangement NSCl and for excluding the arrangement SNCl. The area under a maximum, originating from the contribution of an interatomic interaction, is roughly proportional to $n_{ij}Z_iZ_j/r_{ij}$ (n_{ij} is the number how many times this particular interaction occurs in the molecule, Z_i and Z_j are the atomic numbers of the atoms involved in this interaction and r_{ij} is the interatomic distance). The relative areas of the three maxima on the radial distribution curve are expected to be approximately 2.0 : 3.3 : 1.0 for NSCl and 0.9 : 0.6 : 1.0 for SNCl. It is easily seen in Fig. 59 that the experimental curve corresponds to the former ratio. The molecular parameters including the mean amplitudes of vibration and the bond stretching force constants (Müller, Mohan, Cyvin, Weinstock and Glemser, 1976) are quoted in Table 44. N≡S bonds are only slightly longer than what would correspond to the sum of the covalent radii referring to a triple bond (r_N 55 pm and r_S 87 pm; Pauling, 1960). The S—Cl bond of thiazyl chloride is some 16 pm longer than the calculated single bond: 200 pm according to Pauling (1960) or 199.6 pm according to Glidewell (1976). On the other hand, the S—F bond length determined in thiazyl fluoride is less different from the calculated single bond: 164 pm (Pauling, 1960) or 156.5 pm (Glidewell, 1976).

A recent *ab initio* study of So and Richards (1978) suggested that the SNF molecule is potentially stable although less so than the above discussed NSF molecule.

TABLE 44
Molecular parameters of thiazyl chloride and thiazyl fluoride

Parameter X=Cl or F	NSCl			NSF	
	(a)	*(b)*	*(c)*	*(d)*	*(c)*
r(N≡S), pm	144.8(3)	145.0(3)		144.8(2)	
r(S—X), pm	215.9(3)	216.1(3)		164.3(2)	
r(N...X), pm	310.6(16)				
∢N≡S—X (°)	117.5(10)	117.7(3)		116.91(8)	
l(N≡S), pm	3.8(3)		3.7		
l(S—X), pm	5.3(3)		6.4		
l(N...X), pm	7.7(12)		9.2		
f(N≡S), N m^{-1}			1040(10)		1094(6)
f(S—X), N m^{-1}			165(13)		288(4)

(a) Electron diffraction; Emken and Hedberg (1973)
(b) Microwave spectroscopy; Beppu, Hirota and Morino (1970)
(c) Spectroscopic calculations; Müller, Mohan, Cyvin, Weinstock and Glemser (1976)
(d) Microwave spectroscopy; Cook and Kirchhoff (1967)

Chapter 3

MOLECULES WITH THREE-COORDINATED SULPHUR

The molecular geometry of the following vapour-phase sulphoxides and selenium analogues have been elucidated:

SOF_2 $SOCl_2$ $SOBr_2$ $SeOF_2$ $SeOCl_2$
$(CH_3)_2SO$ $(CF_3)_2SO$ $(CF_3)_2SNCl$ $(C_6H_5)_2SO$
$(CH_3)_2NSON(CH_3)_2$
$(CH_2)_2SO$ $(CH_2)_3SO$

Thus only symmetrically substituted derivatives have been studied so far.

The structure of simple thionyl compounds has been repeatedly investigated employing either microwave spectroscopy or electron diffraction. The accumulated experimental data call for a critical examination. The structural studies of three compounds related to thionyl fluoride, *viz.*

$$S{=}SF_2,\ ClN{=}SF_2,\ \text{and}\ F_3CN{=}SF_2$$

are also discussed in this section.

Recently a few cyclic sulphites (and selenites) have also been investigated by electron diffraction:

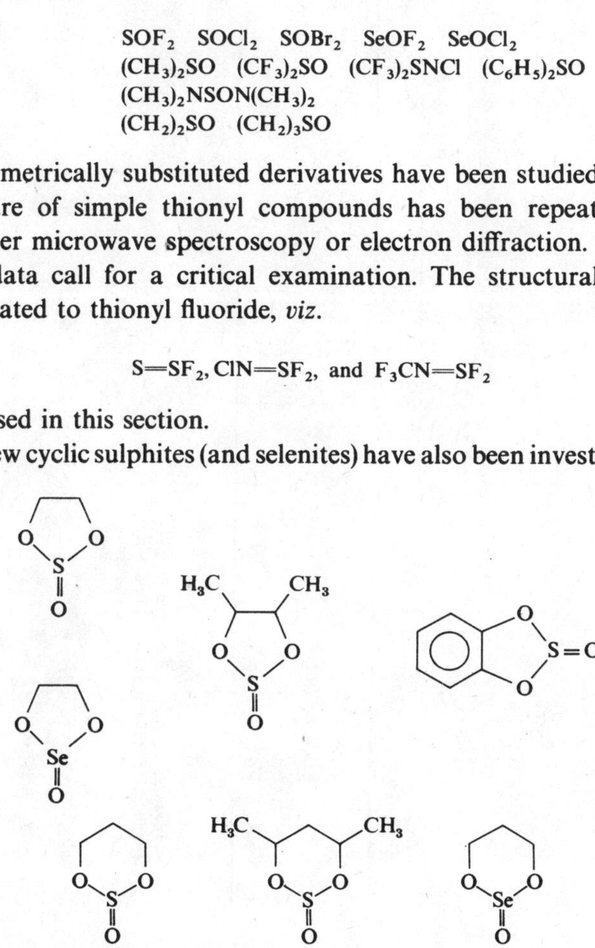

TABLE 45

Bond lengths and bond angles in thionyl halides and analogous selenium derivatives

AOX_2	SOF_2			$SOCl_2$		$SOBr_2$	$SeOF_2$	$SeOCl_2$
	(a1)	*(a2)*	*(b)*	*(c)*	*(d)*	*(e)*	*(f)*	*(g)*
r(A=O), pm	142.1(3)	141.9	141.59(6)	144.3(5)	143.5(11)	144.9(4)	157.6(3)	161.4(5)
r(A—X), pm	158.5(3)	158.3	158.68(4)	207.7(6)	207.2(5)	225.5(5)	172.95(2)	220.5(5)
∢O=A—X (°)		106.2(2)	106.66(4)	106.4(6)	108.0(1)	107.6(2)	104.82(2)	106.0(7)
∢X—A—X (°)		92.3(3)	92.79(4)	96.2(7)	97.2(3)	98.2(2)	92.22(10)	96.9(7)

(a), (c), (e) Electron diffraction

(a) Hargittai and Mijlhoff (1973); *(a1)* r_g; *(a2)* r_α^0

(b) Microwave spectroscopy, r_z; Lucas and Smith (1972)

(c) Hargittai (1969); bond lengths r_g and bond angles r_α

(d) Microwave spectroscopy, r_0; Suzuki, Yamaguchi, Onda, Sakaizumi, Ohashi and Yamaguchi (1981)

(e) Bond lengths r_g and bond angles r_α; Brunvoll, Hargittai and Rozsondai (1982)

(f) Microwave spectroscopy, r_0 except r_s(Se=O); Bowater, Brown and Burden (1967; 1968b)

(g) Bond lengths r_g and bond angles r_α; Gregory, Hargittai and Kolonits (1976)

3.1. THIONYL HALIDES AND RELATED MOLECULES

The bond lengths and bond angles determined in thionyl halides and analogous selenium molecules are compiled in Table 45. The mean amplitudes of vibration are cited in Table 46. The C_s symmetry of the molecules has been established by microwave spectroscopy. The electron diffraction data have also been found to be fully consistent with a C_s symmetry model.

TABLE 46
Mean amplitudes of vibration (pm) in thionyl halides and analogous selenium derivatives

Atomic pair	SOF_2		$SOCl_2$		$SOBr_2$		$SeOF_2$		$SeOCl_2$
	(a1)	*(b1)*	*(a2)*	*(b2)*	*(a3)*	*(b3)*	*(a4)*	*(a5)*	*(b4)*
Temperature (K)	298	298	298	298	298	298	298	298	353
S=O, Se=O	3.2	3.2(1)	3.6	3.1(10)	3.6	2.9(3)	4.2	4.5	5.2(4)
S—X, Se—X	4.5	4.5(1)	5.2	5.1(3)	5.7	5.8(2)	4.5	5.6	5.5(2)
O...X	6.6	6.5(4)	7.7	7.6(6)	8.6	8.2(4)	5.7	7.9	10.9(12)
X...X	6.9	7.4(7)	9.6	10.1(9)	10.6	9.8(3)	6.1	11.6	12.4(16)

(a) Spectroscopic calculations
(b) Electron diffraction
(a1, b1) Hargittai and Mijlhoff (1973)
(a2) Hargittai and Cyvin (1969)
(b2) Hargittai (1969)
(a3, b3) Brunvoll, Hargittai and Rozsondai (1982)
(a4) Müller and Nagarajan (1967a)
(a5) Ramaswamy and Jayaraman (1971a)
(b4) Gregory, Hargittai and Kolonits (1976)

Thionyl fluoride, SOF_2. Following the early work by Ferguson (1954), this molecule was investigated at least four more times by microwave spectroscopy: by Lucas and Smith (1972), Dubrulle (1972), Yamaguchi, Ohashi, Takahashi, Sakaizumi and Onda (1973), and Rimmer, Smith and Whiffen (1973). An important addition in the later studies has been the consideration of the centrifugal distortion effect.

The geometry of the thionyl fluoride molecule possessing C_s symmetry is described by four independent parameters. The rotational spectrum of a given species (e.g. the most abundant $F_2{}^{32}S^{16}O$) yields altogether three rotational constants, i.e. three independent data. Further independent information may be obtained from the rotational spectra of isotopically substituted species. However, a rigorously complete substitution structure (r_s) of this molecule cannot be produced, since the fluorine atom has no isotopes. Lucas and Smith (1972) determined the rotational constants of three isotopic species:

	$F_2{}^{32}S^{16}O$	$F_2{}^{32}S^{18}O$	$F_2{}^{34}S^{16}O$
A_0 (MHz)	8614.80	8582.46	8568.45
B_0 (MHz)	8356.95	7843.39	8304.57
C_0 (MHz)	4952.94	4772.47	4949.63

Fitting the geometrical parameters to these rotational constants, they obtained the following results:

r(S=O)	141.27(3) pm	∢F—S—F	92.83(2)°
r(S—F)	158.54(2) pm	∢F—S=O	106.82(3)°

Of course, these results are not independent of the choice of the isotopic species. There is one geometrical parameter, *viz.* r(F . . . F), however, which can be determined directly from the rotational constants for each isotopic species. If the vibrational contribution (δ) due to the intramolecular motion is negligibly small, the planar moment or second moment of the X_2SO molecule possessing C_s symmetry is determined by the masses and positions of the X atoms situated outside the symmetry plane, and it can be written that

$$I_a + I_c - I_b = \Delta I = m_X r_0^2(\text{X}\ldots\text{X})$$

Naturally, the fluorine–fluorine distance obtained from the planar moment depends on the isotopic composition of the species from whose rotational constants it had been derived.* Thus for example the r_0(F . . . F) values corresponding to the three above-mentioned isotopic species are 229.69, 229.82, and 229.69 pm, respectively. The fitted geometry yields 229.68 pm for the F . . . F distance. The designation "r_0" is used here for the latter in order to distinguish it from the true r_0 parameters obtained directly from the planar moments. Note that the markedly small uncertainties of the "r_0" parameters obtained by the fitting procedure characterize the goodness of fit only and changing the isotopic species may induce differences a whole order of magnitude larger.

Harmonic corrections were applied to the rotational constants A_0, B_0, and C_0 and the average rotational constants A_z, B_z, and C_z were produced, which corresponded to the distance between average nuclear positions (r_z). The distance between the average nuclear positions of the fluorine atoms can be directly obtained from the following relationship:

$$I_a^z + I_c^z - I_b^z = m_F r_z^2(\text{F}\ldots\text{F})$$

and they are 229.81, 229.94, and 229.81 pm for the three isotopic species, respectively. A complete "r_z" structure was also produced by Lucas and Smith (1972) and is given in Table 45. The uncertainties indicated here refer again to the goodness of fit only.** Parameters r_0(F . . . F) were calculated from the data of two other studies as well. In both cases they refer to the $F_2{}^{32}S^{16}O$ species and are in complete agreement with the

* The vibrational contribution is of the order of a few tenths of one per cent and is neglected in our discussion.

** It was noted by Lucas and Smith (1972) that the "r_z" parameters had relatively large standard deviations as compared with those of the "r_0" parameters. In the latter the influences of both the harmonic and anharmonic vibrations, which usually at least partially cancel each other are present, whereas in obtaining the "r_z" parameters harmonic corrections are made, leaving the anharmonic terms uncounteracted.

above results: 229.68 pm from Dubrulle (1972) and 229.69 pm from Yamaguchi, Ohashi, Takahashi, Sakaizumi and Onda (1973).

All the above-mentioned results on thionyl fluoride refer to the vibrational ground state. In two studies rotational transitions belonging to excited vibrational states have also been identified. The data reported for the excited state by Dubrulle (1972) and by Yamaguchi, Ohashi, Takahashi, Sakaizumi and Onda (1973) related to the scissoring F—S—F vibration, $\nu_4(A')$, yielded 232.48 pm for r_0(F . . . F). The excitation of this vibration thus results in an increase of the F . . . F distance by more than 2 pm. The reader is reminded that the interpretation of the microwave spectroscopic data is not hindered by the vibrational excited states being populated (or rather, only to the extent that the intensities will be distributed among all the populated states), since the rotational transitions identified and the geometrical parameters determined refer to individual vibrational states. On the other hand, the parameters emerging from an electron diffraction analysis are averages containing contributions from all species and accordingly, reflect the population distribution among all states characteristic for the sample studied.

The visual technique of electron diffraction was employed to investigate the structure of thionyl fluoride (*cf.* Sutton, 1958). The results of this study became not only obsolete but also proved to be erroneous. The results of a more recent electron diffraction analysis (Hargittai and Mijlhoff, 1973) are cited in Tables 45 and 46. From the average distances (r_g), yielded directly by the electron diffraction structure refinement, the distances between the average nuclear positions (r_α^T) were obtained by applying harmonic corrections

$$r_\alpha^T = r_g - \delta r_{rot} - \frac{\langle(\Delta x)^2 + (\Delta y)^2\rangle}{2r_g}$$

where δr_{rot} refers to the centrifugal distortion, and the second correction term (often denoted also by K) refers to the influence of the perpendicular vibrations. Both r_g and r_α^T refer to the thermal equilibrium defined by the temperature T. The distance between the average nuclear positions in the ground vibrational state can be obtained by the expression

$$r_\alpha^0 = r_\alpha^T - (\langle \Delta z\rangle_T - \langle \Delta z\rangle_0)$$

The best representations of the bond lengths and bond angles are r_g and r_α^0, respectively, and they are cited in Table 45 from the electron diffraction work. For the comparison of electron diffraction and microwave spectroscopic results, r_α^0 and r_z should be selected, since they possess the same physical meaning. The ratios "r_z"/r_α^0 for the distances S═O, S—F, F . . . F, and F . . . O are 0.9980, 1.0019, 1.0069, and 1.0031, respectively. Thus, although the differences are rather large, it is somewhat comforting that for the two best determined bond distances the signs of the differences are different.

The geometrical data on the three molecules related to thionyl fluoride, *viz.* $S{=}SF_2$, $ClN{=}SF_2$, and $F_3CN{=}SF_2$ are collected in Table 47.

TABLE 47
The geometrical parameters characterizing the sulphur bond configuration in the X═SF_2 molecules (X = S, ClN or F_3CN)

Parameter	S═SF_2 *(a)*	ClN═SF_2 *(b)*	F_3CN═SF_2 *(c)*
r(S—F), pm	159.8(12)	159.6(2)	158.3(4)
r(S═S), pm	186.0(20)	–	–
r(N═S), pm	–	147.6(3)	144.7(6)
∢X═S—F (°)	107.5(15)	111.2(1)	112.6(5)
∢F—S—F (°)	92.5(15)	89.8(2)	81.1(16)

(a) Microwave spectroscopy; Kuczkowski (1964)
(b) Electron diffraction; Haase, Oberhammer, Zeil, Glemser and Mews (1970)
(c) Electron diffraction; Karl and Bauer (1975)

Three isotopic species of the so-called *difluorodisulphide*, F_2S_2, were studied by microwave spectroscopy (Kuczkowski, 1964). There are two stable isomers of this compound. One is S═SF_2 (1,1-difluorodisulphide) whose S═S bond has already been mentioned in the discussion of the molecules with one-coordinated sulphur. This molecule is a thionyl fluoride analogue. The other isomer is F—S—S—F, a disulphane derivative whose structure has been discussed among the molecules with two-coordinated sulphur. The sensitivity of the rotational spectra to changes in atom connectivity and geometry is well demonstrated by the rotational constants of the two isomers of difluorodisulphide determined by Kuczkowski (1964):

	$^{32}S{=}^{32}SF_2$	$F^{32}S^{32}SF$
A(MHz)	8179.1(1)	11,273.4(2)
B(MHz)	3968.1(1)	2781.9(2)
C(MHz)	3033.2(1)	2570.0(1)

Of the two isomers of F_2S_2, the thionyl fluoride analogue S═SF_2 is the most stable. Solouki and Bock (1977) calculated the potential energy surface of the FSSF⇄SSF_2 isomerization reaction in the CNDO approximation. They assumed the FSS fragment to be unchanged during isomerization and examined the intramolecular 1,2-fluorine rearrangement. The geometry was optimized in the vicinity of the minima and the multiconfigurational interaction was neglected. Their results on the F_2S_2 system are illustrated by Fig. 60. The potential valley in the potential energy surface corresponding to isomerization is well seen. The minimum near the saddle point is very shallow, and corresponds to an almost planar arrangement of the following configuration

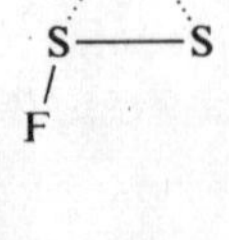

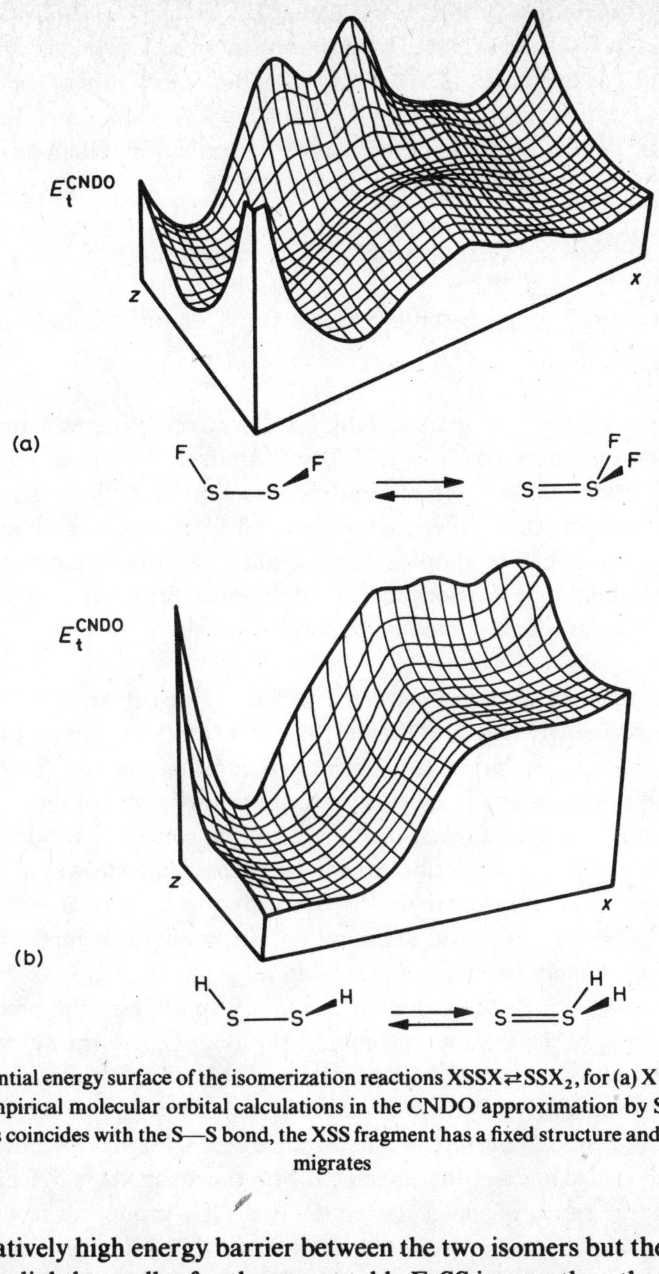

Fig. 60. The potential energy surface of the isomerization reactions $XSSX \rightleftarrows SSX_2$, for (a) X = F and (b) X = H from the semi-empirical molecular orbital calculations in the CNDO approximation by Solouki and Bock (1977). The *x* axis coincides with the S—S bond, the XSS fragment has a fixed structure and the other X atom migrates

There is a relatively high energy barrier between the two isomers but the CNDO total energy is only slightly smaller for the more stable F_2SS isomer than that for the FSSF isomer. Solouki and Bock (1977) carried out similar calculations for the analogous H_2S_2 and Cl_2S_2 systems. According to the potential energy surfaces, the XSSX isomers of C_2 symmetry are the most stable. The results referring to the H_2S_2 system are illustrated by Fig. 60. As has been already mentioned, however, recent *ab initio* calculations of Hinchliffe (1979) have predicted the isomer H_2SS to be as stable with respect to HSSH as F_2SS is to FSSF. Furthermore, infrared spectroscopic evidence was presented for the existence and C_s symmetry pyramidal structure of the isomer Cl_2SS by Chadwick, Grzybowski and Long (1978) and by Feuerhahn and Vahl (1979).

Iminosulphur difluoride molecules were first described by Glemser and Schröder (1956). The molecule with the empirical formula NSF_3 was first thought to have an F—N=SF_2 connectivity but later the $\mathrm{N{\equiv}S}\!\begin{array}{l}\diagup\mathrm{F}\\ -\mathrm{F}\\ \diagdown\mathrm{F}\end{array}$ structure was established unambiguously. Accordingly, its geometry is discussed later, among the molecules with four-coordinated sulphur.

N-Chloro-iminosulphur difluoride, $ClNSF_2$. Its electron diffraction analysis (Haase, Oberhammer, Zeil, Glemser and Mews, 1970) was hindered by the strong overlap of the contributions from the three different bond distances on the radial distribution curve. On the other hand, the four different non-bonded interactions all had individually recognizable features (peak or shoulder) on the radial distribution curve. This greatly increased the reliability of the determination of this structure. The experimental radial distribution and the molecular model are shown in Fig. 61.

Trifluoromethyl-iminosulphur difluoride, F_3CNSF_2. Its electron diffraction structure analysis was performed by Karl and Bauer (1975). The molecule has low symmetry and the apparently strong correlation among the structural parameters (both geometrical and vibrational) hindered a more accurate elucidation of some of them. The shape of the molecule and its *cis* conformation have been unambiguously determined. The molecular model and the deviations from symmetry are shown in Fig. 62. The geometrical parameters are given in this Figure and some of them also appear in Table 47 for comparison with those of other molecules. The most problematic feature of this structure is the extremely small F—S—F bond angle. Since the S—F bond length is normal, it is the F . . . F distance that is strikingly small, *viz.* 206 pm. In all other molecules with SF_2 part of known geometry, the F . . . F distances are considerably larger as illustrated by the examples collected in Table 48. Note also that the F . . . F distance of the trifluoromethyl group was determined to be 216 pm in trifluoromethyl-iminosulphur difluoride. Not only is this distance exactly twice the fluorine 1,3 non-bond radius, but is also in excellent agreement with the mean value of F . . . F distances referring to a large series of molecules containing CF_3 groups as has already been considered in the discussion of the tetrafluoro-1,3-dithietane structure. The contribution from the F . . . F interaction of the SF_2 part is relatively small compared to the total electron scattering of the molecule. A possible erroneous determination of this distance, however, may influence not only the F—S—F bond angle, but also some other parameters correlated with this one. Karl and Bauer (1975) discussed some differences observed between the structural parameters of $ClNSF_2$ and F_3CNSF_2.

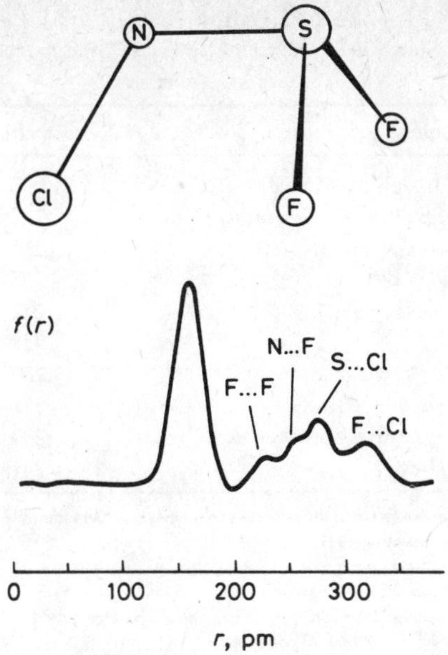

Fig. 61. Molecular model and radial distribution of *N*-chloro-iminosulphur difluoride from the electron diffraction investigation by Haase, Oberhammer, Zeil, Glemser and Mews (1970)

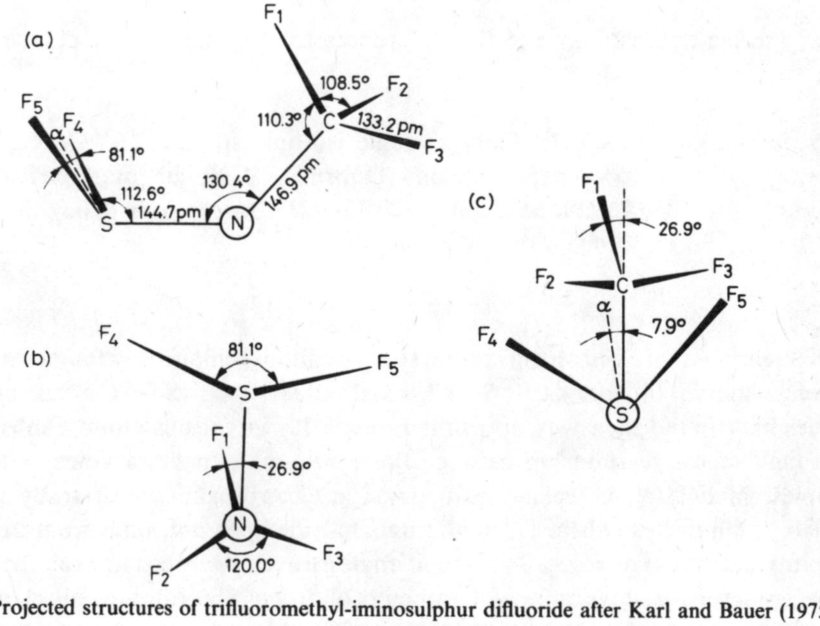

Fig. 62. Projected structures of trifluoromethyl-iminosulphur difluoride after Karl and Bauer (1975)

(a) Side view
(b) View along the CN bond
(c) View along the SN bond. The dashed line α bisects the FSF bond angle.

TABLE 48
Shortest F...F distances of SF_2 fragments of various molecules

Molecule	r(F...F), pm
OSF_4 (axial–equatorial) *(a)*	213–220
F_5SOSF_5 (axial–equatorial) *(b)*	217
$ClSF_5$ (axial–equatorial) *(c)*	221
$ClNSF_2$ *(d)*	225
NSF_3 *(e)*	227
OSF_2 *(f)*	228
O_2SF_2 *(g)*	229
SSF_2 *(h)*	231
SF_4 (equatorial–equatorial) *(i)*	239
SF_2 *(j)*	241

(a) Gundersen and Hedberg (1969), see also Oberhammer and Boggs (1979b); Hargittai (1979c)
(b) Oberhammer and Seppelt (1978)
(c) Marsden and Bartell (1976)
(d) Haase, Oberhammer, Zeil, Glemser and Mews (1970)
(e) Kirchhoff and Wilson (1962)
(f) Hargittai and Mijlhoff (1973)
(g) Hagen, Cross and Hedberg (1978)
(h) Kuczkowski (1964)
(i) Tolles and Gwinn (1962)
(j) Kirchhoff, Johnson and Powell (1973)

They ascribe at least some of the differences to the effect of the chlorine electronegativity.

Thionyl chloride, $SOCl_2$. Only a single isotopic species $^{32}S^{16}O^{35}Cl_2$ has been investigated by microwave spectroscopy (Dubrulle, 1972). The rotational constants A, B, and C are 5086.76, 2822.55, and 1960.32 MHz, respectively. They directly yield r_0(Cl...Cl) 308.5 pm from the expression

$$I_b + I_c - I_a = m_{Cl} r_0^2(\text{Cl} \ldots \text{Cl})$$

The early visual electron diffraction study of thionyl chloride by Palmer (1938) gave, among others, 114° for ∢Cl—S—Cl and 106° for ∢O=S—Cl. These bond angle values have found their way up to the present day in various monographs and texts although such a relationship between these two angles is at variance with even the simplest models of molecular geometry. An electron diffraction study (Hargittai, 1968b; 1969) has established that thionyl chloride does not differ from other X_2SO sulphoxides in that their X—S—X bond angles are relatively constant at about 96–98°. The geometrical and vibrational parameters of thionyl chloride are collected in Tables 45 and 46, respectively. In order to produce the r_α^0 representation of the bond angles, harmonic corrections have been applied (Hargittai, 1974a). The only reliable geometrical parameter from microwave spectroscopy, *viz.* r_0(Cl . . . Cl), is in excellent agreement with the electron diffraction results ($r_a = 308.7$ pm).

A more recent microwave spectroscopic determination of the thionyl chloride structure dealt with the species $SO^{35}Cl_2$ and $SO^{35}Cl^{37}Cl$ (Suzuki, Yamaguchi, Onda, Sakaizumi, Ohashi and Yamaguchi, 1981). The r_s coordinates of the Cl atoms were obtained and the corresponding r_s(Cl . . . Cl) distance was calculated to be 312.8(3) pm. The r_0(Cl . . . Cl) distance for the $SO^{35}Cl_2$ species was obtained as 310.667(1) pm. The authors attributed the rather large difference between r_s and r_0 to large amplitude vibration of the Cl . . . Cl atomic pair. The structural parameters obtained as r_0 structure from separate analyses of the two data sets are cited in Table 45. Assumptions from the electron diffraction study (Hargittai, 1969) were used in the analysis.

Thionyl bromide, $SOBr_2$. The geometrical and vibrational parameters from a recent electron diffraction work (Brunvoll, Hargittai and Rozsondai, 1982) are collected in Tables 45 and 46. The structure is consistent with those of other sulphoxides.

Seleninyl fluoride, $SeOF_2$. The microwave spectra of four isotopic species were studied by Bowater, Brown and Burden (1967; 1968b). The geometrical parameters are given in Table 45. They are independent of the choice of isotopic substitution within the stated error limits. Employing the expression

$$I_a + I_c - I_b = m_F r_0^2(\text{F} \ldots \text{F})$$

in order to directly determine the F . . . F distance, the value of 249.29 pm is obtained for all four isotopic species, while the parameters resulting from the fitting procedure (*cf.* Table 45) correspond to 249.28 pm. Of course, here the isotopic substitution has a limited value only, since there are no isotopes of fluorine, and also because the selenium atom lies near the centre of the mass of the molecule.

Seleninyl chloride, $SeOCl_2$. Two models reproduced equally well the experimental data in the electron diffraction investigation by Gregory, Hargittai and Kolonits (1976). The experimental radial distribution is reproduced in Fig. 63. While the contributions from the two bonds are well resolved, those from the Cl . . . O and Cl . . . Cl interactions strongly overlap. The two models have roughly reversed bond angles. One of them, however, could be excluded on the basis of stereochemical considerations. The geometrical parameters determined for the preferred model are given in Table 45. An independent electron diffraction study by Zharskii, Zasorin, Spiridonov and Novikov (1977) reached similar conclusions regarding both the preferred model and the bond angles. There appears to be, however, some 1% scale difference in the interatomic distances.

Table 46 is a compilation of mean amplitudes of vibration of thionyl halides and analogous selenium derivatives. Experimental electron diffraction results and those calculated from spectroscopic data are presented. In the latter case no attempt at

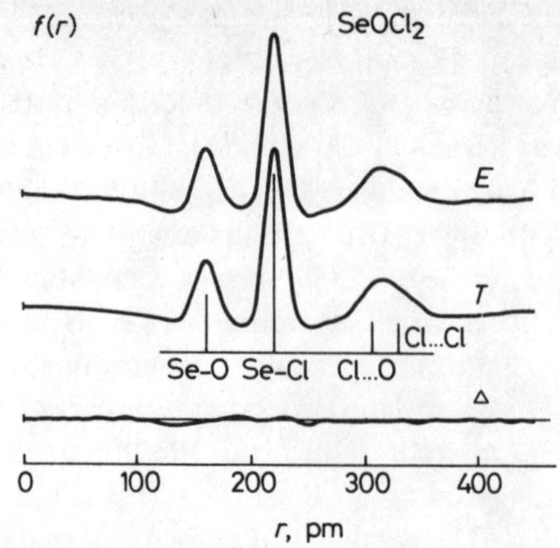

Fig. 63. Radial distribution of seleninyl chloride (Gregory, Hargittai and Kolonits, 1976)

completeness was made. Further references are also given below. They refer to studies devoted primarily to the determination of the molecular force field which is also a central problem of the work aimed at calculating mean amplitudes of vibration. The main difficulty comes from the fact that the SOX_2 and $SeOX_2$ molecules have only six normal frequencies to be determined experimentally. However, their molecular force field is described by 13 force constants which may reproduce the experimental frequencies in numerous combinations. Studies on the molecular force field and/or mean amplitudes of vibration are:

SOF_2	*(a, b, c, d, e, f, g)*
$SOCl_2$	*(h, i, c, d, e, f)*
$SOBr_2$	*(c, e, g, j)*
$SeOF_2$	*(c, g, k)*
$SeOCl_2$	*(c, d)*

(a) Hargittai and Mijlhoff (1973)
(b) Lucas and Smith (1972)
(c) Müller and Nagarajan (1967a)
(d) Siebert (1954)
(e) Venkateswarlu, Thirugnanasambandam and Balasubramanian (1962)
(f) Pichai, Krishna Pillai and Ramaswamy (1967)
(g) Müller, Peacock and Heidborn (1968)
(h) Hargittai and Cyvin (1969)
(i) Ramaswamy and Jayaraman (1971a)
(j) Ramaswamy and Jayaraman (1972)
(k) Ramaswamy and Jayaraman (1970)

3.2. FURTHER OPEN-CHAIN SULPHOXIDES

Dimethyl sulphoxide, $(CH_3)_2SO$. An r_0 structure was determined by microwave spectroscopy by Dreizler and Dendl (1964). Subsequently more and more isotopic species were studied and finally enough data accumulated for a complete substitution structure reported by Feder, Dreizler, Rudolph and Typke (1969). The methyl groups are in staggered position with respect to the other S—C bond, but some deviation from the symmetrical form occurs. Since the methyl group is to a relatively large extent asymmetric in the r_s structure (the parameters are presented in Table 49), Typke (1978) examined whether and to what extent this asymmetry could be ascribed to vibrational effects. This study was made more difficult and more interesting by the fact that three different harmonic force fields were reported for dimethyl sulphoxide, *viz.* those by Horrocks and Cotton (1961), Tranquille, Labarbe, Fouassier and Forel (1971) and by Geiseler and Hanschmann (1971). The differences in the force fields originated at least partly from differences in the assignment of some of the vibrational frequencies. Employing the respective harmonic corrections, however, the resulting r_z structures

TABLE 49
Geometrical parameters of dimethyl sulphoxide from microwave spectroscopy.
The molecular model and labelling of the hydrogen atoms are shown in Fig. 64

Parameters	r_s *(a)*	r_z *(b)*
Bond lengths (pm)		
r(C—S)	179.9(5)	180.8(4)
r(S═O)	148.5(6)	148.5(5)
r(C—H′)	105.4(12)	105.2(39)
r(C—H″)	109.7(8)	109.8(3)
r(C—H‴)	109.3(8)	107.1(8)
Bond angles (°)		
∢C—S—C	96°34′(2′)	96.6(2)
∢C—S═O	106°39′(2′)	
∢S—C—H′	108°16′(6′)	108.3(24)
∢S—C—H″	108°10′(2′)	107.8(2)
∢S—C—H‴	109°33′(3′)	110.6(2)
Dihedral angles (°)		
C—S—C—H′		1.6(9)
C—S—C—H″		123.2(4)
C—S—C—H‴		243.6(3)
CSC/S═O		115.4(1)

(a) Feder, Dreizler, Rudolph and Typke (1969)
(b) Typke (1978)

did not differ significantly. Typke (1978) showed that the asymmetry of the methyl group did not diminish in the r_z structure as compared with the r_s structure. The molecular model and the labelling of the hydrogen atoms are shown in Fig. 64. The r_s and r_z parameters of dimethyl sulphoxide are given in Table 49. The r_z values cited are the mean of the three sets obtained from employing the three different sets of harmonic corrections. The largest changes occur in the S—C bond length and the position of the H‴ atom. However, even these changes are not significant.

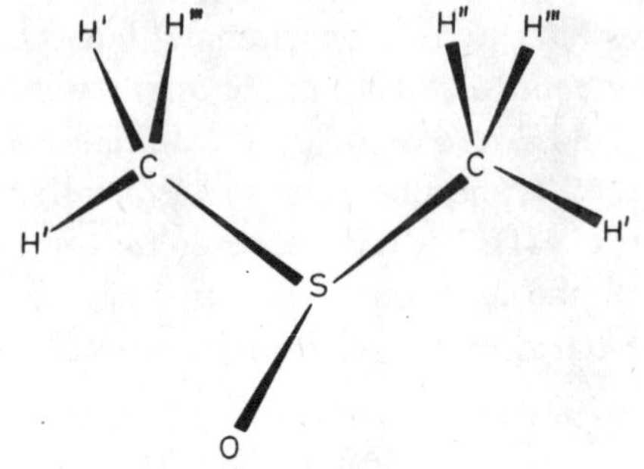

Fig. 64. Molecular model and labelling of the hydrogen atoms of dimethyl sulphoxide

The following geometrical parameters have been communicated from early electron diffraction investigations of dimethyl sulphoxide:

	Rundle (1950)	Bastiansen and Viervoll (1948)
r(C—S), pm	182	184(4)
r(S=O), pm	147	147(4)
∢C—S—C (°)	100(5)	
∢C—S=O (°)	107(5)	106(6)

A modern electron diffraction analysis of the dimethyl sulphoxide structure is desirable in order to determine the average geometrical parameters with a well-defined physical meaning and also to determine mean amplitudes of vibration from experimental diffraction data.

The crystal-phase molecular structure of dimethyl sulphoxide was reported in two independent x-ray diffraction investigations at about the same time. The results of Viswamitara and Kannan (1966) are much closer to the gas-phase structure than those of Thomas, Shoemaker and Eriks (1966). The most striking result of the latter is the 153.1(5) pm S=O bond length.

The molecular structures of two bis(trifluoromethyl)-sulphur(IV) compounds, *viz.* $(CF_3)_2S{=}O$ and $(CF_3)_2S{=}NCl$ have been recently determined by electron diffraction (Oberhammer, Kumar, Knerr and Shreeve, 1981) with the following parameters (r_a):

	$(CF_3)_2S{=}O$	$(CF_3)_2S{=}NCl$
r(C—F), pm	132.8(3)	131.7(2)
r(S=O/S=N), pm	146.9(4)	143.4(8)
r(S—C), pm	188.5(4)	187.8(6)
∢F—C—F (°)	109.0(2)	109.5(3)
∢C—S—C (°)	94.2(8)	99.4(9)
∢C—S=O/C—S=N (°)	104.5(11)	108.8(25)
∢S—N=Cl (°)		138(2)
CF_3 tilt (°)	3.6(5)	2.8(2)
CF_3 torsion (°)	5(4)	6.4(13)

The structures and their variations have been discussed in relation with fluorinated analogues and found to well correlate with the ligand electronegativity changes as accounted for by the VSEPR model.

Diphenyl sulphoxide, $(C_6H_5)_2SO$. The electron diffraction study by Rozsondai, Moore, Gregory and Hargittai (1979) led to a limited amount of structural information. The bond lengths (r_g) could be reliably determined:

r(C—C) (mean value)	139.8(3) pm
r(C—H) (mean value)	110.5(11) pm
r(C—S)	180.4(6) pm
r(S=O)	148.9(5) pm

The experimental radial distribution is reproduced in Fig. 18. The early x-ray diffraction investigation of diphenyl sulphoxide in the crystalline phase by Abrahams (1957) yielded very uncertain bond lengths. The bond angles determined for the crystalline phase, ∢C—S—C = 97.3° and ∢C—S=O = 102.2° are consistent with those found in other sulphoxides.

N,N′-Sulphinylbis(dimethylamine), $(CH_3)_2NSON(CH_3)_2$. The following geometrical parameters (r_a) were determined in the electron diffraction analysis by Hargittai and Vilkov (1970):

r(C—H)	110.3(8) pm	∢H—C—N	110.7(16)°
r(C—N)	146.0(8) pm	∢S—N—C	116.1(5)°
r(S=O)	148.0(9) pm	∢C—N—C	113.9(15)°
r(S—N)	169.3(4) pm	∢N—S—N	96.9(12)°
		∢O=S—N	105.5(8)°

As regards the rotational isomerism about the S—N bond, the conformation similar to those found in the analogous sulphide and sulphone (see later) was established as predominating,

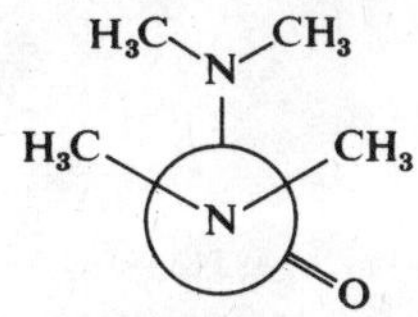

whereas the presence of other forms was indicated.

3.3. CYCLIC SULPHOXIDES

Ethylene sulphoxide, $(CH_2)_2SO$. Its normal isotopic species and also the ^{13}C and ^{34}S isotopic species were investigated by microwave spectroscopy by Saito (1969b) and White and Wollrab (1969). Since the sulphur atom is very near the centre of gravity of the molecule, the sulphur isotopic substitution provided very uncertain additional information. Thus four independent geometrical parameters were fitted to five rotational constants (Saito, 1969b). The parameters characterizing the hydrogen positions were assumed from the structure of ethylene sulphide (Cunningham, Boyd, Myers, Gwinn and Le Van, 1951). The results of the fitting were as follows:

r(S═O)	148.3(3) pm	∢C—S—C	48°46′(10′)
r(C—S)	182.2(3) pm	∢C—S═O	110°1′(10′)

These parameters correspond to a 150.4(3) pm C—C bond distance.

Trimethylene sulphoxide, $(CH_2)_3SO$. In addition to the normal species, eight additional isotopic species were investigated, using microwave spectroscopy, by Bevan, Legon and Millen (1974; 1977). An essentially substitution structure was determined which made it possible to examine the small structural distortions. Trimethylene sulphoxide is the first four-membered ring molecule for which such data became available. The molecular model and the numbering of atoms are given in Fig. 65a. The equatorial position of the S═O bond was unambiguously determined. The equatorial position was indeed found to correspond to a deep potential minimum favourable for the accurate structure determination. The geometrical parameters are cited in Table 50. Bevan, Legon and Millen (1977) also determined the *tilt*, *bend*, and *twist* of the 2-methylene and the *tilt* of the 3-methylene group. The *tilt* and the *bend* are illustrated in Fig. 65b. The interesting features of structural distortions and their stereochemical

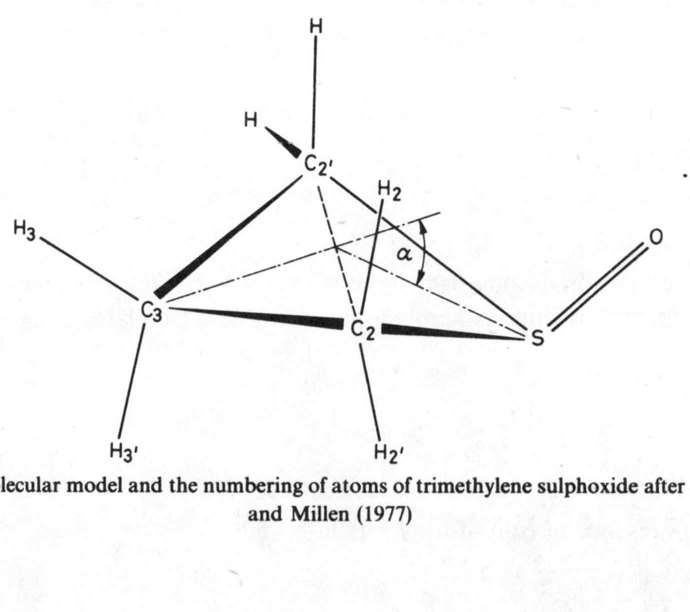

Fig. 65a. The molecular model and the numbering of atoms of trimethylene sulphoxide after Bevan, Legon and Millen (1977)

TABLE 50
Geometrical parameters of trimethylene sulphoxide from microwave spectroscopy by Bevan, Legon and Millen (1977). The molecular model and numbering of atoms are given in Fig. 65a

Bond lengths (pm)	
r(S=O)	147.5(3)
r(C—S)	183.6(3)
r(C—C)	154.2(5)
r(C2—H2)	109.9(10)
r(C2—H2′)	107.5(9)
r(C3—H3)	108.7(7)
r(C3—H3′)	109.8(8)
Bond angles (°)	
∢C—S—C	75.7(3)
∢C—C—C	93.9(4)
∢S—C—C	89.62(8)
∢C—S=O	113.4(5)
∢H2—C2—H2′	110.7(22)
∢H3—C3—H3′	108.8(28)
∢H2—C2—S	106.0(18)
∢H2′—C2—S	116.7(15)
α angle	34.9(6)

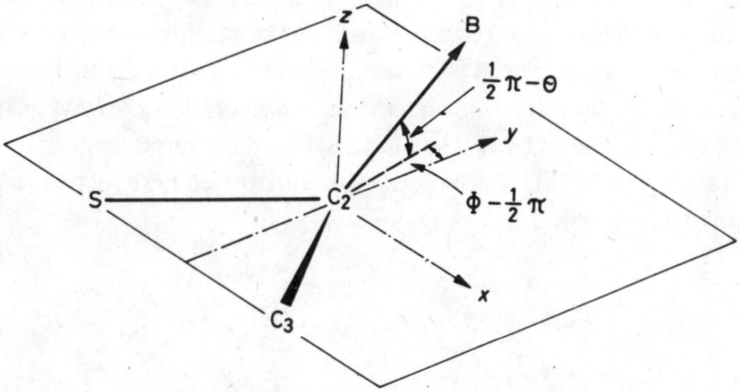

Fig. 65b. The tilt $\left(\frac{1}{2}\pi-\Theta\right)$ and bend $\left(\Phi-\frac{1}{2}\pi\right)$ of the methylene group of trimethylene sulphoxide after Bevan, Legon and Millen (1977). The C_2B vector is the bisector of the 2-methylene group

significance in four-membered ring molecules are discussed by Bevan, Legon and Millen (1974; 1977), Bartell and Andersen (1973), and by Mastryukov, Vilkov and Dorofeeva (1975).

3.4. CYCLIC SULPHITES

The conformational behaviour of cyclic sulphites has been extensively investigated by various techniques and mainly in solution. The relevant references can be found in the papers on electron diffraction structural studies to be cited below. The electron diffraction structure analysis of a cyclic sulphite is hindered by the relatively compact structure producing similar interatomic distances that cause strong correlation among the geometrical and also vibrational parameters. The usual mobility of these rings may cause further difficulties. Combined investigations of such structures involving various physical techniques have yet to appear. The enhanced possibilities of employing more than one physical technique simultaneously were somewhat indicated by the electron diffraction and nuclear magnetic resonance spectroscopic studies of 1,2-dimethyl ethylene sulphites by Geise and Van Laere (1975). The presence of groups adjacent to the cyclic sulphite moiety may make the determination of the ring conformation more reliable as the number of conformation-dependent interactions increases. This was well demonstrated by the conformational analysis of *o*-phenylene sulphite (Schultz, Serke and Kapovits, 1979). The presence of groups adjacent to the cyclic sulphite part probably decreases the ring mobility as well.

The conformational properties of cyclic sulphites and the present status of our knowledge about them are briefly summarized below. The results on *ethylene sulphite* as communicated by Arbuzov, Naumov, Zaripov and Pronicheva (1970) are rather uncertain. The experimental data could be best approximated by models containing planar or nearly planar five-membered rings. Mixtures of conformers could not be excluded and gas-phase vibrational spectroscopic data from Remizov, Kozlova, Vakhrushcheva and Mannafov (1973) also indicated this possibility. While the five-membered ring of *ethylene selenite* was reported to be unambiguously planar, no details on its analysis have been reported (Arbuzov, Naumov, Zaripov and Pronicheva, 1970).

The experimental data on *trans–trans* and *trans–cis 1,2-dimethyl ethylene sulphite* could be best approximated by models having five-membered rings with C_2 symmetry (Geise and Van Laere, 1975). These models are illustrated here by projections showing views along the C—C bond

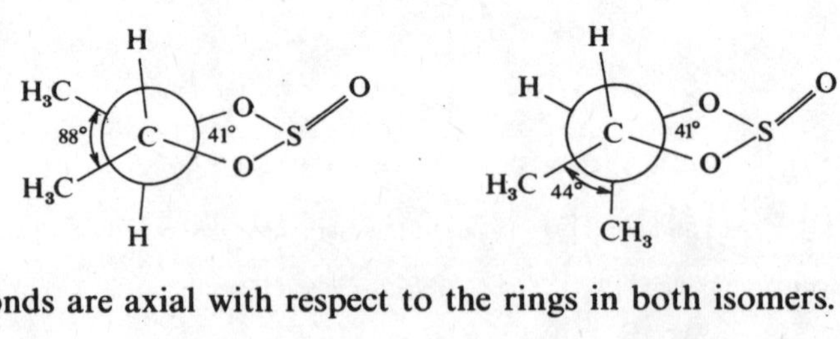

The S=O bonds are axial with respect to the rings in both isomers.

Trimethylene sulphite (Naumov, Zaripov and Shatrukov, 1970) and *trimethylene selenite* (Arbuzov, Naumov and Anonimova, 1970) both take chair forms with axial S=O and Se=O bonds, respectively. Such a structure of the selenite derivative received further support from a gas-phase infrared spectroscopic study by Remizov (1971). Racemic *1,3-dimethyl trimethylene sulphite* was also found to take a chair form with axial S=O bond (Mustoe and Hencher, 1972)

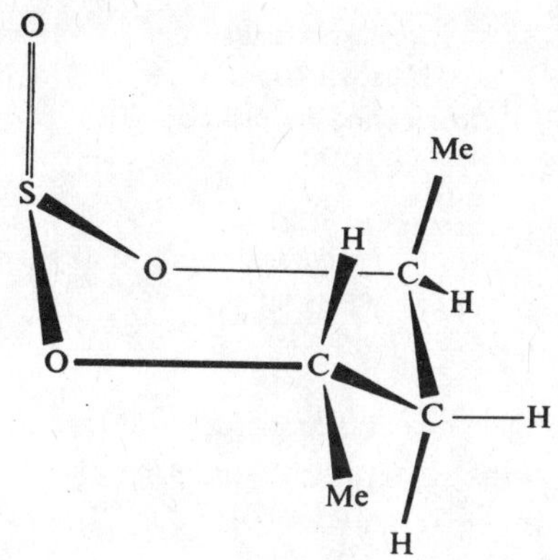

The five-membered ring of *o-phenylene sulphite* has the form of an envelope with an axial S=O bond as depicted in Fig. 66. In addition to the determination of the five-membered ring geometry, Schultz, Serke and Kapovits (1979) also established the elongation of the benzene ring in the direction parallel to the bond common to the two rings. The geometrical parameters characterizing the benzene ring in *o*-phenylene sulphite are shown in Fig. 66.

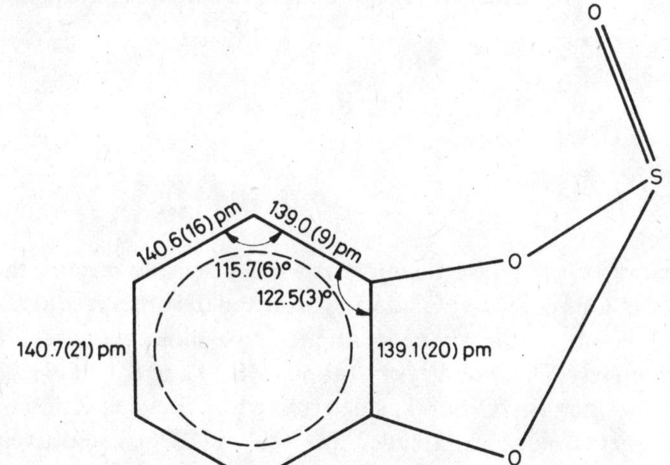

Fig. 66. The geometrical parameters characterizing the benzene ring deformation in *o*-phenylene sulphite after Schultz, Serke and Kapovits (1979)

The most important geometrical parameters of the cyclic sulphites studied to date are compiled in Table 51. The parameters of the *trans–trans* and *trans–cis* isomers of 1,2-dimethyl ethylene sulphite are essentially the same and, accordingly, only one set of data is communicated.

In addition, some parameters characterizing ethylene selenite and trimethylene selenite are cited here

	Ethylene selenite (Arbuzov, Naumov, Zaripov and Pronicheva, 1970)	*Trimethylene selenite* (Arbuzov, Naumov and Anonimova, 1970)
r(Se=O), pm	161.3(11)	160.8(9)
r(Se—O), pm	179.0(8)	177.7(8)
∢O=Se—O (°)	96(3)	115(2)
∢O—Se—O (°)	105(3)	101(1)

Finally some crystal-phase molecular structures are mentioned that may be of interest for comparison with gas-phase structures. 1,1,3-Trimethyl trimethylene

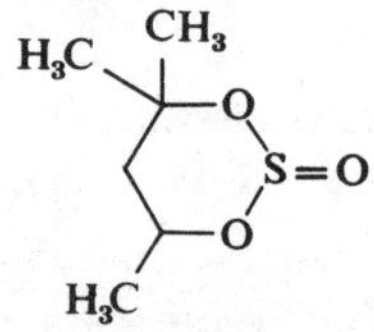

sulphite was studied by Petit, Lenstra, Van de Mieroop, Geise and Hellier (1978). This has been so far the only cyclic sulphite molecule with an S=O bond in equatorial orientation. The skeleton structure of β-thianthrene dioxide (Hosoya, 1966b) is similar

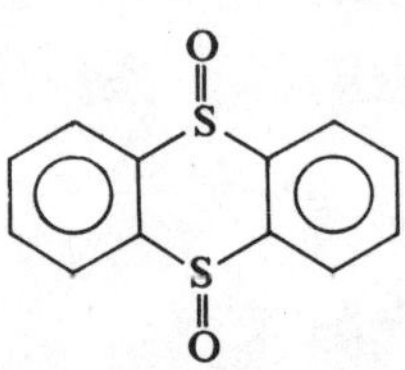

to the structure of gas-phase thianthrene itself discussed among the non-aromatic cyclic sulphides. One of the S=O bonds is axial and the other is equatorial. The lengths of the S=O bonds in the above-mentioned two molecules are 146.0(2) and 148–149 pm, respectively. The similarly crystal-phase S_7O and S_8O molecules have 148.5(3) and 148.3(9) pm long S=O bond as determined by Steudel, Reinhardt and Sandow (1977) and Luger, Bradaczek, Steudel and Rebsch (1976), respectively.

TABLE 51

Bond lengths and bond angles in free sulphite molecules from electron diffraction

Parameter	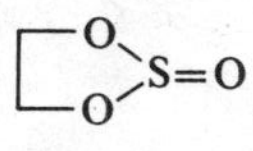 (a)	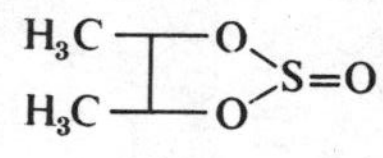 (b)	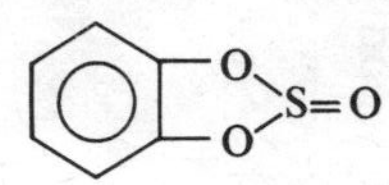 (c)	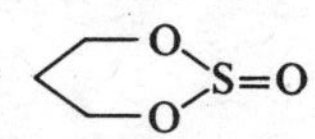 (d)	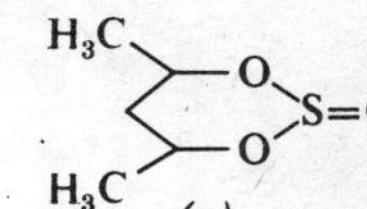 (e)
r(S=O), pm	143.9(15)	143.2(24)	142.7	145(2)	148.0(22)
r(S—O), pm	162.9(10)	162.5(6)	167.0	162(1)	162.2(9)
∢O=S—O (°)	104.5(30)	107.2	107.8	113.5(30)	116.7(14)
∢O—S—O (°)	102.0(3)	93.5(10)	93.0	99(2)	100.7(20)

(a) Arbuzov, Naumov, Zaripov and Pronicheva (1970)

(b) Mean parameters of the *trans–trans* and *trans–cis* isomers. The bond angle O=S—O has been calculated from other parameters. Geise and Van Laere (1975)

(c) Schultz, Serke and Kapovits (1979)

(d) Naumov, Zaripov and Shatrukov (1970)

(e) Mustoe and Hencher (1972)

3.5. CHARACTERISTIC GEOMETRICAL VARIATIONS

The S=O bond lengths and sulphur bond angles determined in free sulphoxide molecules are compiled in Table 52. Two features are noteworthy. One is that the S=O bond considerably shortens on going towards the sulphoxides with chlorine and fluorine ligands. The other is that the X—S—X bond angle is always smaller than the O=S—X bond angle. Both features are well understood on the basis of the VSEPR model. Also, the F—S—F bond angle is smaller than the other X—S—X bond angles and this is also consistent with considerations of the electron pair repulsions. The comparison of the O=S—F and O=S—Cl bond angles, however, leads to less obvious conclusions and here competing effects have to be considered within the VSEPR model. Similar structural variations will be examined in detail for the sulphone molecules invoking comparison with the analogous sulphoxides and sulphides as well.

As has been mentioned before, the geometrical differences between the $AOCl_2$ and AOF_2 molecules (A = S or Se) are well understood with the VSEPR model. The interpretation of the structural changes is more complex in the AOF_2, AO_2, $AOCl_2$ series, e.g. when the S=O bond length variation is considered:

$SOCl_2$	SO_2	SOF_2
144.3(5) pm	143.2(2) pm	142.1(3) pm
$SeOCl_2$	SeO_2	$SeOF_2$
161.4(5) pm	160.8(1) pm	157.6(3) pm

Substituting one of the two A=O bonds of the AO_2 molecule by two A—halogen bonds, the change in the length of the other A=O bond is apparently determined by the electronegativity of the halogen ligand. The shortest A=O bond occurs when the ligand has the largest electronegativity. However, the electron pair repulsions depend also on the double-bond character of the bonds involved. In accordance with the general observation, the A—F bonds have larger double-bond character in the AOX_2 molecules than the A—Cl bonds. This can be seen by comparing the experimental bond lengths with those calculated for single bonds, as follows:

	r_1 (pm) Pauling (1960)	r_2 (pm) Glidewell (1976)	Observed (pm)	Δr_1 (pm)	Δr_2 (pm)
S—F	164	156	158	6	−2
S—Cl	200	200	208	−8	−8
Se—F	176	169	173	3	−4
Se—Cl	214	213	220	−6	−7

Note that the trend of the difference in the double-bond character of the fluorine and chlorine bonds according to the more recent data is consistent with the earlier observation although the actual amount of the difference is smaller. In any case, on the basis of the comparison of the double-bond character, the A=O bond of $AOCl_2$ molecules could be expected to be shorter than that of AOF_2 molecules. However, of

TABLE 52
Characterization of the sulphur bond configuration in X_2SO sulphoxides

X	r(S=O), pm	∢O=S—X (°)	∢X—S—X (°)
F	142.1(3)	106.2(3)	92.2(3)
Cl	144.4(6)	106.4(6)	96.2(7)
$(CH_3)_2N$	148.1(9)	105.5(8)	96.9(12)
C_6H_5	148.9(5)		
CH_3	148.6(7)	106.5(3)	96.6(3)

the two competing effects, *viz.* ligandum-electronegativity *vs.* double-bond character, according to the experimental evidence on the geometries it is the electronegativity effect that appears to be dominating.

As regards the geometrical parameters in the series of the cyclic sulphites no detailed comparison is feasible because of the relatively large experimental errors. Furthermore, it would also be interesting to compare the structures of the cyclic sulphites with those of analogous cyclic phosphites. In both compound series the bond configuration of the central sulphur and phosphorus atoms (A) may be characterized by the formula AX_3E where X is a ligand and E is a lone pair of electrons. The structural information on free cyclic phosphite molecules was reviewed by Naumov (1975). The conformational properties of analogous cyclic sulphites and phosphites are similar indeed. The available geometrical parameters of the cyclic phosphites are also very uncertain. Only one structural feature is chosen here for comparison, namely the angle between the (exocyclic) axial bond and the plane of the (endocyclic) two other bonds of the AX_3E

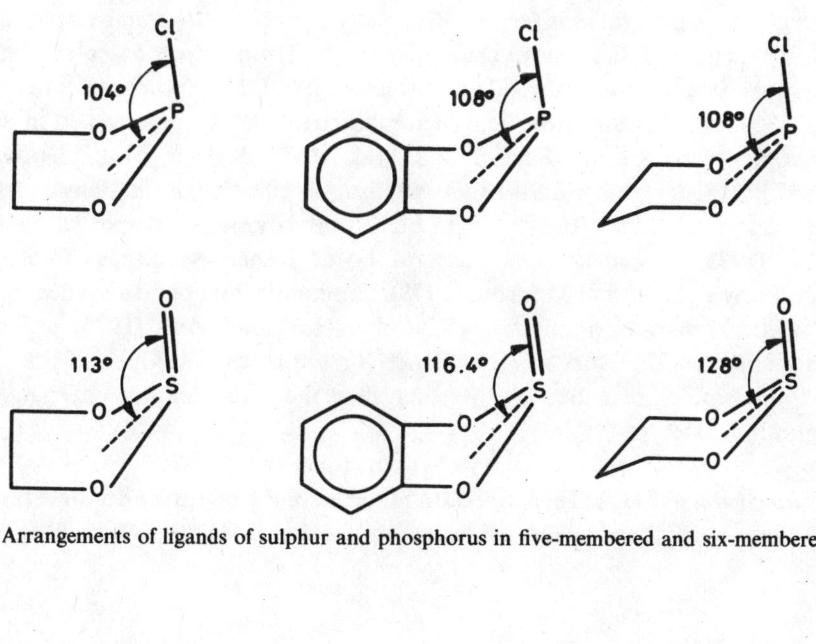

Fig. 67. Arrangements of ligands of sulphur and phosphorus in five-membered and six-membered rings

configuration. According to the data presented in Fig. 67, this angle is larger in the sulphur derivatives than that in the phosphorus compounds. This difference may be interpreted, for example, as the *relatively* larger space requirement of the S=O bond as compared with that of the sulphur lone pair *vs.* the space requirement of the P=O bond (and even more so of the P—Cl bond) as compared with the phosphorus lone pair. This supposition seems to get further support by some structures in which phosphorus has PX_4 configuration rather than PX_3E

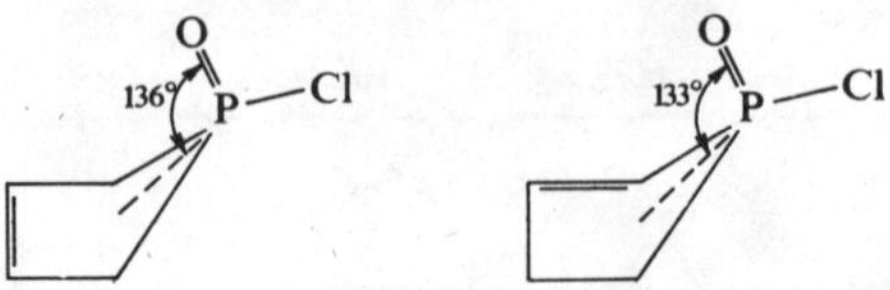

Here the lone pair of electrons is "substituted" by yet another bonding pair which, of course, exercises much less repulsion towards the other bonds than the lone pair. The considerable increase of the angle examined above (from 104–108° to 133–136°) indicates that the electron pair repulsions may indeed be of great importance in these structures.

3.6. OTHER STRUCTURES

Sulphur trioxide, SO_3. The geometrical parameters (r_a) of this D_{3h} symmetry molecule from an electron diffraction investigation (Clark and Beagley, 1971) are as follows:

r(S=O) 141.8(3) pm r(O...O) 245.7(9) pm
∢O=S=O 120.0(6)°

The molecule is quite rigid as is seen by the relatively small vibrational amplitudes and shrinkage effects, and their slow change with the temperature as calculated from spectroscopic data (Cyvin, Brunvoll and Stølevik, 1969). The numerical data are cited in Table 53. The molecular force field of sulphur trioxide was determined in detailed spectroscopic investigations (Kaldor and Maki, 1973; Kaldor, Maki, Dorney and Mills, 1973; Dorney, Hoy and Mills, 1973). An effective bond length (r_0) was determined to be 141.98 ± 0.02 pm from high-resolution infrared spectra by Kaldor and Maki (1973) as well as from pure rotational Raman spectra by Brassington, Edwards, Farwell, Long and Mansour (1978). The equilibrium bond distance, r_e(S=O) $= 141.84 \pm 0.10$ pm, was obtained by Dorney, Hoy and Mills (1973) utilizing an anharmonic force field. Additional evidence for the relative stiffness of the sulphur trioxide molecule is provided by the fact that the bond lengths in r_a, r_0 and r_e representation differ so little.

Selenium trioxide, SeO_3. There are data reported in the literature on the structure of vapour-phase monomeric species of selenium trioxide (Mijlhoff, 1965). They are not

TABLE 53
Mean amplitudes of vibration and shrinkage effects of sulphur trioxide ($^{32}S^{16}O_3$)

T(K)	Mean amplitudes of vibration (pm)		Shrinkage (pm)
	l(S=O)	l(O...O)	
Calculated *(a)*			
0	3.48	5.14	0.20
298	3.49	5.38	0.23
500	3.58	5.96	0.31
1000	4.09	7.60	0.55
1500	4.71	9.09	0.80
2000	5.30	10.41	1.06
Observed *(b)*	4.8(5)	6.7(10)	

(a) Cyvin, Brunvoll and Stølevik (1969)
(b) Clark and Beagley (1971); Electron diffraction
The experimental determination of the mean amplitudes of vibration was somewhat hindered by the presence of a small amount of trimeric species

cited here, however, since subsequent studies of the vapour composition established that only tetrameric species were present in any appreciable amount in the vapour of selenium trioxide in the experimental conditions used (Paetzold and Amouloung, 1965; Bleijerveld and Mijlhoff, 1975).

Dimethyl methylthio aluminium dimer, $[(CH_3)_2AlSCH_3]_2$. Haaland, Stokkeland and Weidlein (1975) determined its molecular structure by electron diffraction. The experimental data could be well approximated by a model of C_{2h} symmetry in which the methyl groups linked to the sulphur atoms take *anti* positions with respect to each other. The presence of a small amount of the *syn* form could not be ruled out. The most important geometrical parameters are given in Fig. 68a. The dimethyl methylthio

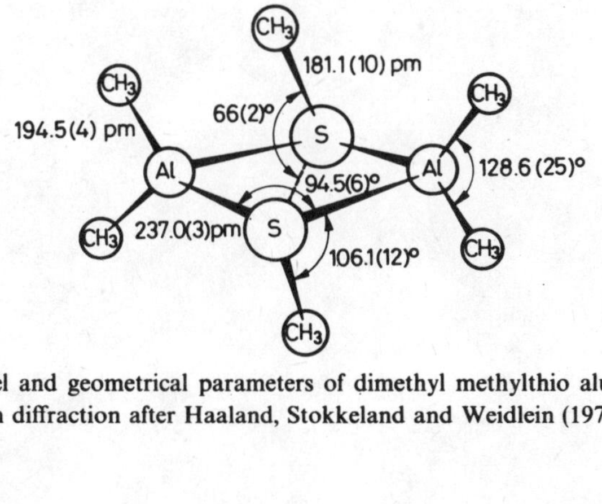

Fig. 68a. Molecular model and geometrical parameters of dimethyl methylthio aluminium dimer from electron diffraction after Haaland, Stokkeland and Weidlein (1975)

aluminium molecules form an infinite chain in the crystalline phase, where the dimethylaluminium groups are bridged by the methylthio groups (Brauer and Stucky, 1969).

Trimethylaluminium dimethylsulphide, $(CH_3)_3AlS(CH_3)_2$. Fernholt, Haaland, Hargittai, Seip and Weidlein (1981) determined its molecular structure by electron diffraction. The molecular model and geometrical parameters are shown in Fig. 68b. The relative orientation of the CSC plane and the Al—S bond is intermediate between those that would be characteristic either for trigonal or for tetrahedral sulphur bond configuration.

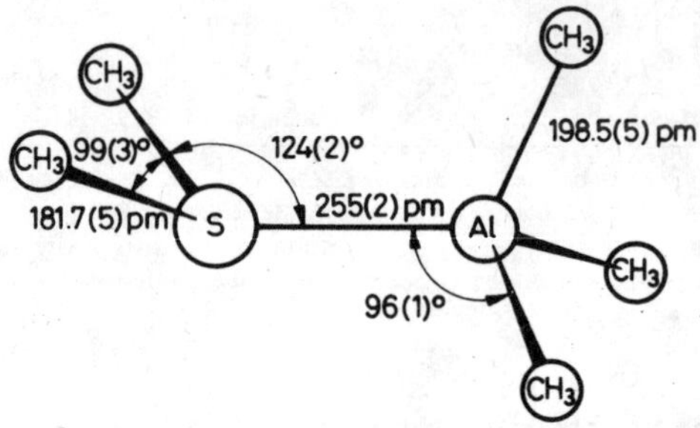

Fig. 68b. Molecular model and geometrical parameters of trimethylaluminium dimethylsulphide from electron diffraction after Fernholt, Haaland, Hargittai, Seip and Weidlein (1981)

Chapter 4

MOLECULES WITH FOUR-COORDINATED SULPHUR

4.1. SULPHUR TETRAFLUORIDE

This molecule possesses C_{2v} symmetry with two different S—F bonds, axial (ax) and equatorial (eq)

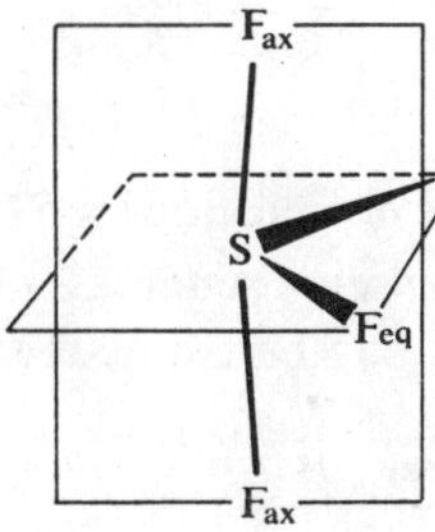

It has a trigonal bipyramidal structure with the sulphur lone electron pair in the equatorial plane. The molecular geometry of sulphur tetrafluoride was investigated by electron diffraction (Kimura and Bauer, 1963; Ewing and Sutton, 1963) and by microwave spectroscopy (Tolles and Gwinn, 1962; Stone, Tigelaar and Flygare, 1970). The bond lengths and bond angles are cited here after Tolles and Gwinn (1962):

$r(S—F_{ax})$	164.6(3) pm	$\measuredangle F_{ax}—S—F_{ax}$	173.1(5)°
$r(S—F_{eq})$	154.5(3) pm	$\measuredangle F_{eq}—S—F_{eq}$	101.6(5)°

with the $S—F_{ax}$ bonds bending away from the sulphur lone electron pair. This information is needed since the direction of the deviation of the $F_{ax}—S—F_{ax}$ chain from linearity is not implied in the value of the bond angle*. The above parameters were obtained in a fitting procedure to the rotational constants of two isotopic species. There is excellent agreement between the above geometry and the F...F distances that can be directly derived from the planar moments.

* In fact, the original paper lists this angle as 186.9(5)°.

The structure of the bis(trifluoromethyl)-substituted sulphur tetrafluoride derivative $(CF_3)_2SF_2$, is also characterized by trigonal bipyramidal configuration with the CF_3 ligands in the equatorial positions. Its geometry was determined by electron diffraction (Oberhammer, Kumar, Knerr and Shreeve, 1981), with the following parameters (r_a):

r(C—F)	131.7(2) pm	∢F—C—F	109.3(2)°
r(S—F_a)	168.1(3) pm	∢C—S—C	97.3(8)°
r(S—C)	188.8(4) pm	∢F_a—S—F_a	173.9(8)°

The structural variations in SF_4 and $(CF_3)_2SF_2$ and their compatibility with the VSEPR model is discussed in detail in the section on thionyl tetrafluoride following the discussion of other trigonal bipyramidal configurations.

Selenium tetrafluoride, SeF_4, was studied both by electron diffraction (Bowen, 1953) and microwave spectroscopy (Bowater, Brown and Burden, 1968a). The latter investigation yielded the following parameters:

r(Se—F_{ax})	177.1(4) pm	∢F_{ax}—S—F_{ax}	169.2(7)°
r(Se—F_{eq})	168.2(4) pm	∢F_{eq}—S—F_{eq}	100.6(7)°

Of the available mean amplitudes of vibration for SF_4 and analogues, some are cited below. For sulphur tetrafluoride, experimental electron diffraction (ED) and spectroscopic calculated (SP) data may be compared:

	SF_4, l(pm)			$TeCl_4$, l(pm)
	ED*(a)*	SP*(b)*		SP*(c)*
S—F_{ax}	4.7(5)	4.6	Te—Cl_{ax}	5.7
S—F_{eq}	4.1(5)	4.3	Te—Cl_{eq}	4.7
$F_{ax}\cdots F_{ax}$	5.9(1)	6.0	$Cl_{ax}\cdots Cl_{ax}$	7.8
$F_{eq}\cdots F_{eq}$	6.8(1)	11.3	$Cl_{eq}\cdots Cl_{eq}$	10.2
$F_{ax}\cdots F_{eq}$	6.7(5)	7.1	$Cl_{ax}\cdots Cl_{eq}$	12.1

(a) Kimura and Bauer (1963)
(b) Cyvin (1969b)
(c) Cyvin, Cyvin, Brockner and Demiray (1978)

The mean amplitudes of vibration of the longer and presumably weaker axial bonds are larger than those of the equatorial bonds. Some other structural features of sulphur tetrafluoride will be discussed later in relation to the thionyl tetrafluoride structure.

4.2. TRIFLUOROSULPHUR NITRIDE

The first step in the elucidation of its structure has been the determination of atom connectivity. Although analogy with $S{=}PF_3$ suggested a similar structural formula*

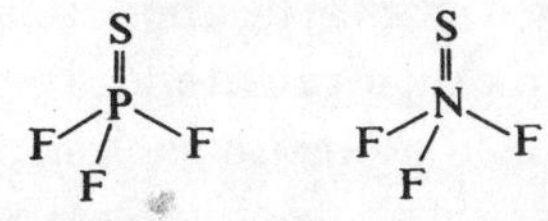

it could be ruled out on the basis of the microwave spectrum by Kirchhoff and Wilson (1962) and the following arrangement was unambiguously established

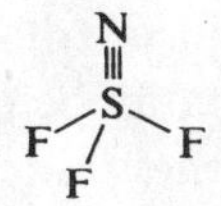

An essentially substitution structure has been derived with the following parameters:

$r(S{\equiv}N)$ 141.6(3) pm $r(S{-}F)$ 155.2(3) pm
$\measuredangle F{-}S{-}F$ 94.0(3)°

4.3. SULPHONE MOLECULES

The sulphones are a group of compounds whose molecular structure has recently become relatively well studied (*cf.* Hargittai, 1978). Before reviewing the structures and their variations, let us consider some of the specifics of the investigation of the sulphone molecules. It is well known that the frequencies of rotational transitions can be determined from the microwave spectrum with great precision and reproducibility. At the same time it is also true that microwave spectroscopy is capable to determine only relatively simple sulphone structures. The rotational spectra of the isotopic species occurring in their natural abundance do not provide sufficient data for a complete structural determination even of the simplest sulphone molecules. One possibility for improving this situation is to synthesize and investigate further isotopic species, which is not always easy and may be in some cases impossible. Another approach is to assume some of the parameters on the basis of other data and then to determine the rest. The influence of such assumptions has to be critically examined in each case.

A very fortunate situation is encountered in the case of sulphones which makes it possible to gain a limited amount of reliable structural information from the microwave spectra. The planar moment of the XSO_2Y sulphone molecules is determined by the masses and positions of the oxygen atoms situated outside the X—S—Y plane, provided that the vibrational contribution (δ) due to the intramolecular

* For example $O{=}PF_3$ has indeed an analogue $O{=}NF_3$ (for its structure see Plato, Hartford and Hedberg, 1970).

motion can be neglected. Thus the distance between the oxygen atoms, $r_0(\mathrm{O}\ldots\mathrm{O})$, can be extracted on the basis of the following expression

$$I_a+I_b-I_c=\Delta I=m_\mathrm{O}r_0^2(\mathrm{O}\ldots\mathrm{O})-\delta$$

where m_O is the mass of the oxygen atom. The above relationship is valid for molecules in which X and Y are, for example, halogen atoms. In the case of dimethyl sulphone, however, only one of the hydrogen atoms in each methyl group is in the X—S—Y plane, while the other two also contribute to the planar moment; accordingly,

$$I_a+I_b-I_c=\Delta I=m_\mathrm{O}r_0^2(\mathrm{O}\ldots\mathrm{O})+2m_\mathrm{H}r_0^2(\mathrm{H}\ldots\mathrm{H})-\delta$$

Note that the planar moment is obtained from the principal moments of inertia not always by the above combination, the latter being determined by the orientation of the principal axes. The orientation of the principal axes is not always established unambiguously. Let us consider, for example, the microwave spectroscopic investigation of sulphuryl fluoride (Lide, Mann and Fristrom, 1957). The rotational spectra of two isotopic species have been studied. Two sets of rotational constants, and accordingly, two sets of principal moments of inertia were produced for each species, differing by some 0.3% depending on the choice of the principal axis orientation. The consequence of this choice is, of course, well seen in the respective O . . . O distances. In the case of sulphuryl fluoride, the F . . . F distances can be obtained and examined in a similar way, so they are also included in the comparison:

	$^{32}SO_2F_2$	$^{34}SO_2F_2$
first version		
$\Delta I\times10^{-4}$ (amu pm^2)	98.147	98.158
$r_0(\mathrm{O}\ldots\mathrm{O})$, pm	247.7	247.7
$r_0(\mathrm{F}\ldots\mathrm{F})$, pm	227.3	227.3
second version		
$\Delta I\times10^{-4}$ (amu pm^2)	98.777	98.786
$r_0(\mathrm{O}\ldots\mathrm{O})$, pm	248.5	248.5
$r_0(\mathrm{F}\ldots\mathrm{F})$, pm	228.0	228.0

The planar moments of the sulphone molecules studied by microwave spectroscopy are collected in Table 54. There may be some small numerical discrepancies between the values given in this table and those given in the original papers since we used the conversion factor 505, 379 × 10^4 MHz amu pm^2 from the IUPAC recommendations (Cohen and Taylor, 1973). The contributions from the methyl and methylene groups were estimated to be uniformly 3.266 and 3.730 × 10^4 amu pm^2, respectively, using 180 and 185 pm for the distances between the methyl hydrogens and methylene hydrogens in this order. The corrected planar moments were obtained by subtracting the contributions of the methyl and methylene groups from the planar moments as calculated from the principal moments of inertia. The corrected planar moments

TABLE 54
Planar moments in sulphone molecules investigated by microwave spectroscopy (in 10^4 amu pm^2)

	Measured *(a)*	Corrected for hydrogen contribution *(b)*
FSO_2F	98.147 (98.777) *(c)*	98.147 (98.777) *(c)*
$ClSO_2Cl$	98.753	98.753
FSO_2Cl	98.700	98.700
FSO_2Br	98.851	98.851
CH_3SO_2F	101.524	98.258
CH_3SO_2Cl	101.904	98.638
CH_3OSO_2F	102.009	98.743
$CH_3SO_2CH_3$	105.628	99.096
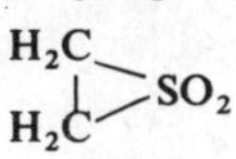	107.507	100.047
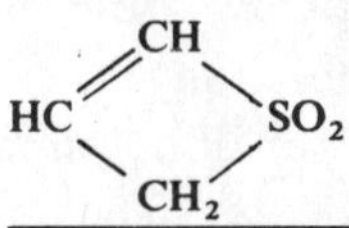	101.547	97.817

(a) For sources see text; Conversion factor 505,379 × 10^4 MHz amu pm^2 (Cohen and Taylor, 1973)
(b) 3.266 × 10^4 and 3.730 × 10^4 amu pm^2 have been subtracted for methyl and methylene contributions, respectively
(c) According to two different orientations of the principal axes

appear to be very similar indicating that the O. . . O distances are very much the same in the various sulphone molecules. The mean value of the planar moments corrected for the contribution of hydrogen atoms is 98.8 $\times 10^4$ amu pm^2 with a standard deviation (σ) of 0.6 $\times 10^4$ amu pm^2. Here the data on sulphuryl fluoride are disregarded (and the data of dimethyl sulphone are considered only once, of course). If, furthermore, the data on the two ring sulphone molecules are also disregarded, the mean value is 98.72 and σ is 0.25 (both 10^4 amu pm^2). These results may or may not suggest that of the two data sets on SO_2F_2 the latter is to be preferred (Lide, Mann and Fristrom, 1957, favoured the former one).

The O. . . O distances determined from the planar moments will be given later, in the discussion of structural variations. However, some comments are made here concerning the estimation of uncertainty of the O. . . O distances obtained in this way. There are two main sources of error: one is the uncertainty in estimating the H. . . H distances and the other is the deviation from zero of the δ vibrational contribution. An error of 5 pm in r(H. . . H) corresponds to an 0.2 pm change in r(O. . . O). The error limit increases then to $\pm$0.3 pm if an uncertainty of 0.2×10^4 amu pm^2 is considered to be introduced by the term δ. The experimental error in the determination of the rotational constants has no appreciable effect in the determination of the O. . . O distance.

Electron diffraction may be successfully applied to the elucidation of more complicated sulphone molecules. This is why the major part of the information on the conformational behaviour of sulphone molecules originates from electron scattering data. It is also true, however, that the electron diffraction data are more subject to misinterpretation than rotational spectroscopic data owing to the possibility of reproducing the experimental data equally well with various models.

The length of the sulphur–oxygen bonds as well as the other bonds can usually be determined with great accuracy from the electron diffraction data. There is, however, a special difficulty in the determination of the sulphur bond angles of the sulphone molecules. The sulphur atom in sulphones has a tetrahedral configuration and, accordingly, there are very similar non-bond distances around the sulphur atom. In the electron diffraction analysis this causes strong correlation among these distances (and naturally also among the distances and the mean vibrational amplitudes associated with them). Since the bond angles are calculated from the experimentally determined non-bond (and bond) distances, the above-mentioned correlation presents a major hindrance to the unambiguous determination of the bond angles.

Of the sulphone molecules so far investigated by electron diffraction, only sulphuryl chloride has a radial distribution (Fig. 69) exhibiting distinguishable features for all different internuclear distances including the three non-bond distances, *viz.* O...O, Cl...O, and Cl...Cl. For the radial distribution of the more compact sulphuryl fluoride (Fig. 70), the contributions from all non-bond distances, *viz.* O...O, F...O, and F...F, appear under the same maximum. The picture is even more complicated with the radial distribution of dimethyl sulphone (Fig. 71) which possesses, by the way, the same C_{2v} symmetry as the previously mentioned sulphuryl chloride and sulphuryl fluoride. The S...H distances in dimethyl sulphone appear together with the non-bond distances determining the bond angles around the sulphur atom. Because of lower symmetry, *viz.* C_s, there are more interaction types close to the same region on the radial distributions of methane sulphonyl fluoride and methane sulphonyl chloride (Fig. 71). Here it is seen that the non-bond distances are again more closely packed on

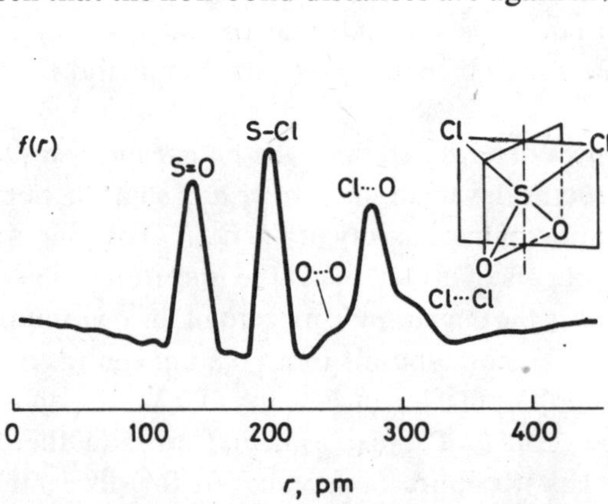

Fig. 69. Radial distribution of sulphuryl chloride (Hargittai and Hargittai, 1981)

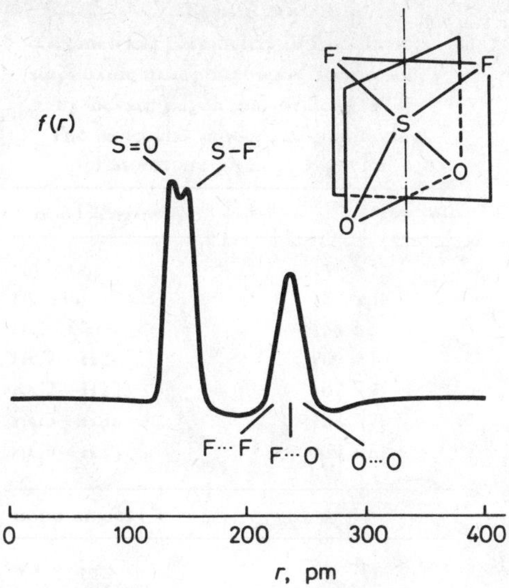

Fig. 70. Radial distribution of sulphuryl fluoride (Hagen, Cross and Hedberg, 1978)

the radial distribution of the more compact methane sulphonyl fluoride than on that of methane sulphonyl chloride.

When there is strong correlation among the parameters, it may easily happen that the experimental data are equally well reproduced by more than one model and a decision among the models can be made only on the basis of further independent evidence.

The difficulties encountered in the electron diffraction determination of the non-bond distances around the sulphur atom on one hand, and the possibility of the unambiguous determination of the O . . . O distance from the microwave spectra on the other hand, led us to recognize the advantages of combining electron diffraction and microwave spectroscopic information in the determination of sulphone molecular geometries. This approach, applied with necessary caution and within certain requirements of accuracy, resulted in the reliable determination of the geometry of SO_2 group in several molecules (see e.g. Hargittai and Hargittai, 1973b; 1974; Hargittai, Seip, Nair, Britt, Boggs and Cyvin, 1977). The above warning for caution refers primarily to the problem of the difference in the physical meaning of parameters from the two techniques (electron diffraction r_a distances and microwave spectroscopic r_0(O. . .O) distance), since an uncritical combination of data could lead to the introduction of systematic errors. It is true, however, that the vibrational effects are expected to be small, since the molecules in question and especially the SO_2 parts are quite rigid. In any case, it is advantageous to remove any constraint on the O. . .O distance when the least-squares procedure converges and obtains a consistent set of parameters. The importance of utilization of rotational spectroscopic information in

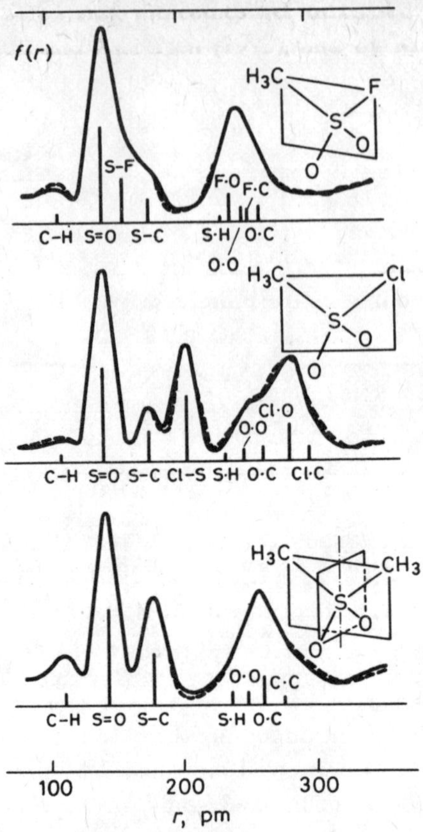

Fig. 71. Radial distribution of methane sulphonyl fluoride (Hargittai and Hargittai, 1973b), methane sulphonyl chloride (Hargittai and Hargittai, 1973), and dimethyl sulphone (Hargittai and Hargittai, 1974).

Dashed lines — experimental
Full lines — theoretical

the electron diffraction structure analysis was emphasized above. Needless to say, the utilization of vibrational spectroscopic information is also of utmost importance in the structural studies on sulphones. In particular, reliable calculated mean amplitudes of vibration may decrease the difficulties presented by the strong correlation among the geometrical and vibrational parameters caused by the tetrahedral bond configuration of sulphur in the electron diffraction analysis.

The molecular geometry of the following sulphones has been determined in the vapour phase by electron diffraction or/and microwave spectroscopy:

FSO_2F	$ClSO_2Cl$	$FO_2SOSO_2OSO_2F$
FSO_2OSO_2F	$ClSO_2CF_3$	$CH_3OSO_2OCH_3$
FSO_2Cl	$ClSO_2CCl_3$	$CH_3SO_2CH_3$
FSO_2Br	$ClSO_2OCH_3$	$CH_2CHSO_2CHCH_2$

FSO_2OCH_3 $ClSO_2CH_3$ $C_6H_5SO_2C_6H_5$
FSO_2CH_3 $ClSO_2CHCH_2$ $(CH_3)_2NSO_2N(CH_3)_2$
$ClSO_2C_6H_5$

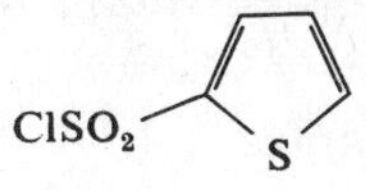
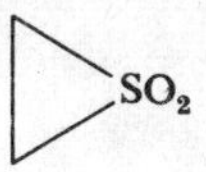

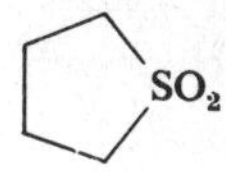

$(ClSO_2)_2NH$
$ClSO_2N(CH_3)_2$
$ClSO_2NCO$

4.4. SULPHURYL DIHALIDES

Sulphuryl fluoride (sulphonyl fluoride), SO_2F_2. This molecule was studied both by microwave spectroscopy (Lide, Mann and Fristrom, 1957) and by electron diffraction (Hagen, Cross and Hedberg, 1978). The bond lengths and bond angles are collected in Table 55. Considering the experimental errors, the two data sets are not in serious disagreement except, perhaps, the S=O bond length. The discrepancies occurring in the S=O bond length and O=S=O bond angle will be discussed in the sections on the characteristic structural variations and on the correlation between geometrical and vibrational parameters.

Sulphuryl chloride, SO_2Cl_2. The bond lengths and bond angles from its most recent electron diffraction reinvestigation (Hargittai and Hargittai, 1981) are given in Table 55. The first electron diffraction study of sulphuryl chloride (Palmer, 1938) used the visual technique. The first sector-microphotometer reinvestigation (Hargittai, 1968a)

TABLE 55
Bond lengths and bond angles in sulphuryl fluoride and sulphuryl chloride

X = F, Cl	SO_2F_2 (a)	SO_2F_2 (b)	SO_2Cl_2 (c)
r(S=O), pm	140.5(3)	139.7(2)	141.8(3)
r(S—X), pm	153.0(3)	153.0(2)	201.2(4)
∡O=S=O (°)	124.0(2)	122.6(12)	123.5(2)
∡X—S—X (°)	96.1(2)	96.7(11)	100.3(2)
∡X—S=O (°)		108.6(2)	108.0(1)

(a) Microwave spectroscopy, r_0 parameters; Lide, Mann and Fristrom (1957)
(b) Electron diffraction, r_a parameters; Hagen, Cross and Hedberg (1978)
(c) Electron diffraction, r_g bond lengths and r_α bond angles; Hargittai and Hargittai (1981)

was based on a poor approximation of the atomic scattering factors. A reanalysis of the same experimental data employing complex electron scattering factors revealed some sensitivity of some of the geometrical parameters to changes in the empirical background (Hargittai, 1969). The overall structure, however, was well determined and was extensively used in discussing sulphone molecular structures (*cf.* Hargittai, 1978). Subsequently, however, with the appearance of revised empirical relationships involving geometrical and vibrational parameters of sulphone molecules (*cf.* Hargittai, 1978; 1979a), the need for more accurate data has arisen, hence the results given in Table 55. In calculating the r_α bond angles, the following perpendicular correction terms (K, pm) were used: S═O 0.35, S—Cl 0.15, O...Cl 0.14, O...O 0.31, and Cl...Cl 0.05 (Hargittai and Brunvoll, 1979b).

Microwave spectroscopic studies of sulphuryl chloride were also performed (Dubrulle and Boucher, 1974; Abbar, 1963; 1965; Abbar, Journel and Moise, 1965) which yielded rotational constants only and no geometrical data. Using the rotational constants communicated by Dubrulle and Boucher (1974) for the ground state of the $SO_2{}^{35}Cl^{37}Cl$,

$$A = 3459.39 \pm 0.02 \text{ MHz}$$
$$B = 2293.83 \pm 0.02 \text{ MHz}$$
$$C = 1888.16 \pm 0.04 \text{ MHz},$$

the O...O distance (r_0) is calculated to be 248.5 pm.

Sulphuryl chloride fluoride, SO_2ClF. The microwave rotational spectra (Holt and Gerry, 1971) of the two most abundant isotopic species allowed an unambiguous determination of the O...O distance only (248.4 pm). Supposing that r(S═O) in sulphuryl chloride fluoride is intermediate (*viz.* 140.8±0.6 pm) between those in sulphuryl fluoride (Lide, Mann and Fristrom, 1957) and sulphuryl chloride (Hargittai, 1968a; 140.9 pm), and that the F—S—Cl bond angle is intermediate (*viz.* 99±3°) between ∢F—S—F of SO_2F_2 and ∢Cl—S—Cl of SO_2Cl_2, the following bond lengths and bond angles were determined by Holt and Gerry (1971):

r(S—F)	155(2) pm	∢O═S═O	123.7(10)°
r(S—Cl)	198.5(15) pm	∢O═S—F	107.5(20)°
	∢O═S—Cl	107.5(25)°	

Sulphuryl bromide fluoride, SO_2BrF. In a microwave spectroscopic study by Raley, Wollrab and Lovejoy (1973) similar to the one on sulphuryl chloride fluoride, using the assumptions: r(S═O) = 140.7 pm, r(S—F) = 156 pm, and ∢O═S═O = 123.7°, the following two parameters were determined:

r(S—Br)	215.5±0.6 pm
∢F—S—Br	100.6±1.5°

The mean amplitudes of vibration of simple sulphuryl halides as determined from electron diffraction and calculated from spectroscopic data are collected in Table 56.

TABLE 56
Mean amplitudes of vibration (pm) of sulphonyl dihalides at 298 K from electron diffraction (ED) and spectroscopic calculations (SP)

Atomic pair	SO_2F_2		SO_2Cl_2		SO_2FCl	SO_2FBr
	SP	ED *(d)*	SP	ED *(e)*	SP	SP
S=O	3.4 *(a)*	3.3(2)	3.5 *(e)*	3.5(7)	3.5 *(f)*	3.5 *(f)*
	4.0 *(b)*		4.0 *(b)*		3.4 *(g)*	3.5 *(g)*
	2.45 *(c)*					
S—F	4.1 *(a)*	4.0(2)			4.3 *(f)*	4.3 *(f)*
	4.3 *(b)*				4.2 *(g)*	4.3 *(g)*
	2.45 *(c)*					
S—Cl			4.75 *(e)*	4.6(4)	4.4 *(f)*	
			5.1 *(b)*		4.9 *(g)*	
S—Br						4.8 *(f)*
						4.9 *(g)*
O...O	5.2 *(a)*	7.1(7)	6.65 *(e)*	[6.1]	5.6 *(f)*	5.5 *(f)*
	6.0 *(a)*		6.1 *(b)*		5.8 *(g)*	5.6 *(g)*
	5.8 *(b)*					
	5.6 *(c)*					
F...F	5.7 *(a)*	7.1				
	6.0 *(a)*					
	6.6 *(b)*					
	8.1 *(c)*					
F...Cl					10.6 *(f)*	
					10.4 *(g)*	
F...Br						8.4 *(f)*
						8.1 *(g)*
Cl...Cl			9.9 *(e)*	10.1(18)		
			8.5 *(b)*			
O...F	6.4 *(a)*	7.1			6.2 *(f)*	6.3 *(f)*
	6.0 *(a)*				6.0 *(g)*	5.9 *(g)*
	5.8 *(b)*					
	6.9 *(c)*					
O...Cl			6.9 *(e)*	7.2(6)	6.9 *(f)*	
			6.4 *(b)*		6.5 *(g)*	
O...Br						7.5 *(f)*
						7.3 *(g)*

(a) Cyvin and Hargittai (1969)
(b) Venkateswarlu and Malathi Devi (1965)
(c) Pichai, Krishna Pillai and Ramaswamy (1967)
(d) Hagen, Cross and Hedberg (1978)
(e) Hargittai and Cyvin (1969)
(f) Cyvin and Cyvin (1972)
(g) Ramaswamy and Jayaraman (1971b)

4.5. FURTHER OPEN-CHAIN SULPHONES

Polysulphuryl fluorides, $S_2O_5F_2$ and $S_3O_8F_2$. Two groups of models were tested in the electron diffraction study of $S_2O_5F_2$ by Hencher and Bauer (1973). In one group of models both SO_2F groups were rotated by the same angle around the (bridging) S—O bonds ensuring at least C_2 symmetry. In the other group of models the rotation of the two SO_2F groups were characterized by two different and independent angles. Hencher and Bauer (1973) noted that all combinations of rotation angles occurring in all possible staggered and eclipsed conformations have been used as starting parameters in the least-squares refinements. The rotation angles are denoted according to Fig. 72. The angle of rotation is 0° when the F—S—O—S—F chain is in one plane and the two S—F bonds are in *syn* orientation. The two angles of rotation have the same sign when the rotations around the bridging S—O bonds have the same direction when looking from the sulphur atoms towards the bridging oxygen atom. The least-squares refinements yielded five rotation angle combinations with local minima of which the combination

τ_1(F4—S2—O1—S3) $54.8 \pm 2.7°$ and
τ_2(F5—S3—O1—S2) $73.7 \pm 2.4°$

was chosen on the basis of statistical criteria. The corresponding form is demonstrated by Fig. 72. An important constraint in this analysis was the assumption that only one conformer was considered to be present. This constraint and the large number of possible conformers make the conclusions as regards the conformational properties of

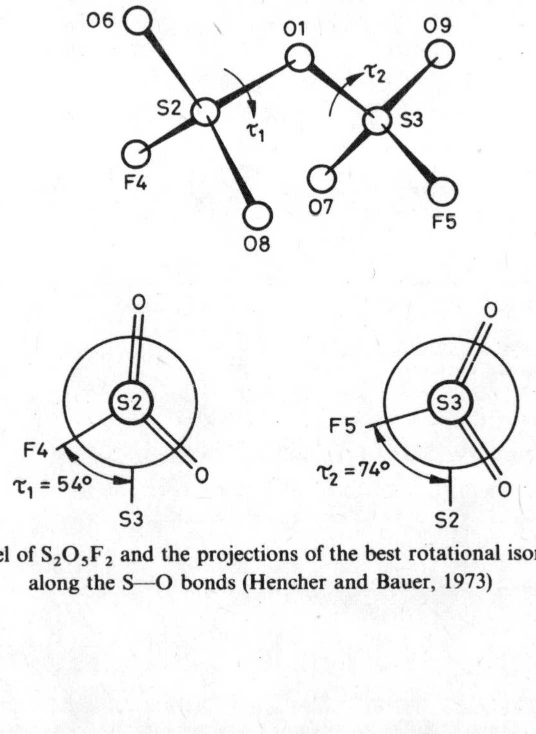

Fig. 72. Molecular model of $S_2O_5F_2$ and the projections of the best rotational isomer representing views along the S—O bonds (Hencher and Bauer, 1973)

$S_2O_5F_2$ rather tentative. At the same time, it is important to note that the geometrical parameters changed very little throughout the structure analysis as various conformations have been tested. Further uncertainty originated, of course, from the application of assumed mean amplitudes of vibration (l values), such as l(S=O) = 4 pm, l(S—O) = l(S—F) = 5 pm, l(O...O) [O—S=O] = l(O...O) [O=S=O] = l(F...O) [F—S=O] = 6 pm. The mean amplitudes of vibration associated with the four-membered chains were assumed to be the same and a value of 8.8 pm was determined for them. All the other l values were also assumed to be the same and 12.3 pm was obtained for them from the refinement. The bond lengths and bond angles are cited in Table 57.

The difficulties mentioned in the general discussion on the electron diffraction investigation of sulphone molecular geometries are of course relevant for the polysulphuryl fluorides as well. It was possible, however, to determine very accurately the bond angle at the bridging oxygen atom.

The conformational properties of the $S_3O_8F_2$ molecule

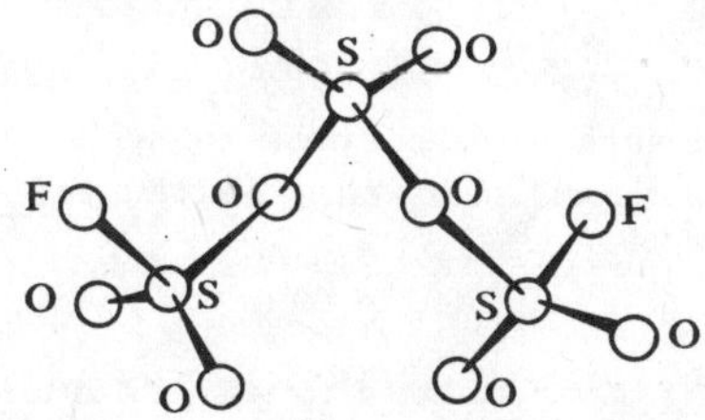

may be even more complicated than those of $S_2O_5F_2$ because of the greater number of possible rotational axes. Thus even more constraints had to be applied in its structure analysis by Hencher and Bauer (1973). No details are given here, only the geometrical parameters reported are reproduced in Table 57.

TABLE 57
Bond lengths and bond angles in two polysulphonyl fluorides from the electron diffraction study of Hencher and Bauer (1973)

Parameter	$S_2O_5F_2$	$S_3O_8F_2$
r_g(S=O), pm	139.8(2)	140.2(3)
r_g(S—F), pm	152.5(5)	152.5(12)
r_g(S—O), pm	161.1(5)	161.3(6)
∢O—S—O (°)		97.8(10)
∢S—O—S (°)	123.6(5)	123.6(12)
∢O—S=O (°)	106.1(9)	106.5(8)
∢F—S=O (°)	106.6(6)	105.5(12)
∢O—S—F (°)	102.4(18)	101.3(15)
∢O=S=O (°)	126.8(12)	128.5(14)
∢O=S=O (bridging group) (°)		128.8(14)

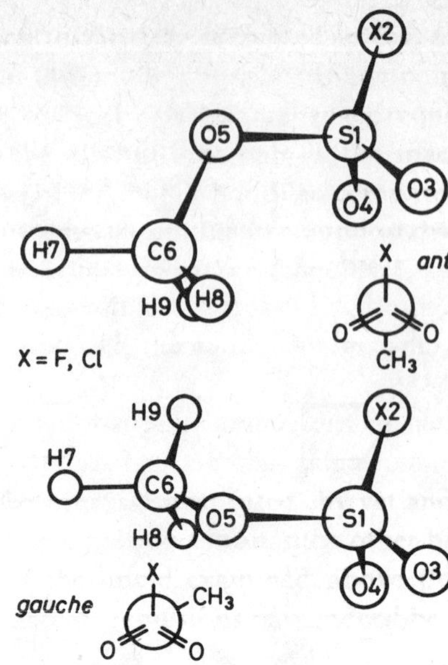

Fig. 73. Molecular models and Newman projections (representing views along the S—O bonds) of two conformers of fluorosulphuric acid methyl ester and chlorosulphuric acid methyl ester

Fluorosulphuric acid methyl ester, CH_3OSO_2F. The microwave spectroscopic and electron diffraction investigations as well as the vibrational spectroscopic calculations were performed concurrently (Hargittai, Seip, Nair, Britt, Boggs and Cyvin, 1977). One conformer possessing a symmetry plane was identified from the microwave spectra. Considering the F—S—O—C chain (*cf.* Fig. 73), this conformer could be the *anti* or the *syn* form but according to the electron diffraction data this is unambiguously the *anti* form. At the same time the electron diffraction data also indicated that a considerable amount of *gauche* form may be present. However, since the rotation-dependent contribution to the total scattering is relatively small, and what is more, the contributions from the *anti* and *gauche* forms are not easily distinguishable (due to the similarity of scattering power of atomic pairs F. . . C and O. . . C), the ratio of the two forms could not be determined reliably. Under the conditions of the microwave spectroscopic experiment, however, no more than a minor amount of the *gauche* form could have been present.

Minimal basis set (STO-6G) *ab initio* molecular orbital calculations by Nair (1978) also indicated that the stable conformer of fluorosulphuric acid methyl ester is the *anti* form. A secondary minimum was found at a *gauche* conformation with an energy higher by an amount in the order of 1 kJ mol^{-1}. The *syn* form was seen to be much less favourable, by an amount in the order of 40 kJ mol^{-1}. Without giving much importance to the numerical values, the general shape of the potential curve established in this investigation is thought to be meaningful.

TABLE 58
Bond lengths and bond angles in fluorosulphuric acid methyl ester and chlorosulphuric acid methyl ester from electron diffraction

X = F, Cl	CH_3OSO_2F *(a)*	CH_3OSO_2Cl *(b)*
r_a(S=O), pm	141.0(2)	141.9(3)
r_a(S—X), pm	154.5(6)	202.3(4)
r_a(S—O), pm	155.8(7)	156.2(4)
∢O=S=O (°)	124.4(7)	122.2(15)
∢X—S=O (°)	106.8(5)	106.4(6)
∢O—S=O (°)	109.5(6)	108.7(8)
∢X—S—O (°)	96.8(6)	102.8(14)
∢S—O—C (°)	116.5(7)	114.4(11)

(a) Hargittai, Seip, Nair, Britt, Boggs and Cyvin (1977)
(b) Hargittai, Schultz and Kolonits (1977)

The geometrical parameters as determined in the electron diffraction analysis are shown in Table 58. The results appeared to be little sensitive to the changing conformational properties of the models used in the structure refinements. It was of great advantage that the O...O distance was available with great accuracy from the microwave spectra. In addition to the requirement of convergency in the least-squares refinement, an important criterion for an acceptable structure was its consistency with the rotational constants derived from the microwave spectra. This criterion, however, was applied in a rather general way for the whole model, or rather acceptable ranges for particular parameters were considered. However, discrepancies between the measured and calculated rotational constants up to about 2% were acceptable, since the parameters were not corrected for the effects of the intramolecular motion, and what may be even more important, the standard deviations were relatively large. The mean amplitudes of vibration from the spectroscopic calculations were very useful in the electron diffraction structure analysis. They are listed in Table 59. They are, however, only of tentative character since, lacking experimental vibrational spectroscopic data, the calculations were based on assumptions and force fields of other molecules constructed again rather tentatively, *viz.* the force fields of fluorosulphuric acid (Cyvin and Hargittai, 1969) and methane sulphonyl fluoride (Cyvin, Dobos, Hargittai, Hargittai and Augdahl, 1973).

Chlorosulphuric acid methyl ester, CH_3OSO_2Cl. This molecule is better suited for the electron diffraction determination than its fluorine analogue, since the contributions of the most important rotation-dependent interactions, *viz.* Cl...C and O...C, differ considerably. On the other hand, attempts for its microwave spectroscopic investigation have been unsuccessful (Boggs, 1975). According to the electron diffraction analysis (Hargittai, Schultz and Kolonits, 1977) based on data taken at

TABLE 59
Calculated mean amplitudes of vibration (pm) for some important distances in fluorosulphuric acid methyl ester and chlorosulphuric acid methyl ester at 298 K (the numbering of atoms is given in Fig. 73)

X=F, Cl	CH_3OSO_2F *(a)*	CH_3OSO_2Cl *(b)*
Rotation-independent		
S=O	3.5	3.5
S—X	4.2	4.3
S—O	4.9	4.9
O3...O4	5.6	5.6
O3...O5	6.5	7.4
O3...X	6.2	7.0
O5...X	9.7	10.5
S...C	6.4	6.4
Rotation-dependent		
X...C *(anti)*	8.8	9.4
C...X *(gauche)*	12.2	13.7
O3...C *(anti)*	9.2	10.0
O3...C *(gauche)*	9.4	10.6
O4...C *(gauche)*	6.8	7.2

(a) Hargittai, Seip, Nair, Britt, Boggs and Cyvin (1977)
(b) Cyvin (1975)

about room temperature of the nozzle, the *gauche* form predominates possibly with a small amount of the *anti* form also present. The *gauche* form was characterized by an angle of rotation of 74° if 0° is assigned to the *syn* form. The rotational forms are shown in Fig. 73. On the basis of dipole moment measurements in solution, Exner, Dembech and Vivarelli (1971) concluded that the angle of rotation in CH_3OSO_2Cl may have values between 0 and 60°. Figure 74 shows the calculated dipole moments after Exner, Dembech and Vivarelli (1971), their experimental dipole moment, and calculated values obtained from the bond moments used by Exner, Dembech and Vivarelli (1971) and the electron diffraction molecular geometry. The agreement is satisfactory when the uncertainty of the angle of rotation obtained for the *gauche* form from electron diffraction, the possibility of a small amount of *anti* form present and the inherent uncertainties of the dipole moment calculations are considered. Also the conformational properties in the vapour and solution need not be the same. The calculated mean amplitudes of vibration used in the structure refinement are listed in Table 59. They were obtained using a similar approach as in the study on fluorosulphuric acid methyl ester. Because of the strong correlation among the parameters, the O...O distance of the SO_2 group was fixed in most of the refinements. The value of 248.5 pm was assumed on the basis of data for a relatively large series of sulphone molecules. The possible effect of this constraint has been considered in the

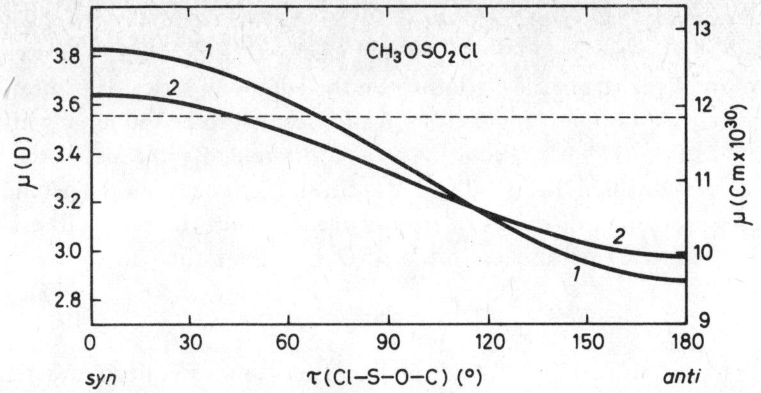

Fig. 74. Experimental (dashed line) and calculated dipole moments of chlorosulphuric acid methyl ester

1 — Bond moments and assumed geometry after Exner, Dembech and Vivarelli (1971)
2 — Bond moments after Exner, Dembech and Vivarelli (1971), geometry from electron diffraction (Hargittai, Schultz and Kolonits, 1977)

error estimations. The bond lengths and bond angles determined in the electron diffraction analysis are collected in Table 58.

Dimethyl sulphate, $CH_3OSO_2OCH_3$. From the electron diffraction data (Brunvoll, Exner and Hargittai, 1981) the prevailing conformer is a C_2 symmetry *gauche–gauche* form in the vapour phase at about 58 °C. The angle of rotation is 132(2)° about the O—S bonds. The second most abundant conformer is the *anti–anti* form (0° angle of rotation)

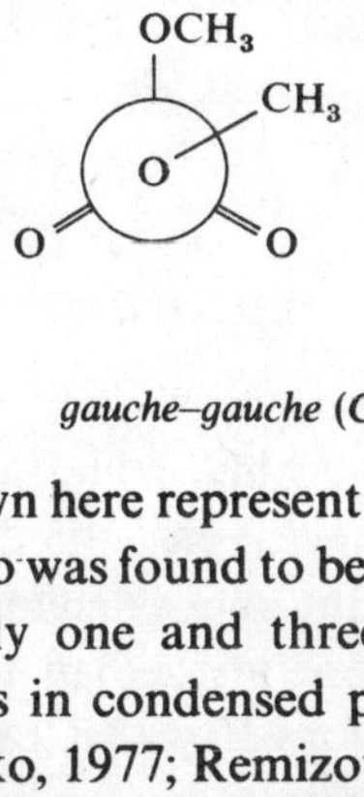

gauche–gauche (C_2)

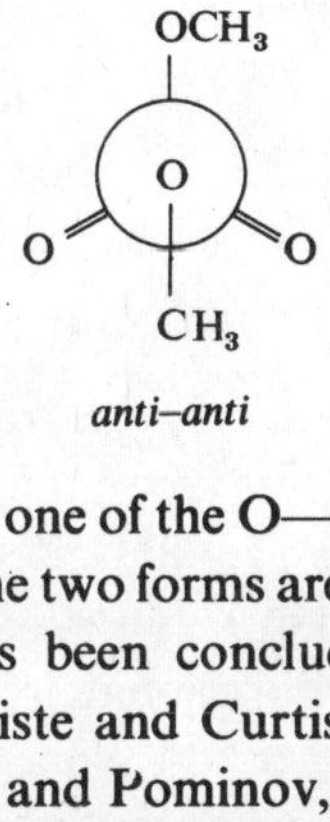

anti–anti

The projections shown here represent view along one of the O—S bonds. The *gauche–gauche*/*anti–anti* ratio was found to be 2:1, thus the two forms are about equally stable. The presence of only one and three forms has been concluded from vibrational spectroscopic studies in condensed phases (Christe and Curtis, 1972 and Remizov, Fishman and Butenko, 1977; Remizov, Fishman and Pominov, 1979, respectively). A dipole moment investigation was consistent with a C_2 conformation (Exner, 1979). The sulphur bond configuration was characterized by the following bond lengths (r_g) and bond angles (r_α) from the electron diffraction analysis:

r(S=O)	141.9(4) pm	∢O=S=O	122.3(8)°
r(S—O)	156.8(3) pm	∢O—S=O	108.4(4)°
		∢O—S—O	98.2(8)°

Further parameters included *r*(C—O) 145.2(10) pm and ∢S—O—C 115.2(7)°.

Trichloromethane sulphonyl chloride, CCl_3SO_2Cl. Its molecular structure was first investigated by Alekseev (1967) using electron diffraction. However, from consideration of the structural variations in the sulphone series (Hargittai, 1978) the communicated S=O bond length (145 ± 1 pm) seemed to be too large while the O=S=O bond angle (111 ± 2°) seemed to be too small. By means of the empirical correlation relationships (Brunvoll and Hargittai, 1978) between the geometrical and vibrational parameters of the SO_2 groups of sulphones (to be discussed in a later section), the following parameters of CCl_3SO_2Cl were estimated

$$r(\mathrm{S{=}O})\ \ 142.2\ \mathrm{pm}\quad \text{and}\quad \sphericalangle\mathrm{O{=}S{=}O}\ \ 121.6°$$

by Hargittai and Brunvoll (1979a) from the vibrational frequencies determined by Nagel (1976). Thus a reinvestigation of the molecular geometry was warranted (Brunvoll, Hargittai and Seip, 1978a) and its results were in complete agreement with the above estimations. The geometrical parameters and the mean amplitudes of vibration are cited in Table 60 and 61, respectively. The same Tables contain the corresponding results from a similar investigation of *trifluoromethane sulphonyl chloride*, CF_3SO_2Cl (Brunvoll, Hargittai and Kolonits, 1978). In both above-mentioned electron diffraction analyses the molecules as a whole were assumed to possess C_s symmetry with C_{3v} local symmetry for the CX_3 groups (X = F or Cl) and the C—S bond coinciding with the three-fold symmetry axis.

According to the experimental data, the staggered conformation is unambiguously established,

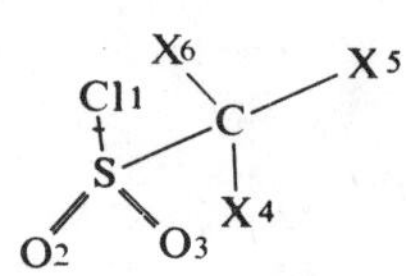

X = Cl or F

as is well demonstrated by the radial distributions shown in Fig. 75, where the contributions of the *anti* type interactions appear individually. Not all of the mean amplitudes of vibration could be determined from the electron diffraction data. Those determined are shown in Table 61 together with the ones calculated from spectroscopic data. Experimental vibrational frequencies were utilized in their calculations, for CF_3SO_2Cl after Lindner and Weber (1967) and for CCl_3SO_2Cl after Nagel (1976). The electron diffraction structure refinements were carried out on both r_a and r_α distances without appreciable differences. The merits of the refinement based on r_α representation will be discussed in connection with the benzene sulphonyl chloride study.

Attempts have been made to estimate the barrier to internal rotation around the C—S bond in the CX_3SO_2Cl molecules. The chlorine derivative was studied in more detail employing two approaches. In one it was supposed that the potential of internal rotation has the form $V = (1/2)\,V_0(1 - \cos 3\varphi)$, where V_0 is the rotational barrier and φ is the angle of rotation. This is, of course, a rather crude assumption for these molecules.

TABLE 60

Bond lengths and bond angles of trifluoromethane sulphonyl chloride and trichloromethane sulphonyl chloride from electron diffraction

X = F, Cl	CF_3SO_2Cl *(a)*	CCl_3SO_2Cl *(b)*
r(C—X), pm	132.4(4)	175.7(10)
r(S=O), pm	141.5(7)	142.0(3)
r(S—Cl), pm	201.5(5)	202.0(5)
r(S—C), pm	185.6(6)	186.5(30)
∢O=S=O (°)	122.4(10)	121.5(9)
∢O=S—Cl (°)	108.3(3)	109.2(6)
∢C—S—Cl (°)	98.7(4)	97.9(8)
∢O=S—C (°)	108.3(7)	108.3(7)
∢S—C—X (°)	109.1(4)	108.8(12)

(a) Brunvoll, Hargittai and Kolonits (1978)
(b) Brunvoll, Hargittai and Seip (1978a)

TABLE 61

Mean (parallel) amplitudes of vibration (*l*, pm) and perpendicular vibrational corrections (*K*, pm) of trifluoromethane sulphonyl chloride and trichloromethane sulphonyl chloride

Atomic pair	CF_3SO_2Cl *(a)*			CCl_3SO_2Cl *(b)*	
	l (c)	*l (d)*	*K (d)*	*l (c)*	*l (d)*
C—X	5.2(5)	4.9	1.49	3.7(14) *(c1)*	5.7
S=O	4.2(3)	3.5	0.66	3.9(4)	3.5
S—Cl	4.5(2)	4.9	0.61	5.1(8)	4.9
S—C	4.6(6)	5.2	0.13	3.9 *(c1)*	5.8
O...O	6.6(18)	6.6	0.97	6.5(23)	5.6
Cl...O	7.0(4) *(c1)*	7.2	0.91		7.1
O...C	7.0(16) *(c2)*	7.0	0.42		7.9
Cl...C	6.7 *(c1)*	6.9	0.35		7.5
S...X	6.8 *(c2)*	6.8	0.72		7.7
X...X	5.5(3)	5.9	2.40		6.9
Cl...X4	8.0(9)	7.0	0.10	8.2(14)	7.9
Cl...X5	17.3(30) *(c3)*	17.5	0.79	19.0(11) *(c2)*	10.4
X4...O2	13.8 *(c3)*	14.0	1.07	19.9 *(c2)*	11.3
X5...O2	6.9(34)	7.0	0.21	6.3(10)	7.8
X6...O2	14.0 *(c3)*	14.3	0.96	20.8 *(c2)*	12.2

(a) Brunvoll, Hargittai and Kolonits (1978)
(b) Brunvoll, Hargittai and Seip (1978a)
(c) Electron diffraction; *(c1)*, *(c2)*, *(c3)* Amplitudes refined in groups
(d) Spectroscopic calculations

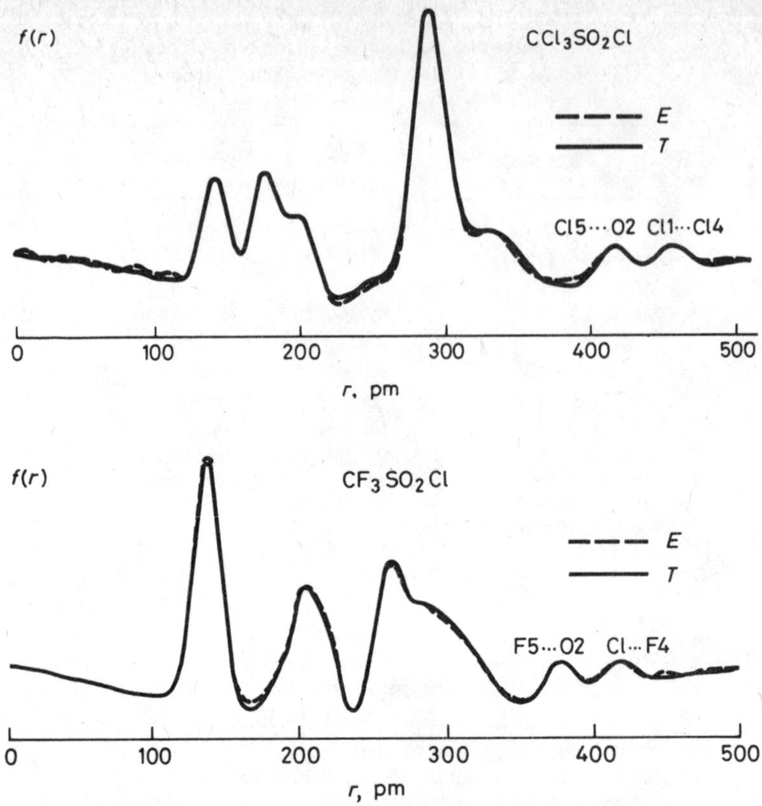

Fig. 75. Radial distributions of trifluoromethane sulphonyl chloride (Brunvoll, Hargittai and Kolonits, 1978) and trichloromethane sulphonyl chloride (Brunvoll, Hargittai and Seip, 1978a)

E — experimental, *T* — theoretical

Following the method proposed by Karle (1966) which assumes that the contributions from the framework and torsional vibrations to the effective vibrational amplitudes may be separated and that the contribution from the framework vibrations is independent of the angle of rotation, the results shown in Fig. 76(a) were obtained. Accordingly, the barrier to internal rotation is between 16 and 24 kJ mol^{-1} for CCl_3SO_2Cl and similar calculation (*cf.* Fig. 76(b)) yielded a barrier between 8 and 20 kJ mol^{-1} for CF_3SO_2Cl. Another approach employed for CCl_3SO_2Cl followed the method proposed by Vilkov, Penionzhkevich, Brunvoll and Hargittai (1978). Here the average torsional angle $\langle|\varphi|\rangle$ with respect to the rotation around the S—C bond was determined from the electron diffraction data assuming spectroscopically-calculated framework vibrational amplitudes in the least-squares refinement. The larger this average torsional angle appears, the lower the barrier to internal rotation is. The results obtained this way are illustrated by Fig. 77. The average torsional angle was 9.6° corresponding to a barrier between 12 and 20 kJ mol^{-1}.

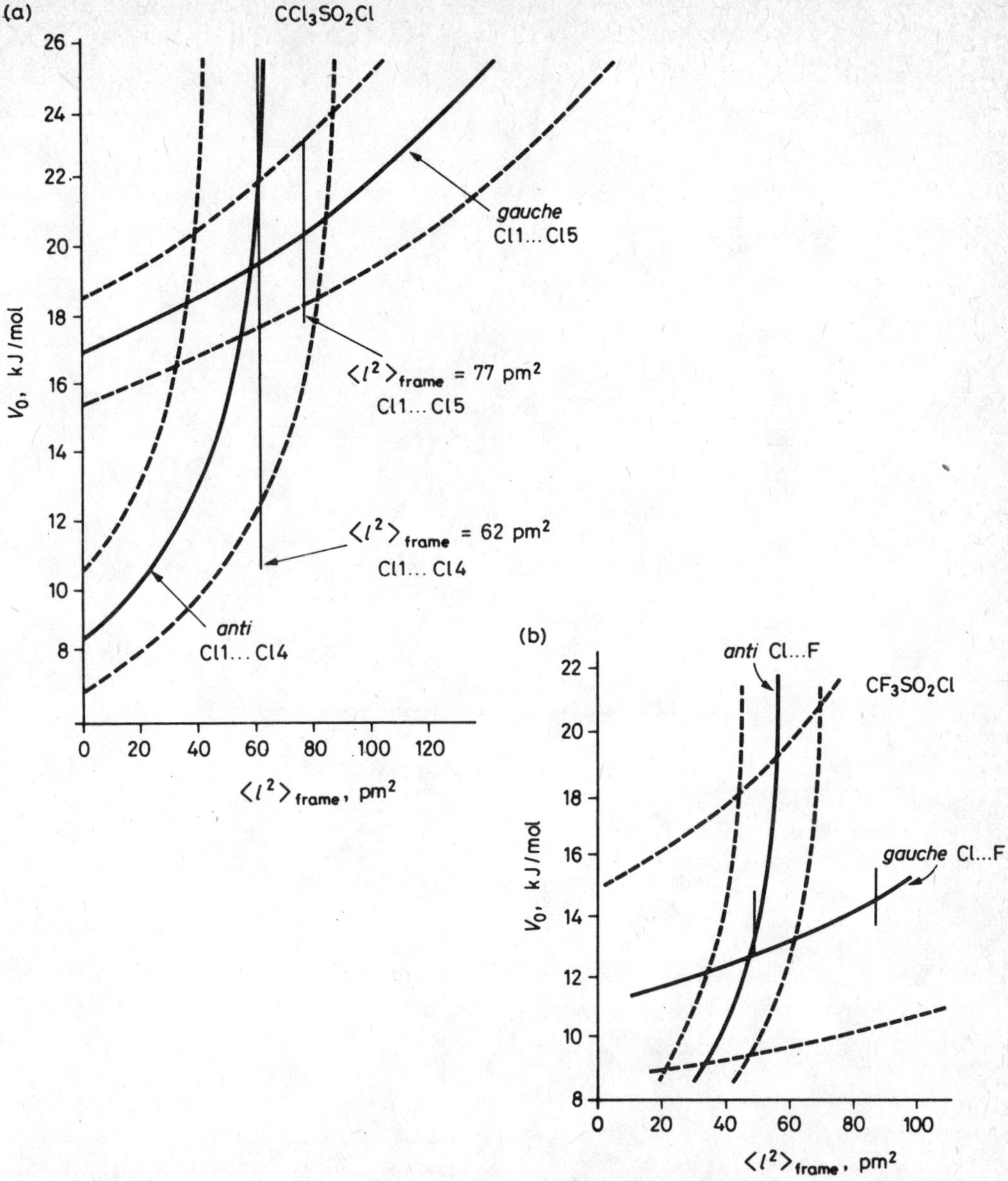

Fig. 76. Graphical solution of the equations determining the potential barrier according to the method of Karle (1966) implying that the framework contributions from the *anti* and *gauche* interactions are assumed to be the same. The continuous lines correspond to the experimental *l*-values and the dashed lines to their error limits. The straight vertical lines correspond to the calculated framework contributions showing that the above assumption is relatively good but also that this assumption results in a little underestimation of the *gauche* framework contribution

(a) Trichloromethane sulphonyl chloride (Brunvoll, Hargittai and Seip, 1978a)
(b) Trifluoromethane sulphonyl chloride (Brunvoll, Hargittai and Kolonits, 1978)

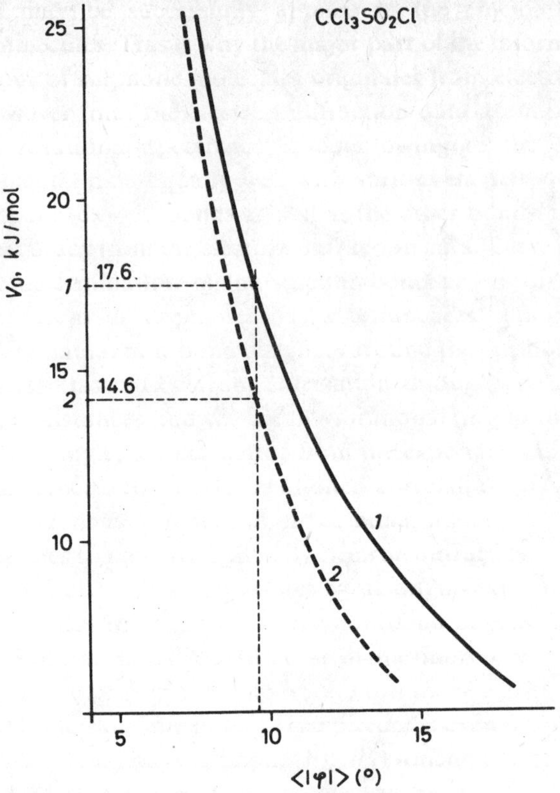

Fig. 77. Estimation of the potential barrier on the basis of the torsional angle, $\langle|\varphi|\rangle$, from the electron diffraction data according to the method of Vilkov, Penionzhkevich, Brunvoll and Hargittai (1978), *1*—by numerical integration, *2*—by approximate formula

Methane sulphonyl fluoride, CH_3SO_2F. The staggered form relative to the rotation around the S—C bond is demonstrated here by a Newman projection representing a view along the S—C bond

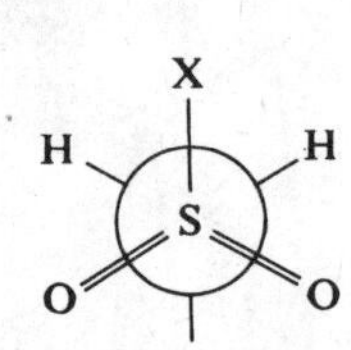

$X = F$, Cl or CH_3

The barrier to internal rotation was determined to be 10.5 ± 1.5 kJ mol^{-1} from microwave spectroscopic data by Jacob and Lide (1971). The O. . .O distance derived from the planar moment was used as a constraint in the electron diffraction analysis by Hargittai and Hargittai (1973b). The geometrical parameters obtained are shown in Table 62. This was, in fact, the first in a series of similar combined analyses of sulphone

TABLE 62
Bond lengths and bond angles of methane sulphonyl fluoride and methane sulphonyl chloride from electron diffraction

X=F, Cl	CH_3SO_2F *(a)*	CH_3SO_2Cl *(b)*
r_a(S=O), pm	141.0(3)	142.4(3)
r_a(S—X), pm	156.1(4)	204.6(4)
r_a(S—C), pm	175.9(6)	176.3(5)
∢O=S=O (°)	123.1(15)	120.8(8)
∢X—S=O (°)	106.2(4)	107.1(7)
∢C—S—X (°)	98.2(15)	101.0(14)

(a) Hargittai and Hargittai (1973b)
(b) Hargittai and Hargittai (1973)

molecular structures. It was realized, of course, that the O...O distance was an r_0 parameter, while the electron diffraction least-squares refinement yielded r_a parameters. In order to estimate the errors, the refinements were repeated utilizing assumed r(O...O) values covering a reasonable range so that the sensitivity of the other parameters to this assumption could be tested. Of the other parameters the bond angle F—S—C proved to be the most sensitive to the changes in the O...O distance. A difference of 1 pm in the latter corresponded to a little more than 1° in ∢F—S—C. All this was taken into account in obtaining the uncertainties cited for the parameters of CH_3SO_2F in Table 62.

Methane sulphonyl chloride, CH_3SO_2Cl. The electron diffraction analysis (Hargittai and Hargittai, 1973) provided three models which reproduced the experimental data equally well. The models differed in the O=S=O bond angle and in some of the vibrational amplitudes. On the basis of the accumulated structural data on sulphone molecules available at the time of the investigation and by selecting the most reasonable mean amplitudes of vibration, it was shown that one of the models was preferable. This model was subsequently unambiguously confirmed by a microwave spectroscopic investigation by van Eijck, Korthoff and Mijlhoff (1975) employing a double resonance technique. The bond lengths and bond angles determined by electron diffraction are given in Table 62. The uncertainty of the bond angle O=S=O was originally given as three times the estimated total error (2.4°). In the light of the microwave spectroscopic investigation, there is no longer any reason to give the uncertainty in this parameter differently from that in the others.

The staggered form of rotation relative to the S—C bond (see above) was assumed in the electron diffraction work. This conformation was confirmed by the microwave spectroscopic study, and the barrier to internal rotation was found to be at least 8 kJ mol^{-1}.

TABLE 63
Mean amplitudes of vibration (pm) of methane sulphonyl fluoride, methane sulphonyl chloride and dimethyl sulphone from spectroscopic calculations (SP) and electron diffraction (ED)

Atomic pair	CH_3SO_2F		CH_3SO_2Cl		$CH_3SO_2CH_3$	
	SP *(a)*	ED *(b)*	SP *(a)*	ED *(c)*	SP *(d)*	ED *(e)*
Temperature (°C)	105	105	105	110	25	130
S═O	3.6	3.9(2)	3.7	4.3(2)	3.7	4.15(10)
S—F	4.8	5.2(3)				
S—Cl			6.1	5.7(2)		
S—C	4.7	5.7(4)	6.5	4.4(4)	4.4	5.7(1)
C—H	7.8	6.3(9)	7.8	7.6(13)	7.8	7.5(2)
O...O	6.9	[5.7]	9.5	6.6(26)	6.3	5.2(14)
C...F	6.9	5.6(24)				
C...Cl			7.9	13.3(50)		
C...C					6.2	5.8(21)
O...F	8.1	5.4(6)				
O...Cl			9.5	8.7(7)		
O...C	6.9	7.7(18)	8.3	7.8(24)	6.1	7.1(8)
S...H	10.6	14.9(43)	11.2	10.6(37)	10.5	12.0(10)
	11.2		12.0			

(a) Cyvin, Dobos, Hargittai, Hargittai and Augdahl (1973)
(b) Hargittai and Hargittai (1973b)
(c) Hargittai and Hargittai (1973)
(d) Cyvin (1973)
(e) Hargittai and Hargittai (1974)

The mean amplitudes of vibration of methane sulphonyl fluoride and methane sulphonyl chloride as determined from electron diffraction and obtained from spectroscopic calculations are compiled in Table 63.

Dimethyl sulphone, $(CH_3)_2SO_2$. Several investigations were performed in the early seventies in order to elucidate its molecular structure. In the models considered the orientation of the methyl group around the S—C bond was such that the C—H bonds staggered the other sulphur bonds. Both the semi-empirical molecular orbital calculations (Robinet, Crasnier, Labarre and Leibovici, 1972) and experimental evidence (Clever and Westrum, 1970) gave support for this model. The rotational barrier for one CH_3 group was found to be around 14–15 kJ mol^{-1} in the cited studies.

The geometrical parameters determined in some of the recent studies are compiled in Table 64. There was yet another electron diffraction investigation (Oberhammer and Zeil, 1970a) which reported $127.1 \pm 2.4°$ for the O═S═O bond angle. As has been shown later for other simple sulphones, electron diffraction alone may not be sufficient to arrive at a unique solution for a sulphone structure. Accordingly, an electron diffraction reinvestigation was carried out by Hargittai and Hargittai (1974) utilizing spectroscopic data. The O...O distance derived from the microwave spectroscopic investigation by Jacob and Lide (1971) was used as a constraint in the first stage of the

TABLE 64
Bond lengths and bond angles of dimethyl sulphone
from vapour-phase microwave spectroscopy *(a, b)* and electron diffraction *(c)*,
and from crystal-phase x-ray diffraction *(d, e)*

	(a1)	*(a2)*	*(b)*	*(c)*	*(d)*	*(e)*
r(S=O), pm	143.3	[142.5]	143.1(3)	143.5(3)	144.5(16)	144.6(3)
r(S—C), pm	[177.0]	178.0	177.7(6)	177.1(4)	177.8(17)	176.5(5)
∡O=S=O (°)	120.8	122.0	121.0(3)	119.7(11)	117.9(8)	117.3(2)
∡C—S—C (°)	103.4	102.7	103.3(2)	102.6(9)	103.0(8)	104.8(2)

(a) Jacob and Lide (1971); r_0; *(a1)* and *(a2)* two sets of results differing in the assumed parameter
(b) Saito and Makino (1972); r_0
(c) Hargittai and Hargittai (1974)
(d) Sands (1963)
(e) Langs, Silverton and Bright (1970)

TABLE 65
Rotational constants* (MHz) measured and calculated for dimethyl sulphone

Rotational constants	*a*	*b*	*a1*	*c*	*g*
A	4638.202 ± 0.006	4638.15 ± 0.02	4639.92	4595	4670.51
B	4295.107 ± 0.006	4295.44 ± 0.02	4295.49	4338	4239.12
C	4177.107 ± 0.007	4177.13 ± 0.02	4178.44	4171	4317.18

* *A*: principal axis perpendicular to the OSO plane; *B*: principal axis C_2 axis; *C*: principal axis perpendicular to the CSC plane.

(a) and *(b)* measured; *(a1)* and *(c)* models; for references see Table 64
(g) As calculated by Jacob and Lide (1971) from the geometry reported by Oberhammer and Zeil (1970a)

structure analysis, but was removed in the final refinement of parameters. This electron diffraction investigation, also utilizing calculated mean amplitudes of vibration (*cf.* Table 63) provided a structure consistent with the microwave spectroscopic data. The calculated and observed rotational constants are compared in Table 65. Here again one must remember that discrepancies up to 1 or 2% in the rotational contants may be due to the difference in the physical meaning of the parameters originating from the two techniques. In any case the uncertainties of the electron diffraction parameters also render a more detailed comparison meaningless. Changes of 0.5 pm and 0.5° in the bond lengths and angles may correspond typically to about 20 and 35 MHz changes in the rotational constants. In the light of these remarks, the agreement between the data from the electron diffraction reinvestigation (Hargittai and Hargittai, 1974) and the microwave spectroscopic results is satisfactory. On the other hand, the *C* rotational constant obtained from the results of the earlier electron diffraction work is certainly too large. Since this rotational constant is primarily determined by the positions of the oxygen atoms, the correlation with the reported large O=S=O angle is obvious. The question naturally arises, how is it possible to arrive at such an erroneous result. For answering this question a test was compiled in which the generalized *R* factor,

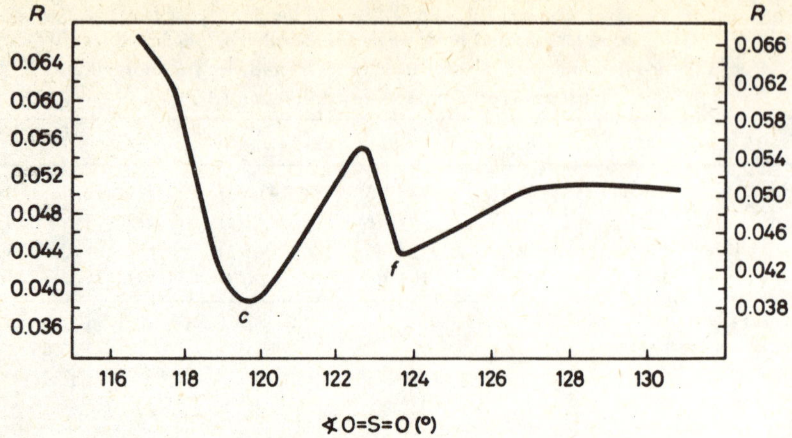

Fig. 78. Generalized R-factor (expressing the "goodness of fit") as plotted against the O=S=O angle values fixed in separate refinements of the electron diffraction analysis of dimethyl sulphone (Hargittai and Hargittai, 1974). The minima c and f correspond to refinement schemes in which the O=S=O bond angle was also refined

$R=\{\Sigma[M^E(s)-M^T(s)]^2W/\Sigma[M^E(s)]^2W\}^{1/2}$, was computed for a series of structure refinements. The bond angle O=S=O was kept fixed at various assumed values in these refinements and all the other parameters were varied. The results are shown in Fig. 78. The minima "c" and "f" correspond to schemes in which the O=S=O angle was refined together with the other parameters. The structure at the minimum "c", corresponding to the results under columns c of Tables 64 and 65, is considered to be the correct model for dimethyl sulphone not only because it is associated with the smaller R value of the two minima. In addition to being a minimum, it is also the model which is consistent with the microwave spectroscopic results. Minimum "f" corresponds to r(O...O) 253 pm and ∢O=S=O 123.7° and is merely another mathematically possible solution. The importance of the utilization of the microwave spectroscopic information in the electron diffraction analysis is that had there been only one starting parameter set, the one somewhere in the vicinity of minimum "f", the true parameter set ("c") would have never been obtained. Similar situations were discussed by Hamilton (1964; 1965).

Ethene sulphonyl chloride, $CH_2{=}CHSO_2Cl$. So far only electron diffraction has been employed to elucidate its molecular structure (Brunvoll and Hargittai, 1977). Spectroscopic calculations based on a force field constructed from those of analogous molecules were used to produce mean amplitudes of vibration to aid the electron diffraction analysis. They are presented in Tables 66 and 67. A number of conformers are *a priori* possible for ethene sulphonyl chloride. They can be characterized by the forms shown in Fig. 79. According to the electron diffraction data, there may be two

TABLE 66
Calculated mean amplitudes of vibration (pm) for bonds and for some important rotation-independent non-bond distances of ethene sulphonyl chloride at 333 K and benzene sulphonyl chloride at 411 K (the numbering of atoms is given in Figs. 79 and 81)

Bonds			Non-bonded atomic pairs			
Atomic pair	CH_2CHSO_2Cl *(a)*	$C_6H_5SO_2Cl$ *(b)*	Atomic pair	CH_2CHSO_2Cl *(a)*	Atomic pair	$C_6H_5SO_2Cl$ *(b)*
S=O	3.6	3.6	S...C(1)	6.1	S...C(2)	7.0
S—Cl	5.1	5.4	Cl...C(2)	6.4	Cl...C(1)	7.0
S—C	4.4	4.5	O...C(2)	6.6	O...C(1)	6.9
C—C	4.7	4.0	O...O	6.9	O...O	7.4
C—H	7.8	7.7	Cl...O	7.6	Cl...O	8.4
					C(1)...C(3)	5.7
					C(1)...C(4)	6.2
					S...C(3)	7.1
					S...C(4)	6.8

(a) Brunvoll and Hargittai (1977)
(b) Brunvoll and Hargittai (1976)

TABLE 67
Calculated mean amplitudes of vibration (pm) for distances dependent on rotation around the S—C bond in ethene sulphonyl chloride at 333 K (the numbering of atoms is given in Fig. 79)*

Atomic pair	Angle of rotation, τ (°)			
	0	60	90	180
Cl...C(1)	6.8	8.4	9.5	10.0
O(8)...C(1)	9.3	10.3	10.4	8.6
O(9)...C(1)	9.3	7.5	6.8	8.6
Cl...H(3)	13.4	17.2	19.8	17.6
Cl...H(4)	11.8	13.3	14.1	13.1
Cl...H(5)	12.9	14.5	14.1	9.7
O(8)...H(3)	18.4	18.0	18.5	16.9
O(8)...H(4)	12.8	12.7	12.9	12.6
O(8)...H(5)	12.3	9.8	10.0	13.1
O(9)...H(3)	18.4	15.5	14.0	16.9
O(9)...H(4)	12.8	11.8	11.3	12.6
O(9)...H(5)	12.3	13.7	13.4	13.1

* After Brunvoll and Hargittai (1977)

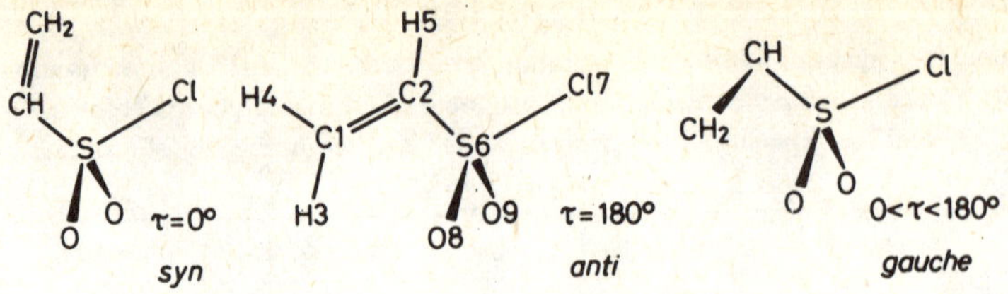

Fig. 79. Some of the *a priori* possible forms of internal rotation for ethene sulphonyl chloride and the numbering of atoms (Brunvoll and Hargittai, 1977)

TABLE 68
Bond lengths and bond angles of ethene sulphonyl chloride and benzene sulphonyl chloride from electron diffraction

Parameter	$CH_2{=}CHSO_2Cl$ (*a*)	$C_6H_5SO_2Cl$ (*b*)
r_a(S=O), pm	142.0(6)	141.7(12)
r_a(S—Cl), pm	203.5(5)	204.7(8)
r_a(S—C), pm	174.4(5)	176.4(9)
r_a(C—C), pm	135.7(18)	140.3(10)
∡C—S—Cl (°)	100.2(6)	100.9(20)
∡C—S=O (°)	109.8(4)	110.0(25)
∡O=S=O (°)	122.0(10)	122.5(36)
∡Cl—S=O (°)	106.4(5)	105.5(18)
∡C—C—S (°)	121.3(17)	

(*a*) Brunvoll and Hargittai (1977)
(*b*) Brunvoll and Hargittai (1976)

conformers present in the vapour of ethene sulphonyl chloride (nozzle temperature 60 °C). The two forms have angles of rotation around the S—C bond of 0 to 20° and 100 to 120° if 0° corresponds to the *syn* form in which the S—Cl bond and the C=C bond are eclipsed. A detailed determination of the conformational properties is hindered by the relatively small contribution of the rotation-dependent interactions to the total electron scattering. Furthermore, this contribution is relatively insensitive to the changes in the angle of rotation in its large intervals. The geometrical parameters given in Table 68 were found to be, however, insensitive to the conformational properties of the models.

Divinyl sulphone, $(CH_2{=}CH)_2SO_2$. The electron diffraction structure analysis was augmented by a detailed vibrational spectroscopic investigation and semi-empirical molecular orbital calculations in the CNDO/2 approximation (Hargittai, Rozsondai, Nagel, Bulcke, Robinet and Labarre, 1978). Two more or less distinct features on the experimental radial distribution could be assigned to the contribution from the terminal carbon–carbon interactions, thus the presence of two rotational forms relative

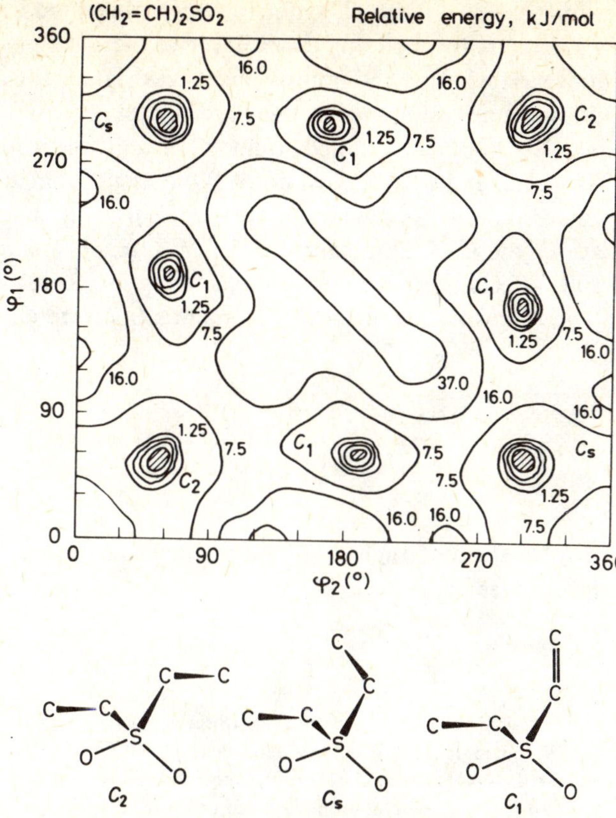

Fig. 80. Some of the *a priori* possible forms of internal rotation for divinyl sulphone and the isoenergetic curves calculated for different combinations of the angles of rotation (Hargittai, Rozsondai, Nagel, Bulcke, Robinet and Labarre, 1978)

to the rotation around the S—C bonds was indicated. Considering the two rotational axes, however, any terminal carbon–carbon distance may correspond to several different rotational forms. Some of the *a priori* conformational possibilities of divinyl sulphone are shown in Fig. 80. According to the CNDO/2 calculations,* as shown by the isoenergy curves of Fig. 80, there are three stable conformers. They are characterized by the following angles of rotation: $C_2(\varphi_1=60°$ and $\varphi_2=60°)$, C_s (60,300°), and C_1 (60, 180°). The energy barriers between the minima are relatively high. Although the electron diffraction experimental data could be rather well

* Semi-empirical CNDO/2 calculations (Pople and Beveridge, 1970) were seen to produce erroneous results for delocalized systems, whereas they proved to be useful for numerous localized systems (see e.g. Gropen and Seip, 1971; Perahia and Pullman, 1973; Corosine, Crasnier, Labarre, Labarre and Leibovici, 1973). The bond lengths determined by electron diffraction indicated no appreciable delocalization in divinyl sulphone.

approximated by a mixture of C_2 and C_s symmetry conformers, adding a relatively small amount of C_1 symmetry form could further improve the agreement. It would be difficult to assess, however, whether the improvement was significant indeed or not. According to the infrared and Raman spectra there are at least two conformers present in the liquid state and also in the supercooled liquid, while only one form occurs in the crystalline phase. The bond lengths and most of the bond angles were not sensitive to the changes in the conformational properties of the models used in the refinements. In most of the calculations the O...O distance of the SO_2 group was assumed to be 248.5 pm. The possible effects of this constraint was taken into consideration in the error estimations. The following geometrical parameters (r_a) were considered to be reliably determined:

r(S=O)	143.8(3) pm	r(C=C)	133.2(4) pm
r(S—C)	176.9(4) pm	r(C—H)	111.8(6) pm
∢O=S=O	119.5(12)°		
∢S—C=C	121.5(3)°		

Unfortunately the C—S—C bond angle could not be determined unambiguously. In various models it has refined to 108° or 95°.

TABLE 69
Calculated mean amplitudes of vibration (pm) for some important distances dependent on rotation around the S—C bond in benzene sulphonyl chloride at 411 K (the numbering of atoms is given in Fig. 81)*

Atomic pair	Angle of rotation, τ (°)			
	0	40	75.3	90
C(6)...Cl	10.5	12.8	13.4	13.0
C(6)...O(15)	10.7	8.1	7.3	7.6
C(6)...O(14)	10.7	12.1	12.4	12.3
C(5)...Cl	11.6	13.9	14.2	13.8
C(5)...O(15)	11.9	9.1	8.3	8.5
C(5)...O(14)	11.9	12.6	12.8	12.8
C(4)...Cl *(a)*	10.2	12.5	13.5	13.6
C(4)...O(15) *(a)*	12.7	10.9	10.6	10.8
C(4)...O(14) *(a)*	12.7	11.9	11.0	10.8
C(3)...Cl	8.6	10.7	13.2	13.8
C(3)...O(15)	12.7	12.4	12.7	12.8
C(3)...O(14)	12.7	11.1	9.1	8.5
C(2)...Cl	8.0	9.6	12.3	13.0
C(2)...O(15)	11.6	11.8	12.1	12.3
C(2)...O(14)	11.6	10.1	8.1	7.6

* After Brunvoll and Hargittai (1976)
(a) The distance itself does not depend on the rotation

Benzene sulphonyl chloride, $C_6H_5SO_2Cl$. The electron diffraction analysis utilized vibrational spectroscopic data (Brunvoll and Hargittai, 1976) extensively. The purpose of the spectroscopic calculations was to provide mean parallel amplitudes of vibration (l values) and correction terms for the perpendicular vibrations (K values). The spectroscopic l values were used in the initial stage of the electron diffraction structure refinement and also for comparison with the electron diffraction results. Some spectroscopic l values were used throughout the refinements for atomic pairs with relatively small contribution to the total scattering. The spectroscopic calculations were based on the force fields of benzene (Brooks and Cyvin, 1962; Brooks, Cyvin and Kvande, 1965) and of the SO_2Cl fragment of methane sulphonyl chloride (Cyvin, Dobos, Hargittai, Hargittai and Augdahl, 1973), and also on the experimental frequencies of benzene sulphonyl chloride (Ham, Hambly and Laby, 1960; Uno, Machida and Hanai, 1968). The l values for rotation-independent and rotation-dependent interactions are listed in Tables 66 and 69. The numbering of atoms is given in Fig. 81. The correction terms for the perpendicular vibrations were used in the (approximate) determination of the r_α average structure, $r_\alpha \approx r_a + l^2/r_a - K$. The conversion was needed primarily to facilitate the conformational analysis in order to distinguish between a symmetrical and an asymmetrical model so as to minimize the effect of intramolecular motion on the results of the structure analysis.

The conformers of benzene sulphonyl chloride, which were studied are shown below as Newman-projections, representing views along the S—C bond:

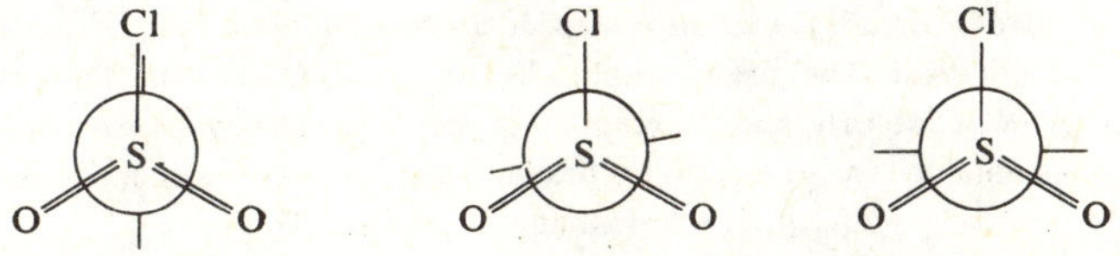

$\tau = 0°$ $\quad 0° < \tau < 90° \quad$ $\tau = 90°$

One of the two symmetrical forms, *viz.* the one with $\tau = 0°$ could be excluded on the basis of the comparison of experimental and theoretical radial distributions. The

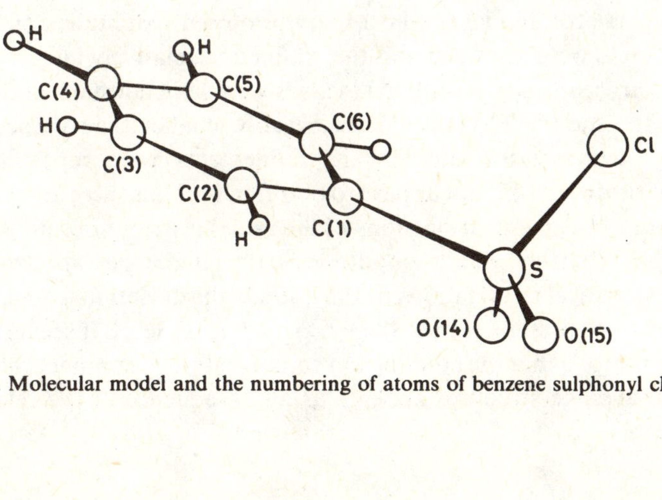

Fig. 81. Molecular model and the numbering of atoms of benzene sulphonyl chloride

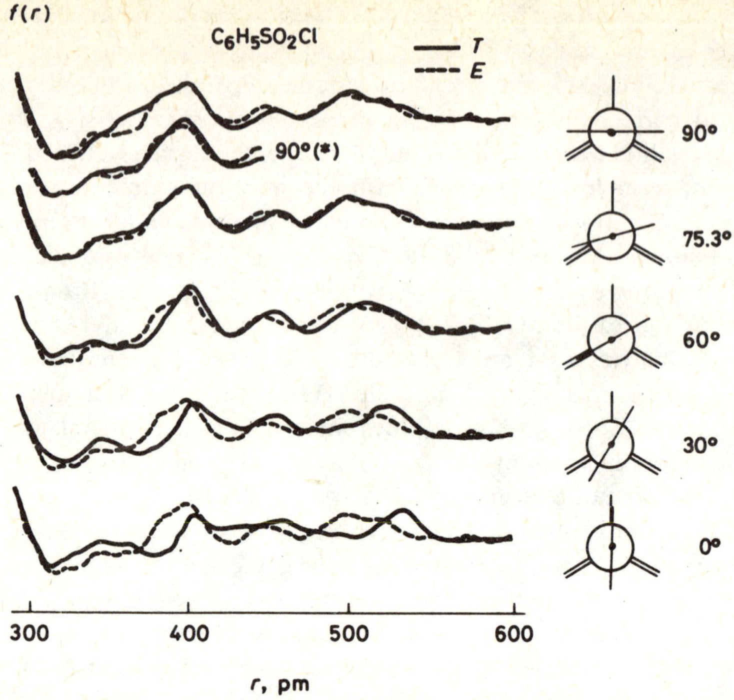

Fig. 82. Newman projections of rotational forms of benzene sulphonyl chloride representing views along the C—S bond and the angles of rotation. The rotation-dependent portions of the radial distributions are shown, *E*—experimental, *T*—theoretical. The mean amplitudes of vibration used in the calculations of the theoretical curves are given in Tables 66 and 69 except for the curve marked by the asterisk (*) for which some larger mean amplitudes obtained in the electron diffraction analysis were used, most notably l(C2. . . Cl) = 25.8 pm (Brunvoll and Hargittai, 1976)

experimental data were reproduced well by two models. One of them was the other symmetrical form with $\tau = 90°$, and the other good model was an asymmetric one with $\tau = 75°$. In the symmetric model ($\tau = 90°$) the vibrational amplitudes of the rotation-dependent distances were not fixed and they refined to relatively large values. On the other hand, spectroscopically-calculated l values were kept constant in the refinement of the asymmetric model. They were considerably smaller than those obtained by refinement in the symmetric model. The refinements were repeated for the r_α parameters, but the results did not appear to be different. Thus there are two acceptable models. The rotation-dependent portions of the radial distributions are shown in Fig. 82, utilizing the relatively small amplitudes calculated from spectroscopic data. According to a statistical test (Hamilton, 1964; 1965), the deviation from the symmetric model appeared to be present even at the 99% significance level. It seems to be equally plausible, if not more so that the equilibrium conformation is symmetrical ($\tau = 90°$) and the asymmetric average structure appears as a consequence of the relatively large-amplitude torsional motion. The corrections, which were based on the available spectroscopic data, probably did not compensate for these vibrations since the spectra of the low-frequency regions were very uncertain or not even known.

The bond lengths and bond angles determined for benzene sulphonyl chloride are given in Table 68.

Recently Boggia, Filgueira, Marañon and Sorarrain (1978) recorded the microwave spectrum of benzene sulphonyl chloride and concluded that the molecule has C_s symmetry with $\tau = 0°$ in direct contradiction with the electron diffraction results. However, van Eijck, Hargittai and Mayer (1980) showed that the very limited amount of microwave spectroscopic data reported were in fact compatible with any τ angle of rotation.

Boggia, Filgueira, Marañon and Sorarrain (1978) also performed some semi-empirical CNDO/2 calculations for the barrier to the rotation about the S—C bond in benzene sulphonyl chloride. They reported an energy minimum at $\tau = 0°$ and a maximum at $\tau = 90°$ with a barrier height of 17 kJ mol^{-1}. It is known, however, that for molecules with second-row atoms the standard CNDO/2 method often fails to predict even qualitatively the relative stability of the conformers. If the molecule contains strongly polarizing groups (Kiss, 1979), the correct conformer may be obtained only by using sp basis instead of the spd one, i.e. by omitting the d orbitals from the standard basis of the CNDO/2 calculations. According to the calculations* of van Eijck, Hargittai and Mayer (1980), the standard CNDO/2 method (spd basis) indeed gives the minimum energy at $\tau = 0°$ and the maximum at $\tau = 90°$ with a barrier height of 14 kJ mol^{-1}. When, however, an sp basis was used, the minimum was at $\tau = 90°$, the maximum at $\tau = 0°$ and the barrier height was 7 kJ mol^{-1}. This was thought to be a consequence of the CNDO/2 approximation strongly exaggerating the population of the d orbitals.

Diphenyl sulphone, $(C_6H_5)_2SO_2$. The electron diffraction analysis (Rozsondai, Moore, Gregory and Hargittai, 1979) resulted in a limited amount of structural information. Only the bond lengths (r_g) could be determined reliably,

r(C—C)	(mean value)	140.0(3) pm
r(C—H)	(mean value)	110.0(11) pm
r(C—S)		177.2(5) pm
r(S═O)		144.0(5) pm

The experimental radial distribution of diphenyl sulphone is shown in Fig. 18 together with the respective curves of its sulphoxide and sulphide analogues. The crystal-phase molecular structure of diphenyl sulphone was determined by Sime and Woodhouse

*Van Eijck, Hargittai and Mayer (1980) assumed the electron diffraction geometry in these calculations. On the other hand, e.g., the OSCl angle was 54.9° in the calculations of Boggia, Filgueira, Marañon and Sorarrain (1978) corresponding to an unrealistically short 165 pm distance between the non-bonded Cl and O atoms.

(1974) using x-ray diffraction. The overall geometries determined in the vapour and crystalline phases are in good general agreement.

2-Thiophene sulphonyl chloride,

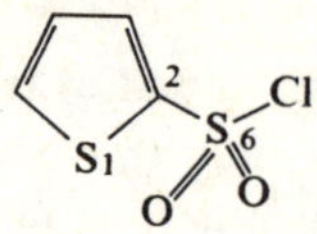

The electron diffraction structure analysis was paralleled by spectroscopic calculations (Brunvoll, Hargittai, Székely and Pappalardo, 1980). These calculations were started with a force field for thiophene (Cyvin and Cyvin, 1969) combined with a force field for the SO_2Cl fragment from calculations on other sulphonyl chloride molecules. This force field was then modified to fit the experimental frequencies (Peron, Saumagne and Lebas, 1970). The geometrical parameters determined by electron diffraction are given in Fig. 83. The data on the size of the thiophene ring are rather uncertain but they are consistent with the thiophene structure itself which was determined accurately by electron diffraction (Bonham and Momany, 1963; Harshbarger and Bauer, 1970) and

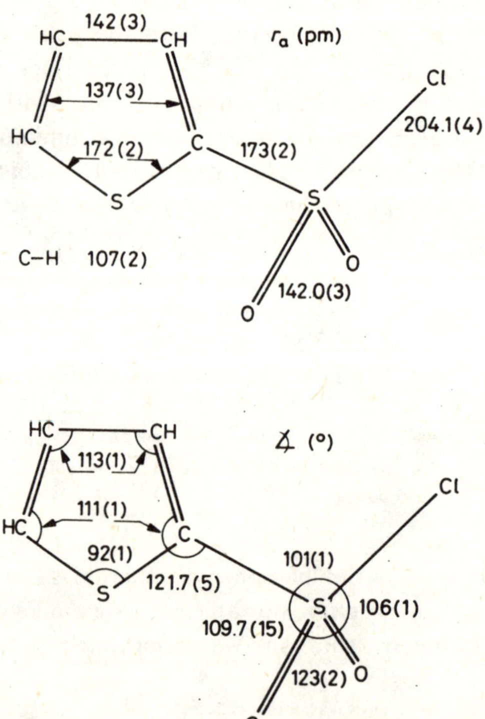

Fig. 83. Bond lengths (r_a) and bond angles (∡) of 2-thiophene sulphonyl chloride from electron diffraction. The parenthesized values are estimated total errors (Brunvoll, Hargittai, Székely and Pappalardo, 1980)

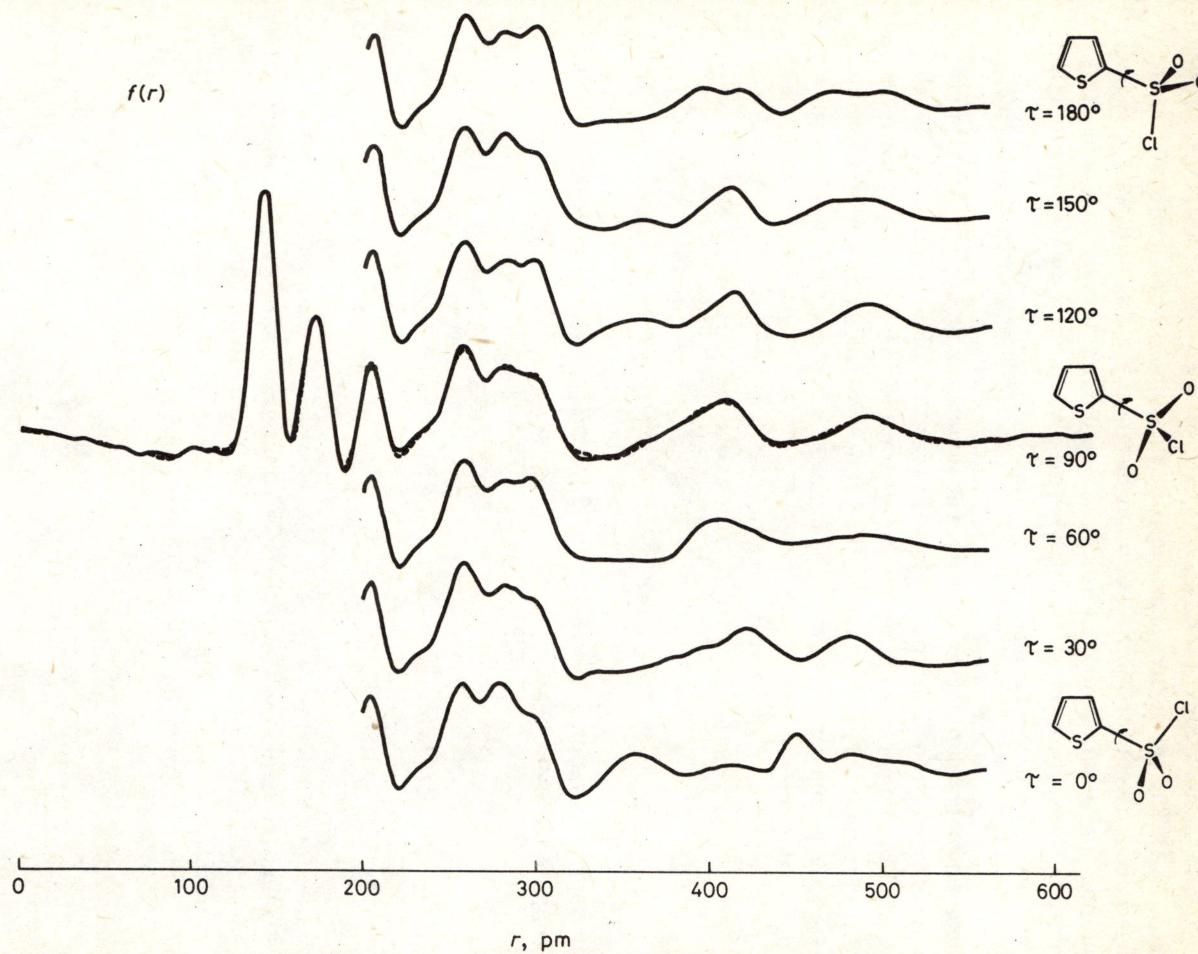

Fig. 84. Experimental (dashed line) and a series of theoretical radial distributions for various conformers of 2-thiophene sulphonyl chloride. τ is the angle of rotation about the C—SO_2Cl bond (Brunvoll, Hargittai, Székely and Pappalardo, 1980)

by microwave spectroscopy (Bak, Christensen, Hansen-Nygaard and Rastrup-Andersen, 1961). The parameters characterizing the sulphur bond configuration are well determined, except for the S6—C bond distance. Figure 84 shows the radial distributions for a series of conformers. The conformation is unambiguously determined by the electron diffraction analysis: the plane of the thiophene ring and the C—S—Cl plane are perpendicular.

Imidodisulphonyl chloride, $(SO_2Cl)_2NH$. The presence of an impurity added to the difficulties of the electron diffraction analysis (Beagley, Moutran, Narula and Ulbrecht, 1979) caused by the correlation among the parameters. Assumptions based on the constancy of sulphone O. . . O distances (Hargittai, 1978) facilitated the determination of a reasonable parameter set. At 86°C the molecule was found to exist as a single C_2 symmetry conformer

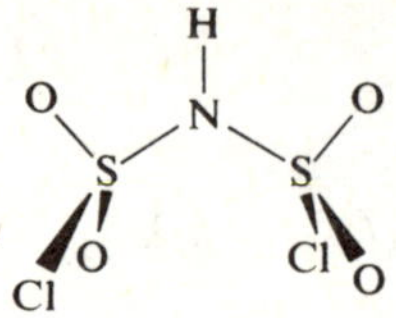

with an angle of rotation H—N—S—Cl 132.3(5)°. The principal bond lengths and bond angles were found to be as follows (r_a):

r(S=O)	141.7(2) pm	∢S—N—S	129(2)°
r(S—N)	162.5(5) pm	∢N—S—Cl	103(1)°
r(S—Cl)	201.6(2) pm	∢N—S=O	109(1)°,

while O=S=O was assumed to be 122.4°.

N,N-Dimethyl sulphamoyl chloride, $(CH_3)_2NSO_2Cl$. A recent electron diffraction reinvestigation (Hargittai and Brunvoll, 1976) confirmed the conformation established by an earlier study (Vilkov and Hargittai, 1967). The molecule has C_s symmetry, the N—C bonds stagger the S—Cl bond and the symmetry plane contains the S—Cl bond and bisects the O=S=O and C—N—C bond angles:

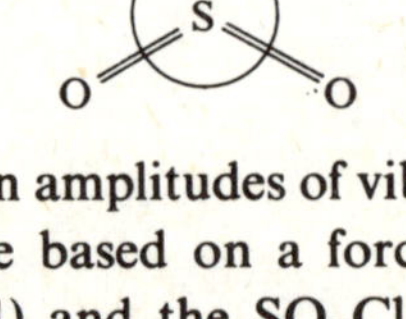

The calculated spectroscopic mean amplitudes of vibration utilized in the more recent electron diffraction analysis were based on a force field constructed from that of trimethyl amine (Gebhardt, 1971) and the SO_2Cl fragment of methane sulphonyl chloride (Cyvin, Dobos, Hargittai, Hargittai and Augdahl, 1973). This force field was modified so as to reproduce the experimental frequencies of *N,N*-dimethyl sulphamoyl chloride (Bürger, Burczyk, Blaschette and Safari, 1971). The geometrical parameters

TABLE 70
Bond lengths, bond angles and mean amplitudes of vibration in *N,N*-dimethyl sulphamoyl chloride from electron diffraction (Hargittai and Brunvoll, 1976)

Parameter	r, ∢	l
r(S=O), pm	142.1(4)	3.8(4)
r(S—Cl), pm	206.4(5)	6.1(3)
r(S—N), pm	161.8(5)	5.1(6)
r(C—N), pm	148.1(12)	5.0*
r(C—H), pm	109.6(10)	7.9(10)
∢N—S—Cl (°)	103.0(5)	
∢N—S=O (°)	108.8(14)	
∢S—N—C (°)	115.8(7)	
∢C—N—C (°)	114.6(22)	
∢O=S=O (°)	122.7(23)	
∢O=S—Cl (°)	105.8(4)	
r(C...Cl), pm		17(6)
r(O...C) (short), pm		15.5*
r(O...C) (long), pm		8(4)

* From spectroscopic calculations

(r_a) and mean amplitudes of vibrations, as determined mainly from electron diffraction, are collected in Table 70.

N,N,N',N'-Tetramethyl sulphonyl diamide, $(CH_3)_2NSO_2N(CH_3)_2$. The principal conformers examined in the electron diffraction analysis (Hargittai, Vajda and Szőke, 1973) are shown below by their Newman projections representing the view along the S—N bond:

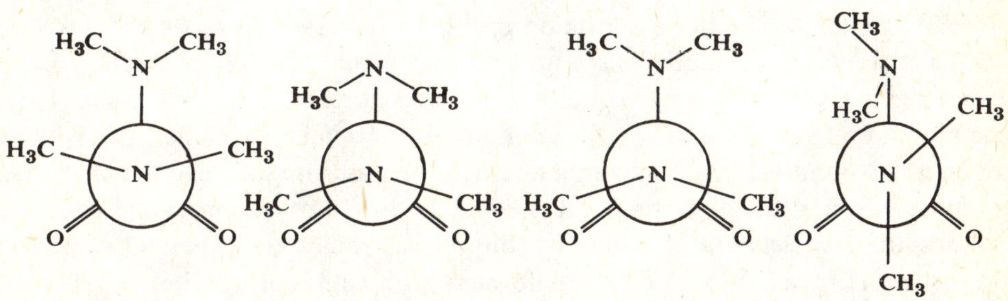

I II III IV

In addition to the form **I** found in the crystalline phase (Jordan, Smith, Lohr and Lipscomb, 1963) the presence of a smaller amount of a second form (**II**) was established in the vapour phase. Both forms have C_{2v} symmetry and their ratio was found to be about 4:1. Table 71 contains the geometrical parameters from the electron diffraction and x-ray diffraction (crystal-phase) studies. Note that the original electron diffraction

TABLE 71
Bond lengths and bond angles in *N,N,N′,N′*-tetramethyl sulphonyl diamide in the vapour phase and in the crystal phase

Parameter	Vapour phase electron diffraction *(a)*	Crystal phase x-ray diffraction *(b)*
r(S=O), pm	143.2(10)	144.5(5)
r(S—N), pm	165.1(3)	162.3(5)
r(C—N), pm	147.5(13)	147.6(7)
∢S—N—C (°)	115.2(11)	118.4(4)
∢C—N—C (°)	118.0(32)	112.9(4)
∢N—S—N (°)	110.5(13)	112.6(4)
∢O=S=O (°)	*(c)*	119.7(4)

(a) Hargittai, Vajda and Szőke (1973)
(b) Jordan, Smith, Lohr and Lipscomb (1963)
(c) See text

report (Hargittai, Vajda and Szőke, 1973) reported an O=S=O bond angle, the corresponding O. . .O distance of which is smaller (242 pm) than the rather constant *r*(O. . .O) observed in the sulphone series (248–249 pm). A relatively large uncertainty was assigned to this parameter, *viz.* 3.8 pm. Later calculations (Vajda and Hargittai, 1975) showed that the other parameters refined practically to the original values when 248 pm was used as constraint for the O. . .O distance.

Sulphonyl chloride isocyanate, $OCNSO_2Cl$. Several assumptions were used in the electron diffraction analysis by Brunvoll, Hargittai and Seip (1978b). The N=C=O chain was supposed to be linear and it was assumed that the geometrical parameters of the individual forms in mixtures of conformers did not differ except, of course, for the angle of rotation. The latter assumption proved to be rather crude for the S—N=C bond angle. As for the relative magnitudes of the C=O and N=C bond lengths, the experimental data could be equally well approximated by two bond length combinations. In one case *r*(C=O) was about 115–116 pm and smaller than *r*(N=C) which was around 122 pm. In the other case the relationship was reversed. According to the electron diffraction data taken with a nozzle at room temperature, the conformational behaviour of sulphonyl chloride isocyanate could be characterized in two ways: *(i)* by a mixture of two conformers with angles of rotation of 109(4) and 70(10)° with 0° corresponding to a form in which the S—Cl bond is *anti* to the O=C=N chain, and the first of the two forms predominating approximately 2 : 1; *(ii)* by one conformer with an angle of rotation of 86(3)° and large-amplitude torsion around the S—N bond. The electron diffraction analysis was augmented by spectroscopic calculations based on experimental frequencies (Kanesaka and Kawai, 1970). Spectroscopic amplitudes of vibration were kept constant for the rotation-dependent distances in the refinement scheme for the mixture of conformers while they were

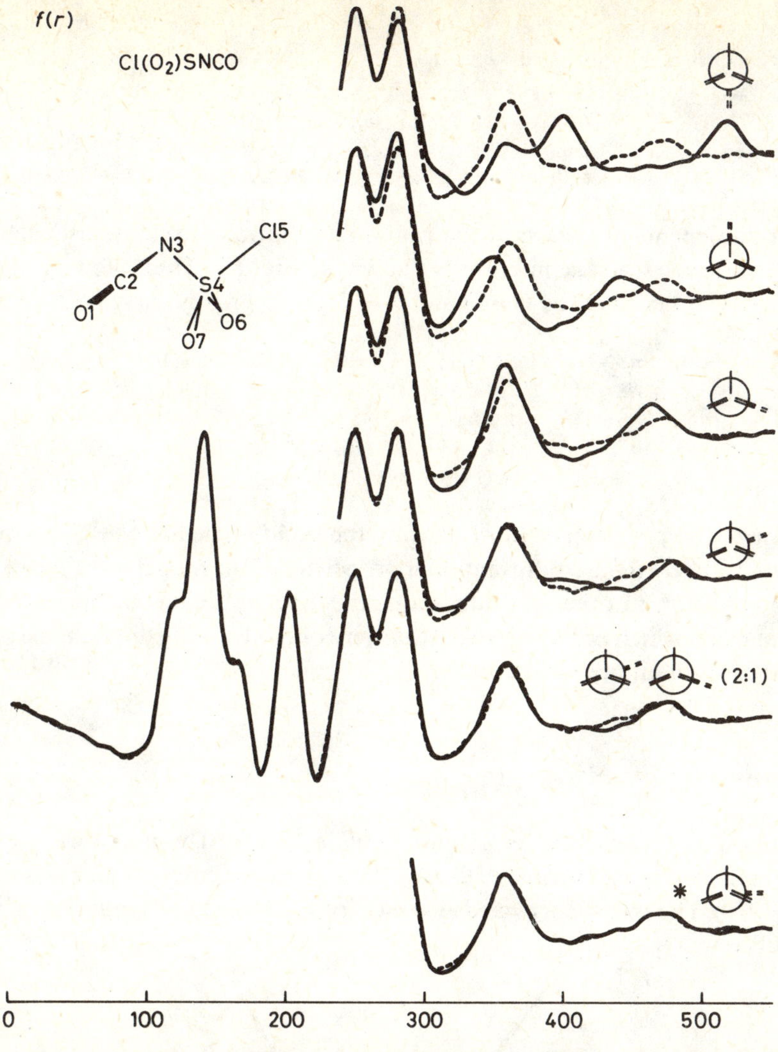

Fig. 85. Experimental (dashed lines) and theoretical (full lines) radial distributions of sulphonyl chloride isocyanate (Brunvoll, Hargittai and Seip, 1978b). The rotational isomers to which the theoretical curves refer are shown by Newman projections representing views along the S—N bond. Spectroscopically calculated mean amplitudes of vibration were used for all the non-bond distances in calculating the upper five theoretical curves. The lowest theoretical curve marked by an asterisk was calculated with larger mean amplitudes for the rotation-dependent distances. The numbering of atoms is also shown

refined in the scheme assuming a single rotational isomer. The refined mean amplitudes of vibration for the rotation-dependent distances are cited below together with the calculated ones for the same rotational form (the numbering of atoms is given in Fig. 85):

	l (refined) pm	l (calculated) pm
O(1)...Cl	27.3(28)	22.0
O(1)...O(6)	10.8(13)	7.8
O(1)...O(7)	29.4 } (14)	19.2
C...Cl	25.4 } (14)	15.4
C...O(6)	13.4 } (14)	6.9
C...O(7)	25.4(53)	13.0

The rotation-dependent region of the radial distribution curves as calculated for a series of conformers and a mixture is shown in Fig. 85. The following molecular parameters (r_a) other than the C═O and N═C bond lengths, have been determined:

r(S—N)	165.6(4) pm	∢S—N═C	124(4)°
r(S—Cl)	201.9(3) pm	∢N—S═O	108.3(22)°
r(S═O)	141.7(3) pm	∢N—S—Cl	98(3)°
	∢Cl—S═O	107.8(7)°	
	∢O═S═O	122.8(24)°	

Since some of the parameters, most notably the bond angle S—N═C, proved to be somewhat sensitive to the conformational properties of the model for which they were refined, in addition to other considerations the error estimates included changes in these parameters which occurred when they were obtained using different assumptions about the conformational properties.

4.6. CYCLIC SULPHONES

Thiirane-1,1-dioxide (ethylene episulphone), $(CH_2)_2SO_2$. The microwave spectra of three isotopic species, *viz.* normal, ^{13}C, and ^{34}S were investigated by Nakano, Saito and Morino (1970). The spectra provided evidence for C_{2v} molecular symmetry. The model is shown below,

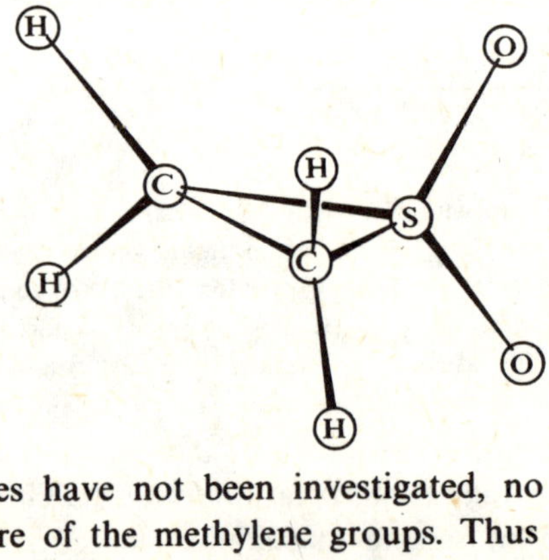

Since deuterated species have not been investigated, no information is available concerning the structure of the methylene groups. Thus it was assumed that the methylene groups of thiirane-1,1-dioxide and thiirane (ethylene episulphide) have the same geometry. Furthermore, the rotational spectrum of the ^{34}S isotopic species added practically nothing to the structural information since the sulphur atom is situated near the centre of gravity of the molecule. The following four parameters (r_0) were determined for the C_{2v} symmetry heavy-atom skeleton:

r(S=O)	143.9(6) pm	∢O=S=O	121° 26′±30′
r(C—S)	173.1(6) pm	∢C—S—C	54° 40′±15′

From the above independent parameters the C—C bond length was calculated to be 159.0±1.1 pm (Nakano, Saito and Morino, 1970).

Thiete-1,1-dioxide (thiete sulphone), $\underline{CH{=}CH{-}CH_2{-}SO_2}$. In order to determine the O. . .O distance from the rotational constants, Ralowski, Ljunggren and Mjöberg (1973) assumed a planar ring configuration. This assumption was consistent with the results of an x-ray diffraction study of this compound in the crystalline phase by Lowenstein (1965). The latter work yielded the following parameters (bond lengths in pm, angles in degrees):

r(C—C)	152(4)	∢C—S—C	80.5
r(C=C)	139(3)	∢S—C—C	85.1
r(—C—S)	179(4)	∢S—C=C	90.0
r(=C—S)	177(3)	∢C—C=C	104.5
r(S=O)	143(2)	∢O=S=O	115.5

The O. . .O distance (247.8 pm) determined unambiguously from the microwave spectrum is not consistent with the S=O bond length and O=S=O bond angle reported from the x-ray diffraction work. Either the S=O distance or the O=S=O angle is too small. The latter seems to be probable while the O. . .O distance from the microwave spectroscopic data and the S=O bond length determined by x-ray diffraction give a plausible value of 120.1°.

Tetrahydrothiophene-1,1-dioxide (tetramethylene sulphone), $(CH_2)_4SO_2$. An electron diffraction investigation by Naumov, Semashko and Shaidulin (1973) showed that an envelope and a half-chair model reproduces the experimental data well. These two conformers are illustrated in Fig. 44. A model with planar ring skeleton could be rejected. However, the conformational properties could not be established unambiguously. Fortunately, the bond lengths and bond angles depended very little on the conformations assumed in the refinements of the parameters. The results obtained for a half-chair model are cited below (bond lengths in pm, angles in degrees):

r(CC—CC)	153.3(15)	∢O=S=O	114.7(30)
r(SC—CC)	154.0(14)	∢C—S—C	101.1(15)
r(S=O)	144.9(10)	∢C—C—S	104.3(10)
r(S—C)	179.8(8)	∢C—C—C	112.1(15)

The cited paper communicated mean amplitudes of vibration for the bonds only.

4.7. IMINOSULPHUR DERIVATIVES

Iminosulphuryl difluoride,

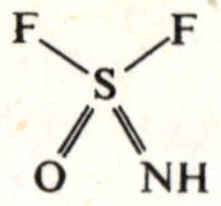

The microwave spectra of four isotopic species were analyzed by Cassoux, Kuczkowski and Creswell (1977). As regards the rotation around the S═N bond, the *trans* orientation of the O═S═N—H chain was established

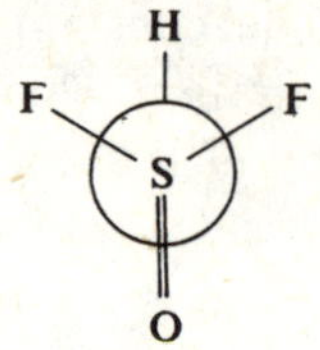

The geometrical parameters (r_0, and some of them r_s) are given in Table 72 including some non-bond distances.

N-chloro-iminosulphuryl difluoride, $ClNSOF_2$. The geometrical parameters determined by an electron diffraction investigation of Oberhammer, Glemser and

TABLE 72
Geometrical parameters of iminosulphuryl difluoride and *N*-chloro-iminosulphuryl difluoride

Parameter	$HNSOF_2$ *(a)*	$ClNSOF_2$ *(b)*
r(S—F), pm	154.9(2)	154.8(3)
r(S═O), pm	142.0(5)	139.4(3)
r(S═N), pm	146.6(3)	148.4(7)
r(N—X), pm	102.3(7)	171.5(5)
∢N═S═O (°)	119.5(2)	117.4(31)
∢F—S—F (°)	93.7(1)	92.6(8)
∢O═S—F (°)	107.3	108.6(8)
∢N═S—F (°)	112.9(1)	111.8(9)
∢X—N═S (°)	115.5(5)	114.7(8)
r(O. . .N), pm	249.3	246.0
r(O. . .F), pm	239.2	239.3
r(N. . .F), pm	251.3	251.1

(a) Microwave spectroscopy; Cassoux, Kuczkowski and Creswell (1977). The cited uncertainties were based on the experimental uncertainties of the rotational constants. The estimated errors taking the vibrational effects also into account were 1 pm and 0.5° for the bond lengths and angles, respectively
(b) Electron diffraction; Oberhammer, Glemser and Klüver (1974)

Klüver (1974) are cited also in Table 72. The rotational form here is characterized also by the *trans* orientation of the O═S═N—Cl chain

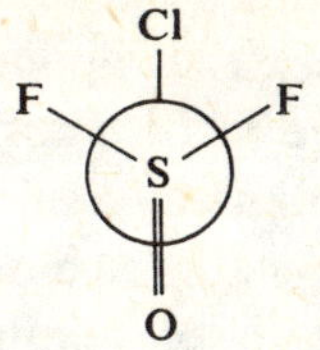

The vapour-phase molecular geometry of *S,S-dimethyl-sulphoximide*, $HN(O)S(CH_3)_2$ and *diimino dimethyl sulphur*, $(HN)_2S(CH_3)_2$ was investigated by electron diffraction by Oberhammer and Zeil (1970b and 1969, respectively).

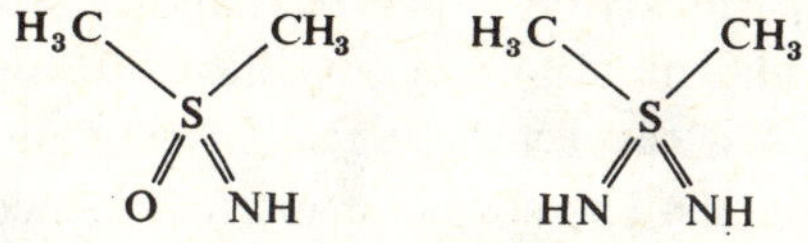

The reported results are cited in Table 73. x-Ray diffraction structural data on diimino dimethyl sulphur have also been published (Webb and Gloss, 1967), and they are

TABLE 73
Geometrical parameters of *S,S*-dimethyl-sulphoximide and diimino dimethyl sulphur

Parameter	$HN(O)S(CH_3)_2$	$(HN)_2S(CH_3)_2$	
	Vapour phase *(a)*	Vapour phase *(b)*	Crystal phase *(c)*
r(S—C), pm	178.1(3)	179.1(2)	176(5)
r(S═O), pm	143.7(3)		
r(S═N), pm	152.1(3)	153.3(2)	153(4)
r(N—H), pm	102.2(18)	102.2(18)	
r(C—H), pm	108.3(7)	108.5(8)	
∢N═S═O (°)	132.6(9) *(d)*		
∢N═S═N (°)		135.0(12) *(d)*	122(3)
∢C—S—C (°)	101.6(13)	101.2(14)	103(3)
∢O═S—C (°)	106.7(10)		
∢N═S—C (°)	102.7(13)	104.0	103(2); 113(2)
∢H—N═S (°)		114.1(24)	
r(N...N), pm		283.3	268
r(O...N), pm	270.9		
r(C...C), pm	259		
r(N...C), pm	258.4	262.4	258; 275

(a) Electron diffraction; Oberhammer and Zeil (1970b)
(b) Electron diffraction; Oberhammer and Zeil (1969)
(c) x-ray diffraction; Webb and Gloss (1967)
(d) Warning: these bond angles may be erroneously too large, see discussion in the text

included in Table 73. There is a large (13°) difference in the N=S=N bond angle of $(HN)_2S(CH_3)_2$ as reported in the two studies and this difference cannot possibly be accounted for by the large uncertainties of the x-ray diffraction results or by any expected structural difference between the vapour and crystalline phases. The latter statement is made in the light of the relative stiffness of the bond configuration of diimino dimethyl sulphur. Also, dimethyl sulphone has similar sulphur bond configurations in the vapour and crystalline phases (*cf.* Table 64). Unfortunately, an unambiguous structure determination of diimino dimethyl sulphur by electron diffraction is hindered by the strong correlation among parameters such as non-bond distances invoking distances between atoms separated by a bond angle and the mean amplitudes of vibration associated with these atomic pairs. The contributions from all such interactions appear under the same peak of the experimental radial distribution. The dangers in the electron diffraction structure analysis of such a structure, especially if additional independent information is lacking, have already been seen on the example of dimethyl sulphone. It is suspected that the investigation of both molecules has suffered from similar difficulties. This suspicion is strengthened by a closer examination of the variations of the N...N and N...O non-bond distances. The N...O distance of $HNSOF_2$ determined reliably by microwave spectroscopy is consistent with the N...O distance of $ClNSOF_2$ yielded by the electron diffraction analysis (see data in Table 72). Note that the rotation-dependent portion of the radial distribution of $ClNSOF_2$ is richer in various characteristic features than the analogous curves of the other two molecules also investigated by electron diffraction. The N...O distance of $HN(O)S(CH_3)_2$ was reported to be 20 pm longer than the N...O distances of $HN(O)SF_2$ and $ClN(O)SF_2$ pointing to the former being probably erroneous. The structures or at least some of their features of the two iminosulphur derivatives, *viz.* $HN(O)S(CH_3)_2$ and $(HN)_2S(CH_3)_2$ are suspect and they certainly call for a reinvestigation possibly involving both electron diffraction and spectroscopic data.

4.8. ON THE CONFORMATIONAL PROPERTIES OF SULPHONES

The conformers prevailing in the vapours of sulphones are summarized in Fig. 86. It is seen that staggered forms predominate while eclipsed orientations appear when two double bonds (e.g. C=C and S=O) may take a *syn* arrangement. However, this is a rather qualitative observation. Also, there remain some important questions unanswered. Some of them are listed below with some additional comments.

It is unexplained why the conformational behaviour of fluorosulphuric acid methyl ester and chlorosulphuric acid methyl ester is different. The van der Waals interactions were estimated and no appreciable difference was discovered (Schultz, 1976).

An unsettled question is the conformation of benzene sulphonyl chloride. Apart from the apparent controversy between the electron diffraction and microwave

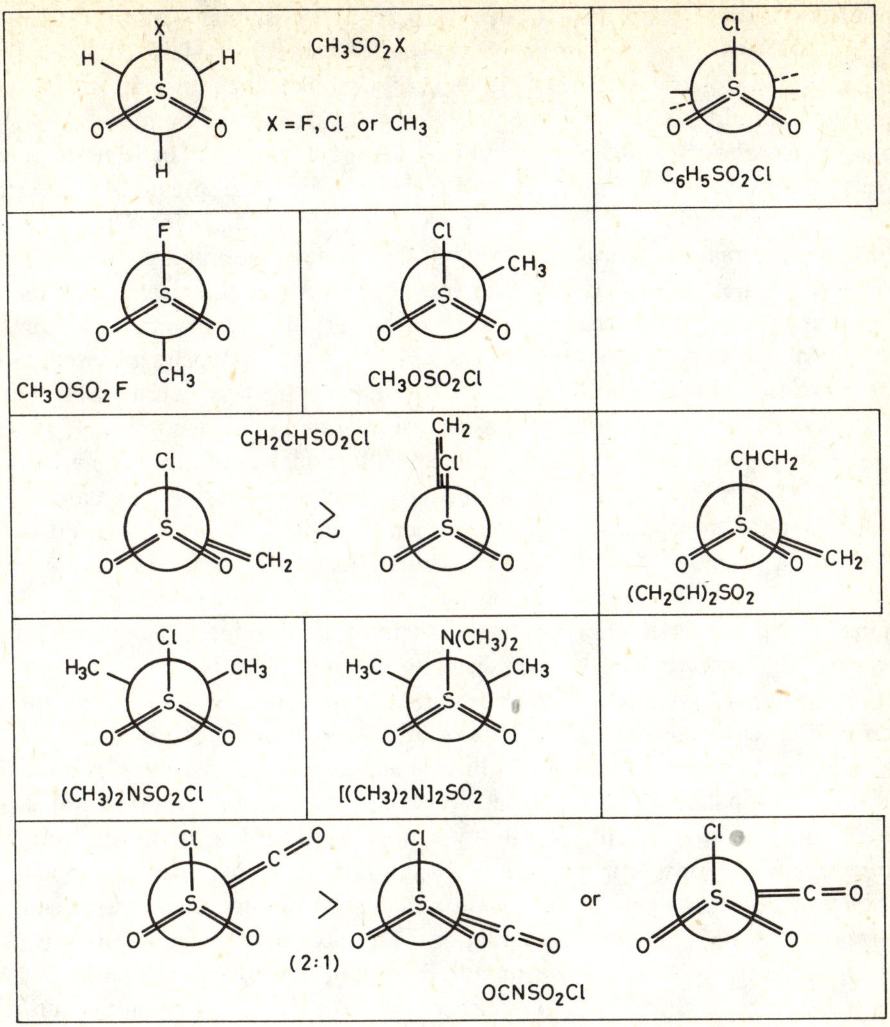

Fig. 86. Summary of characteristic conformations of sulphone molecules

spectroscopic reports, it is not clear whether the symmetric and asymmetric forms found by electron diffraction could be distinguished or indeed which of them should be preferred. Additional information on the low-frequency region of the vibrational spectrum would probably be very useful. From stereochemical point of view it may be argued that the stabilization of the asymmetric structure might be facilitated by possible intramolecular H...O interaction. However, such interaction obviously cannot be present in the molecule of benzyl chloride, $C_6H_5CH_2Cl$ for which a similar asymmetric conformation was observed in its electron diffraction investigation (Sadova, Vilkov, Hargittai and Brunvoll, 1976). It is also of interest that even more pronounced asymmetries have been observed in similar systems, e.g., in $C_6H_5SO_2NHSO_2C_6H_5$ (Cotton and Stokely, 1970) in the crystalline phase. Here, however, packing forces may strongly influence the molecular conformation. The root-mean-square amplitude of the torsional motion around the S—C bond, $\langle \Delta\tau^2 \rangle^{1/2}$, was estimated to be about 8 to 8.5° for the symmetric model ($\tau = 90°$) of benzene sulphonyl chloride, using the same force field as that on which the previously described spectroscopic calculations were based (Brunvoll and Hargittai, 1976). This amplitude seems to be too small to indicate that the intramolecular motion may account for the appearance of the strongly asymmetric rotational form in the electron diffraction structure analysis. It should be emphasized, however, that the asymmetric structure was associated with relatively small l values for the rotation-dependent interactions. These amplitudes for rotation-dependent distances refined to considerably higher values when allowed to vary under the assumption of symmetrical conformation. Had a force field reproducing these larger mean vibrational amplitudes been used, a considerably larger value of $\langle \Delta\tau^2 \rangle^{1/2}$ would have been most certainly calculated. If it is assumed that the appearance of the asymmetric model in the electron diffraction analysis is a consequence of the intramolecular (torsional) vibrations, a barrier to internal rotation of 16 kJ mol^{-1} could be estimated by the method of Vilkov, Penionzhkevich, Brunvoll and Hargittai (1978) from the amount of the deviation from the symmetric structure. This barrier seems to be quite plausible.

The perpendicular form of 2-thiophene sulphonyl chloride was unambiguously determined by electron diffraction. This result was in complete agreement with the dipole moment measurements in solution as communicated by Gruttadauria, Maccarone, Pappalardo and Tomaselli (1976). However, the experimental dipole moment could be interpreted not only by a single conformer but alternatively by a mixture of several forms, or free rotation. The authors cited suggested that in fact free rotation was more likely to occur. They based their discussion on supposing the absence of steric and conjugative interactions. The electron diffraction analysis ruled out such a hypothesis. Moreover, Brunvoll, Hargittai, Székely and Pappalardo (1980) estimated the barrier to internal rotation to be some 22(2) kJ mol^{-1} utilizing calculated framework vibrational amplitudes and assuming a form of $V = \frac{1}{2} V_0[1 - \cos(2\varphi)]$ for the potential of internal rotation with $\varphi = 0$ for the 90° angle of rotation. This relatively high rotational barrier and also the very short S—C bond all may indicate steric hindrance as well as considerable conjugation.

The predominating conformer of *N,N,N′,N′*-tetramethyl sulphonyl diamide is the same in the vapour and in the crystalline phases. One more form has also been detected, however, in the vapour phase, which can be obtained by a 180° rotation around the S—N bond or by inversion of the nitrogen bond configuration of the predominant form. The thiobis(dimethylamine) molecules were found to display similar conformational properties in the vapour phase. As for the rotational behaviour of *N,N′*-sulphinyl-bis(dimethylamine), the most important form was found to be analogous to the forms

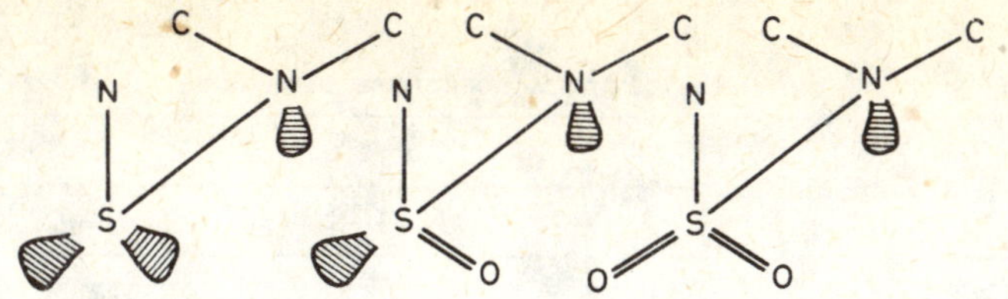

Fig. 87. The ethane-like rotational forms of the $NSNC_2$, $NSONC_2$, and NSO_2NC_2 skeletons in $(CH_3)_2NSN(CH_3)_2$, $(CH_3)_2NSON(CH_3)_2$, and $(CH_3)_2NSO_2N(CH_3)_2$

mentioned above although other forms have also been indicated. Considering also the sulphur and nitrogen lone pairs of electrons, all these prevailing forms may be looked upon as ethane-like staggered conformers as depicted in Fig. 87. The single conformation existing in the vapour of *N,N*-dimethyl sulphamoyl chloride is completely analogous to the prevailing forms of the other S—N derivatives.

The two conformational models found for sulphonyl chloride isocyanate differ only in the character of the motion relative to the N—S bond. The model of the mixture with two forms assumes a relatively high rotational barrier. These two forms are characterized by planar five-membered chains

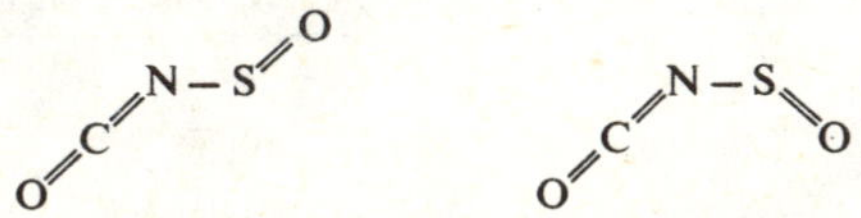

The other model containing a single form is associated with large-amplitude torsional vibrations around the N—S bond and thus is supposed to have a relatively low barrier to internal rotation.

4.9. VARIATIONS OF THE SULPHUR BOND CONFIGURATION IN SULPHONES

A general and very characteristic structural feature of the XSO_2Y sulphones is the following relationship between the bond angles

$$\measuredangle X\text{—}S\text{—}Y < \measuredangle X(Y)\text{—}S{=}O < \measuredangle O{=}S{=}O$$

This relationship is invariant to the nature of the X and Y ligands. The relevant data are collected in Table 74. The relative magnitude of the bond angles around the sulphur atom in sulphones is an excellent example for the multiple bond rule of the VSEPR model. The largest bond angle is formed between the two S=O bonds, i.e. between the two double bonds, and the smallest between the two single bonds. The former is considerably larger while the latter is considerably smaller than the ideal tetrahedral angle. The structure of *N,N,N′,N′*-tetramethyl sulphonyl diamide seems to be the only exception in which the N—S—N bond angle was found to be larger than the

TABLE 74
Sulphur bond angles in sulphone molecules

Compound	∢(°)		
	X—S—Y	X(Y)—S=O	O=S=O
FSO_2F	96	108	124
CH_3OSO_2F	97	107, 109	124
CF_3SO_2Cl	99	108, 108	122
CCl_3SO_2Cl	98	108, 109	122
CH_3SO_2F	98	106, 110	123
$ClSO_2Cl$	100	108	123
CH_3OSO_2Cl	103	106, 109	122
CH_3SO_2Cl	101	107, 109	121
$CH_3SO_2CH_3$	103	108	120
$CH_2{=}CHSO_2Cl$	100	106, 110	122
$C_6H_5SO_2Cl$	101	106, 110	122
$OCNSO_2Cl$	98	108, 108	123
$(CH_3)_2NSO_2Cl$	103	106, 109	123
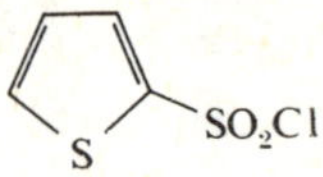	101	106, 110	123

tetrahedral value, *viz.* 110.5±0.4° in the vapour phase by electron diffraction (Hargittai, Vajda and Szőke, 1973) and 112.6±0.4° in the crystalline phase by x-ray diffraction (Jordan, Smith, Lohr and Lipscomb, 1963). There were two models of the divinyl sulphone structure differing in the C—S—C bond angle, 108° *vs.* 95° (Hargittai, Rozsondai, Nagel, Bulcke, Robinet and Labarre, 1978) between which it was not possible to distinguish on the basis of the available data. Neither of them is in serious discrepancy with the generally observed structural features of the sulphone molecules although 95° seems to be a little too small compared with other X—S—Y angles, while 108° is a little too large compared with ∢C—S=O 107° of divinyl sulphone itself. The bond angle relationships in the iminosulphur derivatives are also in complete agreement with the VSEPR model.

In the subsequent comparisons the structural variations will be examined primarily in the light of the changing X and Y ligands. Sulphur–carbon and sulphur–halogen bond lengths as well as a series of bond angles are collected in Table 75, while S=O bond lengths and O=S=O bond angles are compiled in Table 76. Some non-bond distances, *viz.* O. . .Cl and O. . .O are also listed in these Tables. The data presented could be grouped in other ways and some other comparisons could also be made, but those given were thought to be sufficient to characterize the observations made on the sulphur bond configurations in the sulphone series. Before inspecting and commenting upon the tendencies in the individual series of data, the qualitative character of these observations is emphasized. The ligand electronegativities for groups are for example

TABLE 75

(a) The S—C bond lengths and C—S—X bond angles in CH_3SO_2X molecules, with X = F, Cl, CH_3

Compound	r(S—C), pm	∢C—S—X (°)
CH_3SO_2F	175.9(6)	98.2(15)
CH_3SO_2Cl	176.3(5)	101.0(14)
$CH_3SO_2CH_3$	177.1(4)	102.6(9)

(b) The S—C bond lengths and C—S—Cl bond angles in CX_3SO_2Cl molecules, with X = H, F, Cl

Compound	r(S—C), pm	∢C—S—Cl (°)
CH_3SO_2Cl	176.3(5)	101.0(14)
CF_3SO_2Cl	185.6(6)	98.7(4)
CCl_3SO_2Cl	186.5(30)	97.9(8)

(c) The S—F bond lengths and X—S—F bond angles in XSO_2F molecules, with X = F, CH_3O, CH_3

Compound	r(S—F), pm	∢X—S—F (°)
FSO_2F	153.0(3)	96.1(2)
CH_3OSO_2F	154.5(6)	96.8(6)
CH_3SO_2F	156.1(4)	98.2(15)

(d) The S—Cl bond lengths, O=S—Cl bond angles and O...Cl interatomic distances in a series of sulphonyl chloride molecules

Compound	r(S—Cl) (pm)	∢O=S—Cl (°)	Δ∢* (°)	r(O...Cl) (pm)	Δr* (pm)
FSO_2Cl	198.5	107.5	−0.5	275.7	3.5
$ClSO_2Cl$	201.1	107.7	−0.7	278.1	1.1
CF_3SO_2Cl	201.6	108.3	−1.3	280.2	−1.0
$OCNSO_2Cl$	201.8	107.8	−0.8	279.8	−0.6
CCl_3SO_2Cl	202.1	109.2	−2.2	282.5	−3.3
CH_3OSO_2Cl	202.3	106.3	0.7	277.9	1.3
$CH_2{=}CHSO_2Cl$	203.5	106.3	0.7	278.9	0.3
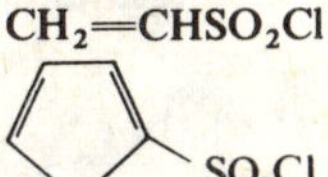	204.1	105.4	1.6	277.9	1.3
$C_6H_5SO_2Cl$	204.7	105.5	1.5	278.0	1.2
CH_3SO_2Cl	204.6	107.1	−0.1	281.6	−2.4
$(CH_3)_2NSO_2Cl$	206.4	105.8	1.2	280.6	−1.4

* Deviations from the mean

TABLE 76

S=O bond lengths, O=S=O bond angles and O...O interatomic distances in vapour-phase sulphone molecules *(a)*

Compound	r(S=O), pm	∢O=S=O (°)	r(O...O), pm
FSO_2F	139.8(2) ED	125.1 *(c)*	248.1(6) MW
FSO_2OCH_3	141.0(3) ED	124.4(7) ED	248.4(3) MW
FSO_2Cl	[140.8] *(b)*	(123.7) *(d)*	248.4(3) MW
FSO_2Br	[140.7] *(b)*	(123.7) *(d)*	248.6(3) MW
FSO_2CH_3	141.1(3) ED	123.1 *(c)*	248.0(3) MW
$ClSO_2Cl$	140.5(4) ED	123.5(10) ED	248.5(3) MW
$ClSO_2CF_3$	141.6(7) ED	122.4(10) ED	248.0(14) ED
$ClSO_2CCl_3$	142.1(3) ED	121.5(9) ED	247.8(15) ED
$ClSO_2OCH_3$	142.0(4) ED	(122.2) *(d)*	[248.5] *(b)*
$ClSO_2CH{=}CH_2$	142.1(6) ED	122.0(10) ED	248.4(1)) ED
$ClSO_2C_6H_5$	141.8(12) ED	122.3(36) ED	248.2(39) ED
$ClSO_2NCO$	141.7(3) ED	(122.8) *(d)*	[248.4] *(b)*
$ClSO_2N(CH_3)_2$	142.2(4) ED	122.7(23) ED	249.4(26) ED
$ClSO_2CH_3$	142.5(3) ED	120.8(8) ED	248.3(3) MW
$CH_2{=}CHSO_2CH{=}CH_2$	144.0(4) ED	(119.5) *(d)*	[248.4] *(b)*
$(CH_3)_2NSO_2N(CH_3)_2$	143.3(10) ED	(119.8) *(d)*	[248] *(b)*
$CH_3SO_2CH_3$	143.6(3) ED	119.7(11) ED	249.3(3) MW
SO_2	143.2(2) ED		247.6(3) MW
$(CH_2)_4SO_2$	144.9(10) ED	115(3) ED	244 ED
$(CH_2)_2SO_2$	143.9(6) MW	121.4(5) MW	250.1(3) MW

(a) For sources see earlier; ED — electron diffraction; S=O bond length r_g parameter; MW — microwave spectroscopy; O...O distance r_0 parameter; the mean of the MW O...O distances is 248.5 pm (σ 0.7 pm); ignoring the datum for $(CH_2)_2SO_2$, the mean value is 248.4 pm (σ 0.4 pm)

(b) Assumed parameter

(c) Calculated from the electron diffraction r(S=O) and microwave spectroscopic r(O...O)

(d) One of the distances has been assumed in calculating this bond angle

not well-defined*, and the changes in the geometrical parameters from compound to compound may be comparable with or even smaller than the experimental uncertainties. Thus it is important to discuss the structural variations in relatively large series of compounds.

Keeping in mind the above warnings, let us now consider the tendencies in the changes of the geometrical parameters accompanying an increase of the electronegativities of the ligands linked to the SO_2 group. It is observed that

(i) the S—X bonds shorten, X = O, C, F or Cl;

(ii) the O=S=O angle opens;

(iii) the C—S—X (X = F, Cl or CH_3) and X—S—F (X = F, CH_3O or CH_3) angles decrease.

Special comments are made on the variations of the O=S—Cl bond angles.

* It will be attempted later to estimate group electronegativities on the basis of the geometrical variations observed in the sulphone series.

According to the VSEPR model, the more electronegative ligand draws away electron density from the central sulphur atom and, accordingly, the space requirement of the bonding pair between the sulphur atom and the ligand decreases ("electronegativity rule"). Since the neighbouring bonds suffer less repulsion from this bonding pair, they may shorten. At the same time, the bond angle, in whose formation the sulphur–ligand bond participates, may decrease. All this accounts for tendencies *(i)* and *(iii)*.

According to the "multiple bond rule" of the VSEPR model, the shortening of the S═O bonds, i.e. the increase in their multiple bond character, results in an opening of the O═S═O bond angle, hence tendency *(ii)*.

Note also, however, that according to the "electronegativity rule", the presence of a more electronegative ligand attached to the SO_2 group tends to shorten all the other sulphur bonds. This, in turn, should, according to the "multiple bond rule", open the bond angles which is not observed; tendency *(iii)* shows the opposite. Thus of the two effects competing here, the one described by the electronegativity rule seems to predominate in determining the changes in the C—S—X and X—S—F bond angles (*cf.* the data of Tables 75(a), 75(c)).

Table 75(b) contains some X_3C—S bond lengths. There is a strikingly large difference of almost 10 pm between the S—C bond lengths of the methyl and trifluoromethyl derivatives. Both parameters are well determined. On the other hand, *r*(S—C) of CCl_3SO_2Cl was reported with large uncertainty. The S—C bond of trifluoromethane sulphonyl chloride is longer than that in some other molecules containing the F_3C—SO_2 part. Some data are cited here from crystal-phase x-ray diffraction studies showing only fragments of the molecules studied

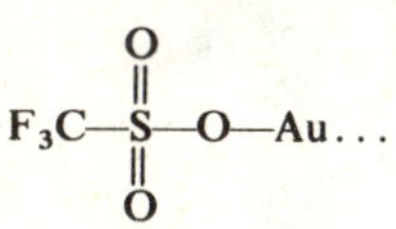

r(C—S) 181.6(9) pm, after Komiya, Huffman and Kochi (1977),

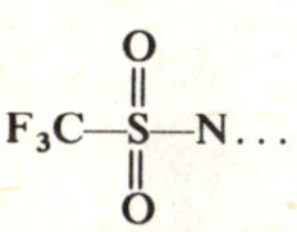

r(C—S) 182 pm, after Roesky, Diehl, Fuess and Bats (1978).

In any case, the difference in the C—S bonds of the molecule pair CH_3SO_2Cl/CF_3SO_2Cl is larger than the other known structural changes caused by CH_3/CF_3 substitution (see Yokozeki and Bauer, 1975). In considering the short C—S bond of CH_3SO_2F, Robinson and Aroca (1978) called attention to the hyperconjugative release of electrons from the C—H bonds into the C—S bond. Some semiempirical CNDO/2 calculations were carried out (Mayer and Hargittai, 1979) to further examine the origin of the large variation in the C—S bond length upon CH_3/CF_3 substitution. Although the semi-empirical character of the CNDO/2 procedure makes it difficult to assess the reliability of the quantitative results obtained for a given individual molecule, useful qualitative information may be anticipated

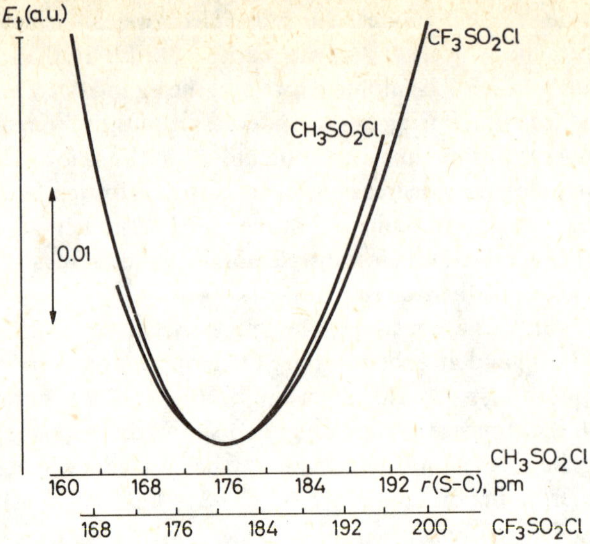

Fig. 88. Calculated total CNDO/2 molecular energy *vs.* bond length curves for CH_3SO_2Cl and CF_3SO_2Cl molecules (after Mayer and Hargittai, 1979)

concerning differences between two similar systems. The electron diffraction geometries were assumed in the calculations and the C—S bonds were calculated. Although the difference in the bond lengths was smaller from the calculations than from the experimental data, the effect was considered to be large enough (5 pm) to expect qualitative information as to its origin. The data indicated that the C—S bond of the fluorinated derivative became poorer in electrons due to the loss of electrons on the carbon atom. Accordingly, the C—S bond became also more compliant as is well seen by the difference in curvature of the energy *vs.* bond length plots of Fig. 88. This was also consistent with the values of the Wiberg (1968) indices giving a measure of the bond order

$$H_3C\overset{1.089}{\text{——}}SO_2Cl \qquad F_3C\overset{0.925}{\text{——}}SO_2Cl$$

Similar effects may be expected for trichloromethane sulphonyl chloride. Without going into the details of the calculations, they have supported the conclusion according to which the C—S bond length variation upon the CH_3/CF_3 substitution may be attributed in simple terms to the difference between the electron releasing ability of the CH_3 group and the electron withdrawing ability of the CF_3 group.

The smaller bond angle C—S—Cl of CF_3SO_2Cl as compared with the analogous parameter of CH_3SO_2Cl is in agreement with the expectation based on the electronegativity rule of the VSEPR model and also with the increased double bond character of the C—S bond in the methyl derivative as compared to the fluorinated molecule. Thus in determining the actual C—S—Cl angles, the effects described by the electronegativity and multiple bond rules of the VSEPR model strengthen each other.

Consider now the changes in the O=S—X bond angles. When the electronegativity of ligand X changes, there are two competing effects influencing the O=S—X bond angle. On the other hand, when changing the electronegativity of ligand Y in the XSO_2Y sulphones, both the electronegativity and the multiple bond effects act in the same direction from the point of view of the O=S—X angle, e.g., they both tend to increase this angle with increasing Y electronegativity. Thus the bond angle O=S—F increases in the CH_3SO_2F, CH_3OSO_2F, FSO_2F series (the respective bond angles are 106.2, 106.8 and 108°). The same tendency would be expected for the O=S—Cl bond angles in the sulphonyl chloride series. It is not obvious from the data of Table 75(d) that such a tendency is present, although there is a qualitative pattern suggesting that with increasing Y electronegativity in the YSO_2Cl molecules the O=S—Cl bond angle increases. Indeed, the tendency is somewhat better seen when the deviations from the mean of the O=S—Cl bond angles are considered instead of the actual angles. It is also noteworthy that the X—S—Cl bond angles of the XSO_2Cl sulphonyl chloride molecules do not really follow the changes of the X—S—F angles of the XSO_2F sulphonyl fluoride molecules. Here again the two rules of the VSEPR model cited represent competing effects so the situation is rather complicated. The suspicion also arises, however, that the VSEPR model may be less applicable for sulphonyl chlorides than for sulphonyl fluorides. This would be consistent with the general observation according to which the importance of non-bonded interactions *vs.* the electron pair repulsions is higher as the size of the ligands increases relative to the size of the central atom.

Attention has been drawn repeatedly to the importance of non-bonded interactions in the sulphone molecular geometries, and in particular in the variations of the geometry of the SO_2 group (Hargittai and Hargittai, 1974; Hargittai, 1978). The oxygen–oxygen distance is strikingly constant at about 248–249 pm according to the data of Table 76. The S=O bond lengths (r, pm) and O=S=O bond angles ($\measuredangle$ in degrees) determined by electron diffraction are demonstrated in Fig. 89. The straight line represents the empirical expression

$$\measuredangle = -1.477r + 331.7 \quad (\sigma = 0.6°)$$

resulting from a least-squares fit to the experimental data. As the observation on the constancy of the O . . . O distances is utilized, the relationship between the S=O bond lengths and O=S=O bond angles can be given by the following trigonometrical expression,

$$\measuredangle = 2 \operatorname{arc\,sin}\left(\frac{248.4}{2r}\right),$$

and the corresponding curve indeed barely deviates from the straight line obtained from the least-squares fit.

The constancy of the oxygen–oxygen distance of the SO_2 group indicates that the non-bonded atom–atom interactions may be at least as important in determining the

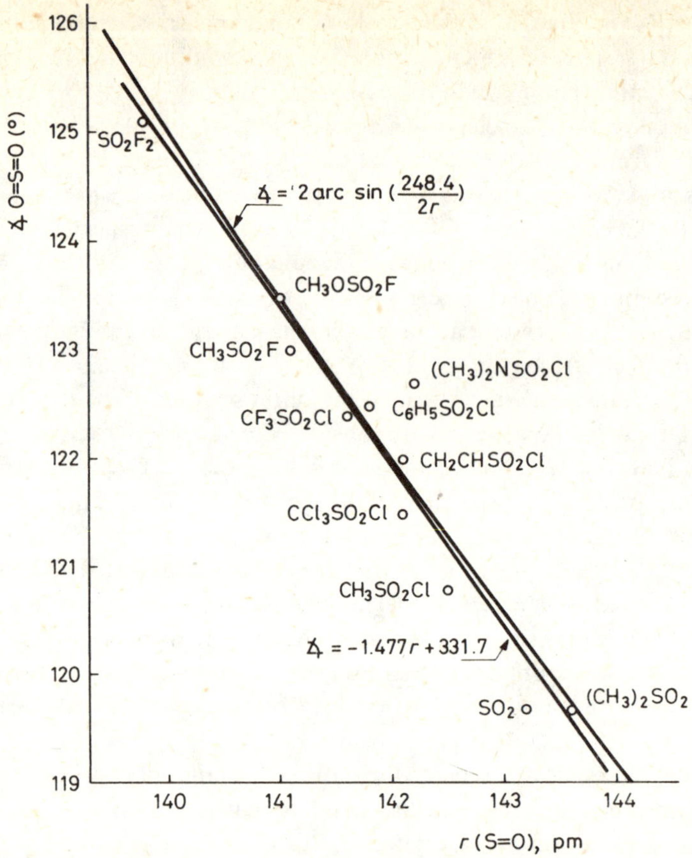

Fig. 89. The O═S═O bond angle (∡) as a function of the S═O bond length (r). The circles represent the experimental data. The empirical relationship $\measuredangle = -1.477r + 331.7$ was found by a least-squares procedure ($\sigma = 0.6°$). The trigonometrical expression utilizes the remarkable phenomenon of the O. . .O distance being constant (248.4 pm) in a large series of sulphone molecules (Hargittai, 1978; 1979a)

geometry of the sulphone group as the electron pair repulsions considered in the VSEPR model. The emerging pattern may be visualized by a tetrahedral arrangement of the XSO_2Y molecule with the sulphur in the centre of the tetrahedron and the two oxygen atoms in two of its apexes. As the two ligands X and Y are changed, the two oxygen atoms remain in their position, while the sulphur atom may move up and down along the direction as a continuation of the bisector of the O═S═O angle.

It is hoped that this discussion will give further emphasis to the importance of considering the non-bond distances along with the more easily conceivable bond lengths and bond angles in discussing molecular geometries, their variations and interpretation.

The oxygen–chlorine non-bond distances are given in Table 75(d) for a relatively large series of sulphonyl chloride molecules. They scatter considerably around their mean value of 279 pm ($\sigma = 2$ pm) but there is definitely no tendency in their changes. At the same time a gradual increase is observed in the S—Cl bond length with decreasing ligand electronegativity.

Some other non-bond distances are listed below, providing further examples of their constancy among the sulphone derivatives:

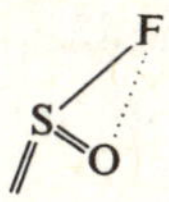

r(F. . .O), pm

238 SO_2F_2
237 CH_3OSO_2F
238 CH_3SO_2F
239 $HNSOF_2$
239 $ClNSOF_2$

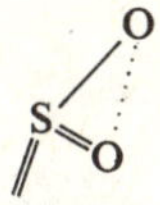

r(O. . .O), pm

241 $S_2O_5F_2$
242 $S_3O_8F_2$
242 CH_3OSO_2F
242 CH_3OSO_2Cl
242 $CH_3OSO_2OCH_3$

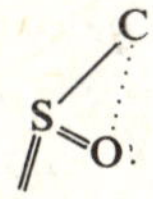

r(O. . .C), pm

261 $(CH_3)_2SO_2$
260 CH_3SO_2F
261 CH_3SO_2Cl
261 $C_6H_5SO_2Cl$
260 $CH_2{=}CHSO_2Cl$

When emphasizing the importance of the non-bonded interactions in the sulphone sulphur bond configurations, it is also stressed that they cannot be considered by far to be exclusively determining these structures. The non-bond distances themselves give a good indication of this as they are considerably larger than the respective sums of the 1,3 non-bond radii (*cf.* Table 2), *viz.*

	Actual atom–atom distance (pm)	$\Sigma r(1,3)$, pm
O. . .O(=S=)	248.4	226
O. . .O(=S—)	242	226
Cl. . .O	279	257
F. . .O	238	221
O. . .C	261	238

The deviations probably express the great importance of the electron pair repulsions. Furthermore, the difference in the actual O. . .O(=S=) and O. . .O(=S—) distances may be looked upon as direct evidence of the role of the electron pair repulsions in these structures (*cf.* the multiple bond rule of the VSEPR model). The difference between the SO_2 oxygen "non-bond radius" deduced from these structures and that postulated by

Bartell (1960) is 11 pm (i.e. 248/2 *vs.* 113 pm). Considering in addition the O...Cl distances, the chlorine "non-bond radius" appears to be also 11 pm larger than that postulated by Bartell (1960) (i.e. 279–124 *vs.* 144 pm). The coincidence of the differences is fortuitous but their relatively large value is certainly of interest. The "non-bond radii" for fluorine and single-bonded oxygen deduced from the sulphone structures are larger than the radii of Bartell (1960) the differences being half of what is observed for double-bonded oxygen or the chlorine atom.

To summarize, both the electron pair repulsions and the non-bonded atom–atom interactions are important in determining the sulphone molecular geometries. The actual structures obviously depend on the relative magnitudes of the various interactions involved. It is hoped that the relative abundance of the experimental data accumulated for this class of compounds will stimulate theoretical work aiming at establishing potentials quantitatively describing these structures and the intramolecular interactions involved in them.

4.10. CRYSTALLINE SULPHONES

Characteristic variations in the S=O bond lengths and O=S=O bond angles were observed in vapour-phase sulphone molecules depending on the nature (electronegativity) of the ligands attached to the SO_2 group. Data on molecules whose structure has been studied both in the vapour and in the crystalline phase are very scarce. The relevant data have been collected (*cf.* those on dimethyl sulphone or *N,N,N′,N′*-tetramethyl sulphonyl diamide). They do not warrant, however, a detailed comparison. It is only hoped that the amount of structural information concerning molecules examined in both phases will gradually grow in order to enable us to examine finer details of changes upon transition from one phase into another. Since there are already numerous crystal-phase structure determinations of sulphone derivatives by x-ray diffraction (in fact much more than vapour-phase studies), it is of considerable interest to see whether the most characteristic structural variations observed for the free molecules are also present in the crystalline phase.

While a complete coverage was attempted for the vapour-phase structures, it is stressed that no such aim was pursued in the case of the crystal-phase structures. Instead, the literature was scanned in a casual manner and no critical approach applied in selecting and presenting the data. Those given in Tables 77 and 78 refer to organic and inorganic crystalline sulphones and contain S=O bond lengths, O=S=O bond angles and the corresponding oxygen–oxygen non-bond distances. The arbitrary criterium of a sulphone to be called inorganic here is that neither atom directly linked to the SO_2 group be carbon. The inorganic sulphones are considered separately because of their structural peculiarities and the greater variety of their parameters. It appears that it is the organic sulphones which can be considered primarily to be representative sulphones. The crystal-phase data, originating almost exclusively from x-ray diffraction, are not analysed as regards the application of thermal corrections, although these are of great importance. This is illustrated by the example of the uncorrected and corrected parameters of the SO_2 group of 3,4-epoxy-sulpholane (Sands, 1972):

$C_4H_6OSO_2$	Uncorrected	Corrected
r(S=O), pm	144.4(2)	145.5
∢O=S=O (°)	117.7(2)	117.8
r(O...O), pm	247.0(4)	249.2

According to the data of Tables 77 and 78, the geometries of the SO_2 groups in crystalline sulphones are very similar to those in the free sulphone molecules. The O...O distances themselves, however, show a greater variation in the crystalline

TABLE 77
The geometry of the SO_2 group in crystal-phase sulphones (organic)

Number	r(S=O) pm	∢O=S=O (°)	r(O...O) pm	Reference
1a	144.5	117.9	248	*(a)*
b	144.6	117.3	247	*(b)*
2	143.6	119.2	248	*(c)*
3	143.7	117.3	245	*(d)*
4	143.5	118.3	246	*(e)*
5	143.4	118.5	246	*(f)*
6	143.2	118.1	246	*(g)*
7	144.0	118.4	248	*(h)*
8	144.4	117.7	247	*(i)*
9	145.6	117.8	249	*(j)*
10	143.5	117.5	245	*(k)*
11	149	120	258	*(l)*
12	144.8	116.5	247	*(m)*
13	143.9	116.5	245	*(m)*
14	143.3	116.8	244	*(n)*
15	144.0	117.0	246	*(o)*
16	142.4	117.6	244	*(p)*
17	142.8	120.3	248	*(q)*
18	144.6	119.7	250	*(r)*
19α	144	119	248	*(s)*
19β	145.1	118.2	249	*(t)*
19γ	144.5	117.3	245	*(u)*
20	145.6	118.3	250	*(v)*
21	143.4	118.6	247	*(w)*
22	143.2	118.1	246	*(x)*
23	141.2	120.6	245	*(y)*
24	143.2	120.0	248	*(z)*
*25**	142.4	118.3	244	*(aa)*
**	141.7	119.1	244	

TABLE 77 (cont.)

Number	r(S=O) pm	∢O=S=O (°)	r(O...O) pm	Reference
26	142.7	118.9	246	*(bb)*
27	143.4			*(cc)*
28	142.6	121.2	248	*(dd)*
29	143.5	120.0	249	*(ee)*
30	144	119.0	248	*(ee)*
31	144	120.0	249	*(ee)*
32	142	119.0	245	*(ee)*
33	144.6	116.0	245	*(ff)*
34	142.8	120.3	248	*(gg)*
35	142.8	119.5	247	*(hh)*
36	143.4	120.5	249	*(hh)*
Mean	143.7	118.6	247.1	
σ	1.3	1.3	2.5	
Mean	143.6	118.5	246.8	(without the data of *11*)
σ	1.0	1.3	1.8	
Interval	3.9	5.2	6	

(a) Sands (1963)
(b) Langs, Silverton and Bright (1970)
(c) Sime and Woodhouse (1974)
(d) Chawdhury (1976)
(e) Harlow, Sammes and Simonsen (1974)
(f) Harlow, Simonsen, Pfluger and Sames (1974)
(g) Loghry and Simonsen (1975)
(h) Wadhawan (1976)
(i) Sands (1972)
(j) Towns and Simonsen (1974)
(k) Towns and Simonsen (1975)
(l) Kronfeld and Sass (1968)
(m) Jandacek (1975)
(n) Guin (1969)
(o) Sands and Day (1967)
(p) Beall, Herdklotz and Sass (1970)
(q) Kiers and Vos (1972)
(r) Noordik and Vos (1967)
(s) O'Connor and Maslen (1965)
(t) O'Connell and Maslen (1967); Alléaume and Decap (1965a)
(u) Alléaume and Decap (1965b)
(v) Alléaume and Decap (1968)
(w) Klug (1968)
(x) Kálmán, Duffin and Kucsman (1971)
(y) Cotton and Stokely (1970)
(z) Karle (1973)
(aa) Dupont and Dideberg (1972)
(bb) Gogoi and Hargraves (1970)
(cc) Meinwald, Knapp, Obendorf and Hughes (1976)
(dd) Mathew and Palenik (1974)
(ee) Rérat (1969)
(ff) Lindley, Mahmoud, Dodd, Smith, Boyd and Norris (1977)
(gg) Calleri, Chiari, Chiesi Villa and Guastini (1976)
(hh) Calleri, Chiari, Chiesi Villa, Gaetani Manfredotti, Guastini and Viterbo (1976)

The numbers in the above Table refer to the following compounds:

1
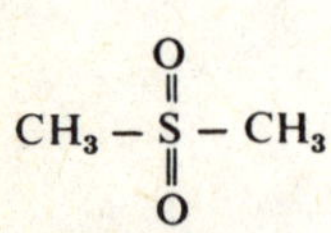

2
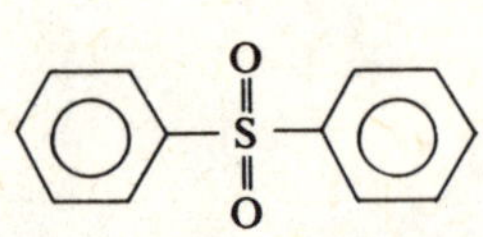

3
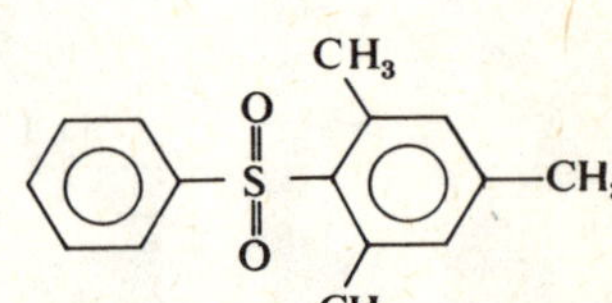

4
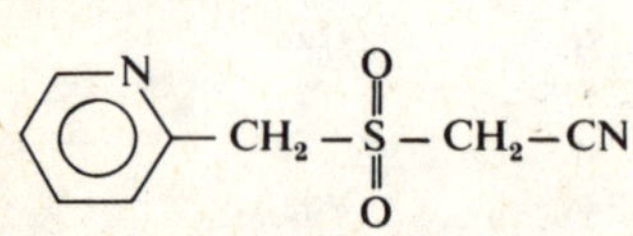

5
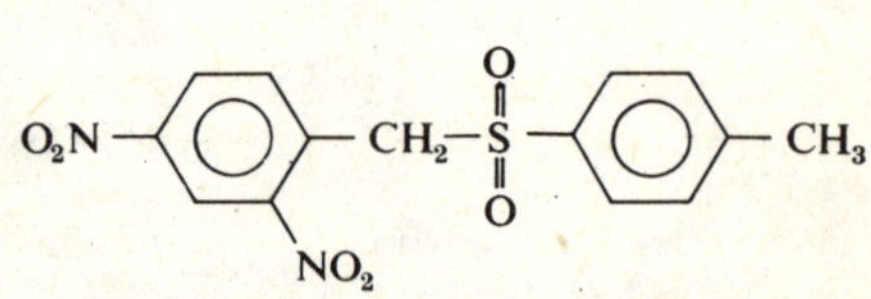

6 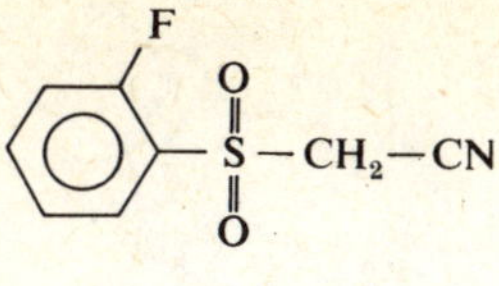

7 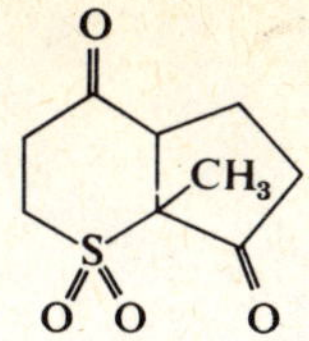

8 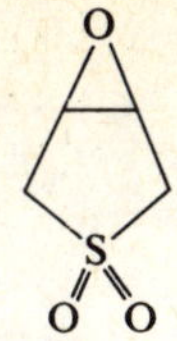

9 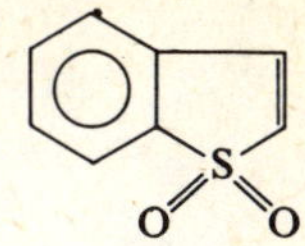

10 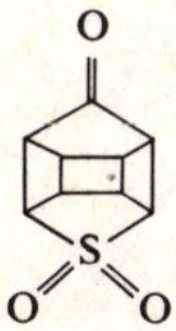

11 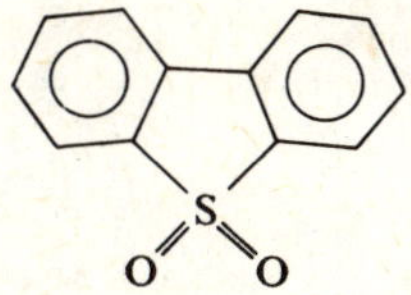

12 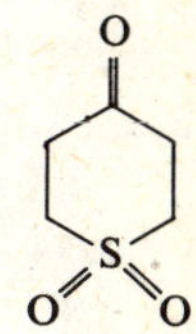

13 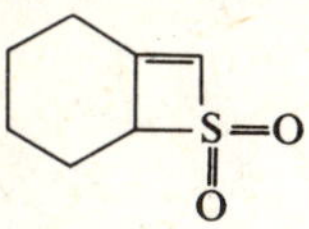

14 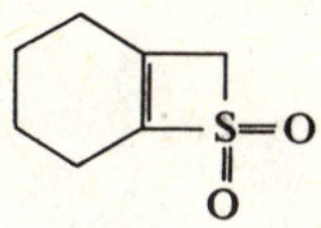

15 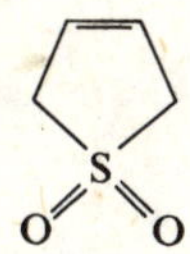

16 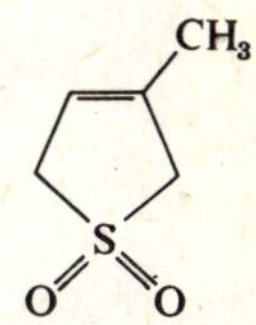

17 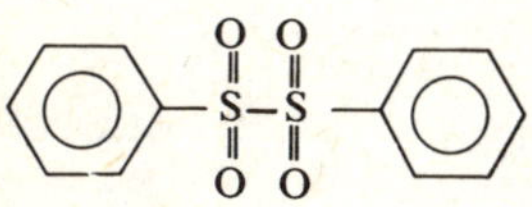

18 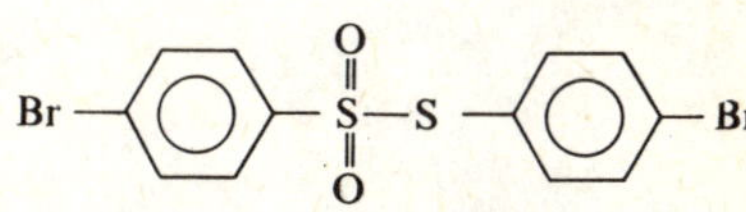

19 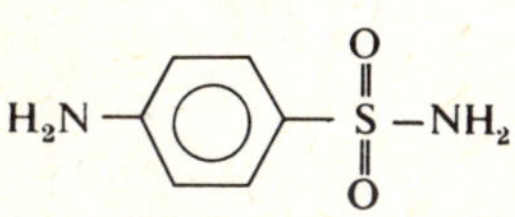

20 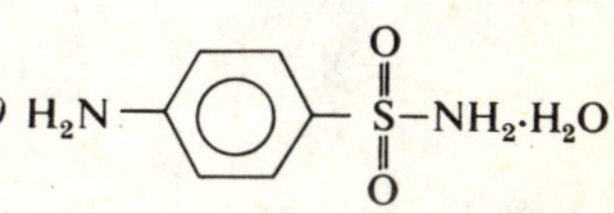

21 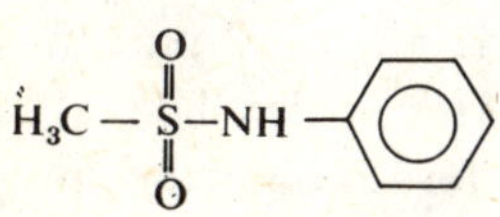

22 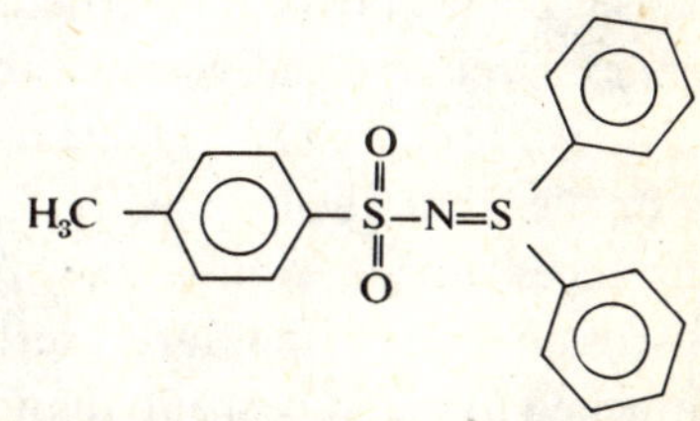

23 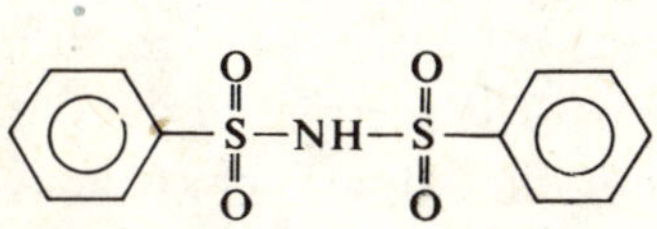

24 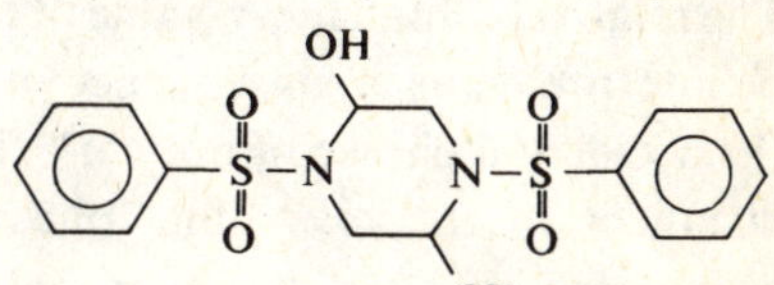

25 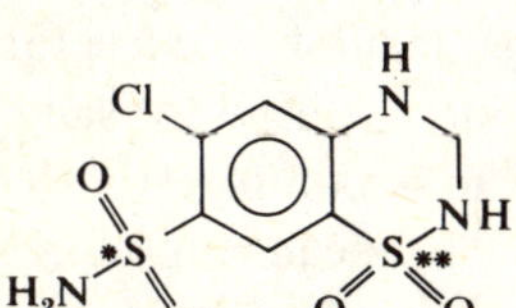

26 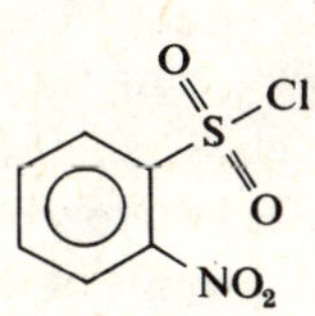

27 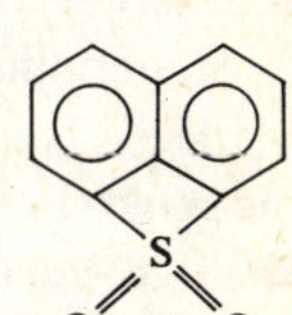

28 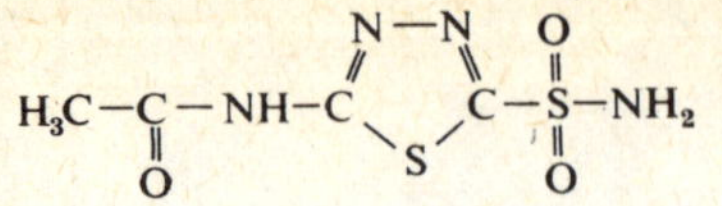

29 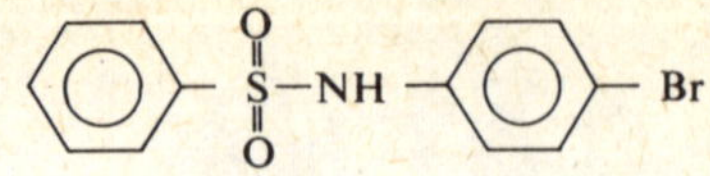

30 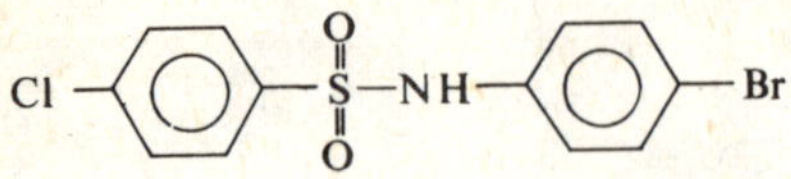

31 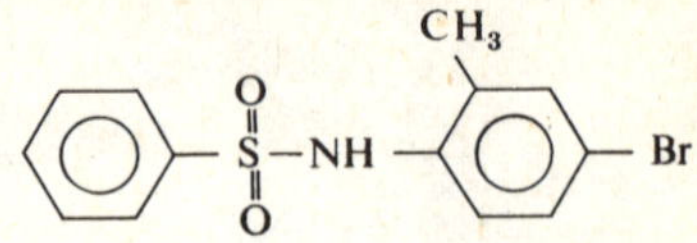

32 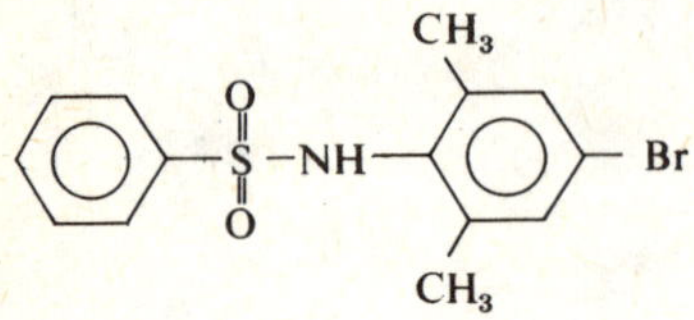

33 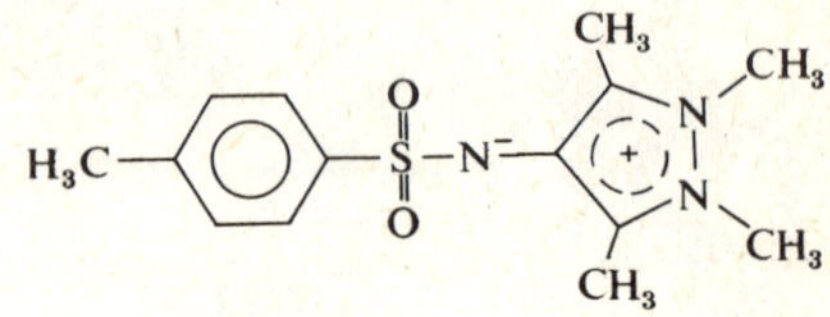

34 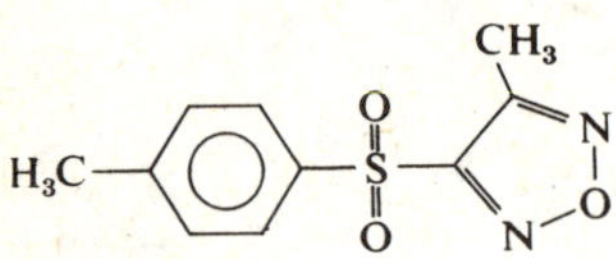

35 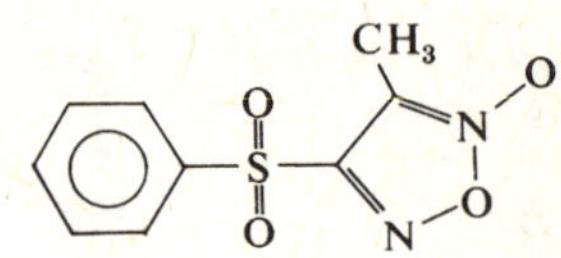

36 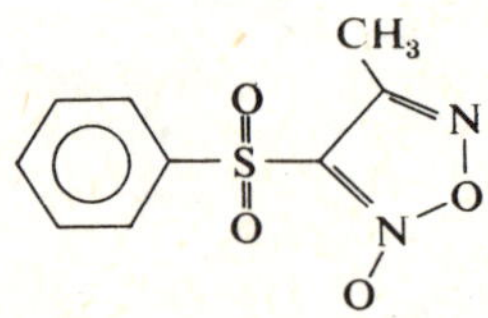

phase. There may be real structural causes for this (intermolecular interactions?) apart from the relatively large experimental errors. It seems to be of special interest that the mean value of the O. . . O distances appearing in Table 77 is 247 pm ($\sigma = 2$ pm), thus it is smaller than the mean O. . . O distance in the vapour-phase molecules. The O. . . O distances are, of course, the most appropriate parameters for comparison in our discussion, since they are presumably less influenced by the nature of the ligands attached to the SO_2 group than the S═O bond lengths or the O═S═O bond angles. The relatively small mean value of the O═S═O bond angles (118.6°, σ1.3°), e.g., could be interpreted as a consequence of intermolecular interactions, but could result from the accidental choice of not very electronegative ligands. The decrease of the O. . . O distances in the crystalline phase as compared with the vapour phase may be considered to be a more probable indication of the influence of intermolecular interactions.

Some comments are made concerning the crystalline inorganic sulphone structures. In the S_3O_9 molecule the endocyclic O. . . O distance is 246 pm, i.e. almost the same as the exocyclic O. . . O distance. The latter is, of course, the one to be compared with the analogous parameter of the SO_2 groups of other sulphones. The mean lengths of the endocyclic S—O bonds is 162.6(7) pm and the mean endocyclic sulphur bond angle is

TABLE 78

The geometry of the SO_2 group in crystal-phase sulphones (inorganic)

Compound	Reference	r(S=O) pm	∢O=S=O (°)	r(O...O) pm
$H_2NSO_2NH_2$	*(a)*	139.1(8)	119.4(8)	240
$(CH_3)_2NSO_2N(CH_3)_2$	*(b)*	144.9 and 144.1(5)	119.7(4)	250
S_3O_9	*(c)*	142.8 and 136.7(25)	127.9(15)	250
$FXeOSO_2F$	*(d)*	141.5 and 143.0(9)	119.6(6)	247
$[Fe(CO)_2C_5H_5]_2SO_2$	*(e)*	148.0 and 147.6(1)	112.9(8)	246
$(CH_3)_3N \cdot SO_2$	*(f)*	137.7 and 141.6(8)	114.8(5)	235
$IrCl(CO)[P(C_6H_5)_3]_2(SO_2)$	*(g)*	141.1 and 146.7(25)	117.1(15)	246
$[RuCl(NH_3)_4SO_2]Cl$	*(h)*	139.4 and 146.2(10)	113.8(6)	239

(a) Trueblood and Mayer (1956)
(b) Jordan, Smith, Lohr and Lipscomb (1963)
(c) McDonald and Cruickshank (1967)
(d) Bartlett, Wechsberg, Jones and Burbank (1972)
(e) Churchill, DeBoer, Kalra, Reich-Rohrwig and Wojcicki (1972)
(f) van der Helm, Childs and Christian (1969)
(g) La Placa and Ibers (1966)
(h) Vogt, Katz and Wiblerley (1965)

The above compounds are characterized by the following structures:

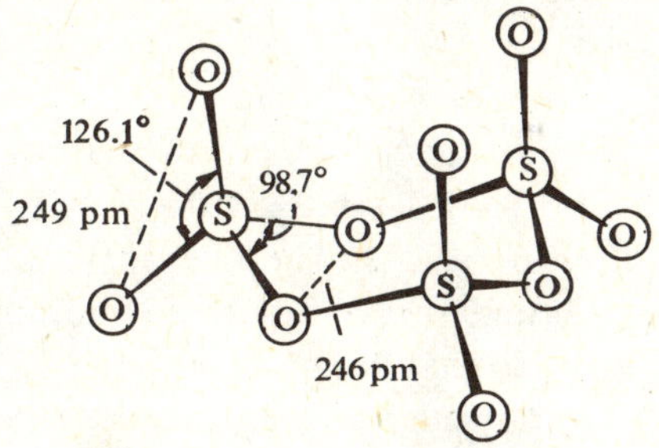

S_3O_9

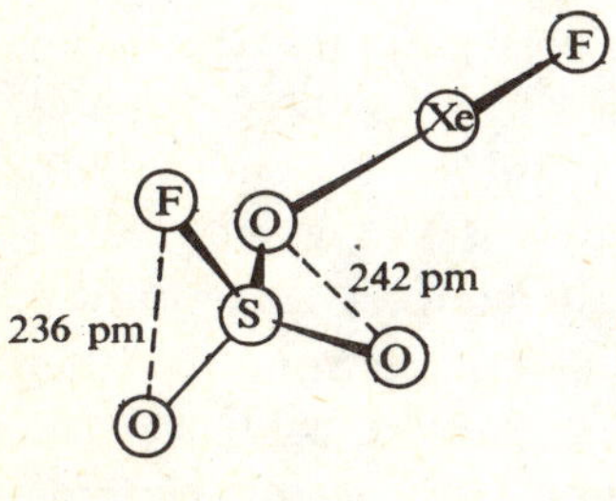

$FXeOSO_2F$

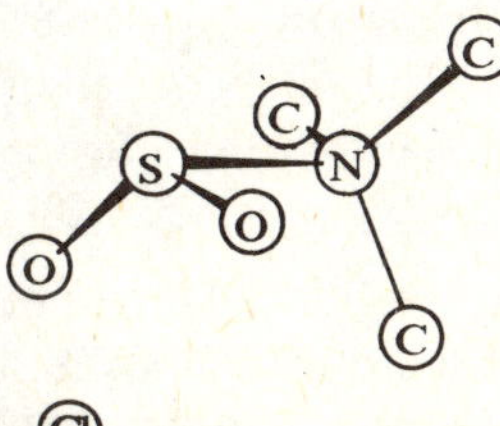

C_3NSO_2 skeleton
$(CH_3)_3N{\cdot}SO_2$

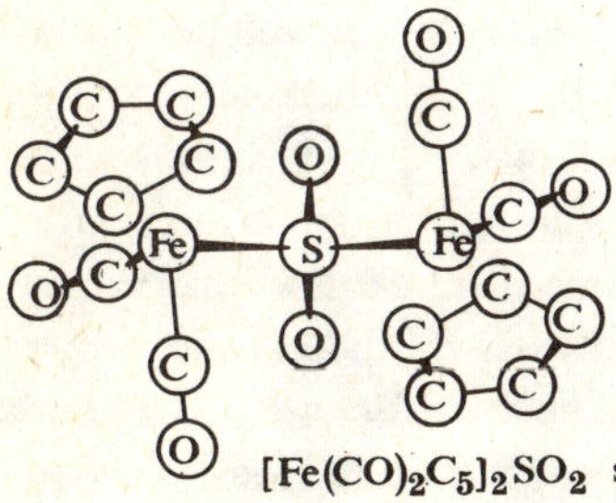

$[Fe(CO)_2C_5]_2SO_2$ skeleton
$[Fe(CO)_2C_5H_5]_2SO_2$

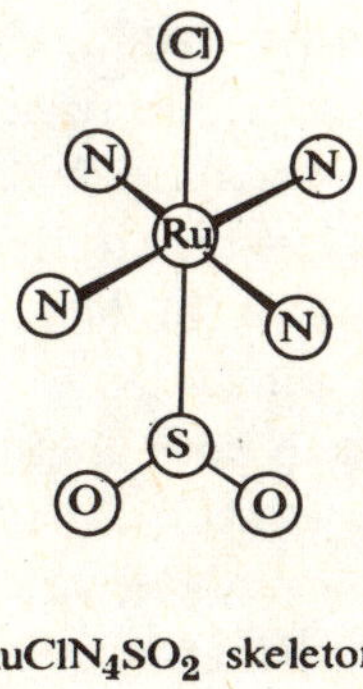

$RuClN_4SO_2$ skeleton
$[RuCl(NH_3)_4SO_2]Cl$

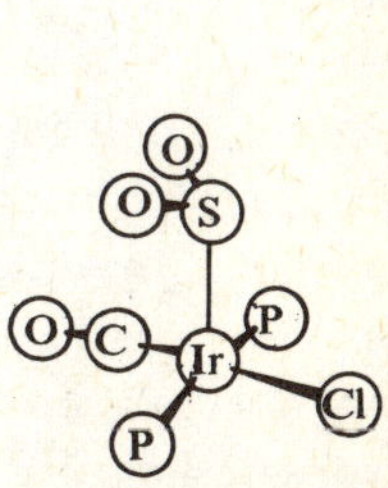

$Ir(CO)P_2ClSO_2$ skeleton
$Ir(CO)[P(C_6H_5)_3]_2ClSO_2$

98.7° (McDonald and Cruickshank, 1967). Some of the non-bond distances of the $FXeOSO_2F$ molecule are also given as their agreement with analogous vapour-phase parameters is noteworthy (after Bartlett, Wechsberg, Jones and Burbank, 1972).

The SO_2 parts are considered to be Lewis acid in both the sulphur dioxide trimethyl amine complex (van der Helm, Childs and Christian, 1969) and the iridium compound (La Placa and Ibers, 1966). The SO_2 group as Lewis acid is supposed to draw away electron density from the nitrogen and iridium atoms, respectively. This is consistent with the observation of the distorted tetrahedral configuration of the sulphur bonds. In order to account for the structure of the ruthenium compound, it was supposed that the sulphur dioxide part acts as Lewis base donating electron density to the ruthenium atom, hence the planar sulphur bond configuration (Vogt, Katz and Wiblerley, 1965). Thus the bonding peculiarities of sulphur are very diverse in these substances (*cf.* Shriver, 1970).

4.11. COMPARISON OF ANALOGOUS SULPHONE, SULPHOXIDE, AND SULPHIDE GEOMETRIES

There are four different ligands X for which complete structural comparison is possible in the series SO_2X_2, SOX_2 and SX_2. These ligands are the dimethylamino and methyl groups and the chlorine and fluorine atoms. The relevant bond angles and bond lengths are compiled in Table 79, and the variations in the bond angles are also depicted in Fig. 90 because of their special importance.

As the structural changes are inspected going from the sulphones toward the sulphides, what is seen is indeed the consequence of the bonding pairs being replaced by lone pairs of electrons in the series SO_2X_2, SOX_2E, SX_2E_2, where E denotes the lone pair of electrons in the valence shell of the central sulphur atom. Going from the four-ligand case to the three-ligand plus one lone electron pair case, it is observed that the bond angles decrease and the bonds lengthen. This is simple to interpret by the VSEPR model. As the ligand bonding pairs are replaced by a lone pair of electrons, the latter has larger space requirement around the sulphur atom and, accordingly, exercises more repulsion toward the other bonding electron pairs. As a second ligand is replaced by an other lone pair, a further decrease of the X—S—X bond angle and more lengthening of the sulphur bonds could be expected. However, this is not observed. Instead, the bond angles are larger and the bonds are shorter in sulphides than in the analogous sulphoxides. However, in the systems discussed above double bonds were "substituted" by lone pairs and it has to be recognized that the application of the VSEPR model for systems with both double bonds and lone pairs around the central atom is not really straightforward. The model does not indicate whether a lone pair of electrons or a double bond has more space requirement. The data for the SO_2X_2 and SOX_2 molecules may indicate that in case of sulphur as central atom, the lone pair exercises more repulsion than the double bond. Similar conclusion can be made for

TABLE 79
Summary of bond angles and bond lengths
in analogous sulphone, sulphoxide and sulphide molecules
(sources in the respective discussions)

Compounds				
$[(CH_3)_2N]_2SO_2$		$[(CH_3)_2N]_2SO$		$[(CH_3)_2N]_2S$
$(CH_3)_2SO_2$		$(CH_3)_2SO$		$(CH_3)_2S$
SO_2Cl_2		$SOCl_2$		SCl_2
SO_2F_2		SOF_2		SF_2
∢X—S—X Bond angles (°)				
110.5		96.9		114.5
102.6		96.6		99.2
100.0		96.1		102.8; 103.0
96.1; 96.7		92.2; 92.8		98.2
Bond lengths, pm				
S—X	S=O	S—X	S=O	S—X
165.1	143.2	169.3	148.0	168.8
177.1	143.5	179.9	148.5	181.0
201.1	140.4	207.6	144.3	201.4; 200.6
153.0; 153.0	140.5; 139.9	158.3; 158.5	142.0; 141.3	159.2

phosphorus as central atom on the basis of the data shown in Table 80 on X—P—X bond angles in POX_3 and PX_3E molecules. On the other hand, the data on the nitrogen bond angles in NOX_3 and NX_3E molecules (Table 80) show the opposite trend.

It is of interest to examine the geometrical changes in other AX_4, BX_3E, CX_2E_2 series of molecules in which there are no double bonds. Gillespie (1972) noted that "... in the series CH_4, NH_3, H_2O the bond angle decreases from 109.5° to 107.3 and to 104.5° as the number of non-bonding pairs increases." Data on bond angles for some simple molecules are shown in Table 81 and Fig. 91. Again, going from the four-ligand case to the three-ligand plus one lone electron pair case is in agreement with the above statement. As the "substitution" is further continued, *viz.* as the second bonding pair is replaced by an other lone pair, a further decrease of the bond angle would be expected to occur according to the above statement. This is observed, however, for NH_3 and H_2O, and PH_3 and SH_2 only. For other molecule pairs with chlorine or fluorine ligands, the opposite trend is realized. Note also the larger than tetrahedral angle in OCl_2, which may be a consequence of the relatively large size of the ligands.

The interpretation of the changes in the bond angles, as going from the three-ligand plus one lone pair case to the two-ligand plus two lone pair case, is rather complicated

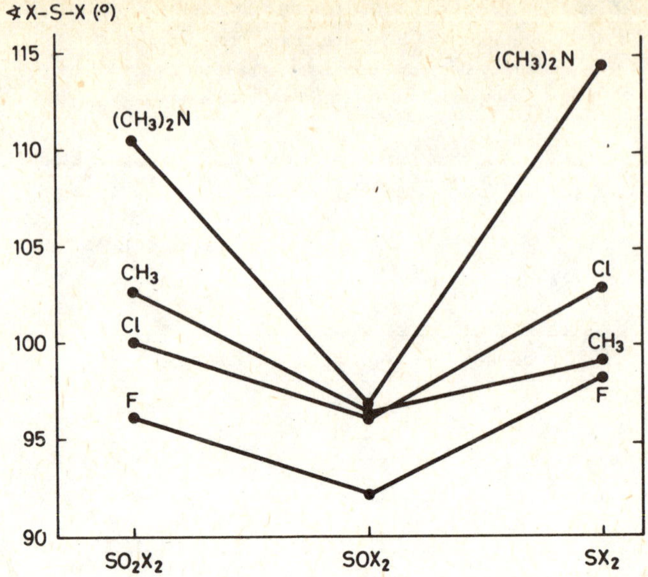

Fig. 90. The variations of the bond angles X—S—X in analogous sulphones, sulphoxides and sulphides (Hargittai, 1974b; 1978)

TABLE 80

X—P—X and X—N—X bond angles (X = F, Cl, Br or CH_3) in simple AOX_3 and AX_3 molecules (A = P or N)

AOX_3		∢X—A—X (°)	AX_3		∢X—A—X (°)
POF_3	*(a)*	101.3	PF_3	*(f)*	97.8
$POCl_3$	*(a)*	103.3	PCl_3	*(g)*	100.3
$POBr_3$	*(b)*	105.6	PBr_3	*(h)*	101.0
$PO(CH_3)_3$	*(c)*	106	$P(CH_3)_3$	*(i)*	98.6
NOF_3	*(d)*	100.8	NF_3	*(j)*	102.4
$NO(CH_3)_3$	*(e)*	109.0	$N(CH_3)_3$	*(k)*	110.6

(a) Moritani, Kuchitsu and Morino (1971)
(b) Olie and Mijlhoff (1969)
(c) Wang (1965)
(d) Plato, Hartford and Hedberg (1970)
(e) Caron, Palenik, Goldish and Donohue (1964)
(f) Morino, Kuchitsu and Moritani (1969)
(g) Iwasaki and Hedberg (1962)
(h) Kuchitsu, Shibata, Yokozeki and Matsumura (1971)
(i) Bartell and Brockway (1960)
(j) Otake, Matsumura and Morino (1968)
(k) Beagley and Hewitt (1968)

TABLE 81
Bond angles in AX_3E and AX_2E_2 molecules

AX_3E		∢X—A—X (°)	AX_2E_2		∢X—A—X (°)
NH_3	*(a)*	108.2	OH_2	*(g)*	107.2
NCl_3	*(b)*	107.1	OCl_2	*(h)*	111.2
NF_3	*(c)*	102.4	OF_2	*(i)*	103.2
PH_3	*(d)*	93.5	SH_2	*(j)*	92.2
PCl_3	*(e)*	100.3	SCl_2	*(k)*	103.0
PF_3	*(f)*	97.8	SF_2	*(l)*	98.2

(a) Kuchitsu, Guillory and Bartell (1968)
(b) Bürgi, Stedman and Bartell (1971)
(c) Otake, Matsumura and Morino (1968)
(d) Bartell and Hirst (1959)
(e) Iwasaki and Hedberg (1962)
(f) Morino, Kuchitsu and Moritani (1969)
(g) Shibata and Bartell (1965)
(h) Beagley, Clark and Hewitt (1968)
(i) Morino and Saito (1966)
(j) Landolt–Börnstein (1976)
(k) Morino, Murata, Ito and Nakamura (1962)
(l) Kirchhoff, Johnson and Powell (1973)

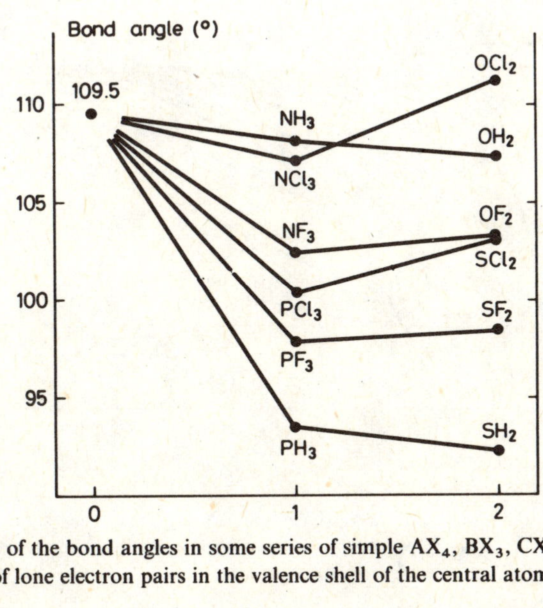

Fig. 91. The variations of the bond angles in some series of simple AX_4, BX_3, CX_2 molecules; 0, 1, and 2 indicates the number of lone electron pairs in the valence shell of the central atom (Hargittai, 1973; 1978)

since in addition to the bonding pair–bonding pair and bonding pair–lone pair repulsions, there are also lone pair–lone pair repulsions present. The resulting configuration depends, in the final account, on the relative magnitudes of the three different types of interactions.

To further examine the above changes, a simple point-charges-on-the-sphere model was constructed (Hargittai and Baranyi, 1977) in which bonding and non-bonding electron pairs are represented by smaller (q_X) and larger (q_E) charges, respectively. The configuration was then determined in which only radial forces acted on the charges. At the same time it was strongly emphasized that using the charges q_X and q_E by no means implied that the origin of repulsion in the systems under discussion could be considered to be simply electrostatic. In the expression of the force affecting the charges, the force was, of course, inversely proportional to a power of the distance between the charges, and the exponent $p+1$ varied over a wide range (between 1 and 15) in the calculations.

The results of these calculations are illustrated in Fig. 92. The variations of the bond angles are shown at different values of the repulsion exponent p against the ratio of the two different charges. It is seen that the bond angle Θ of the AX_3E configuration is always smaller than the regular tetrahedral angle, while the bond angle β of the AX_2E_2 configuration may be smaller, as well as larger than Θ, depending on the repulsion exponent. Thus the direction of the changes in the bond angles as going from AX_4 to AX_3E is well understood and is invariant to the choice of the repulsion exponent. On the other hand, the relationship between the bond angles of molecule pairs AX_3E and AX_2E_2 strongly depends on the choice of the repulsion exponent. Accordingly, the rules constituting the VSEPR model seem to lose here the usefulness in predicting the trends in the structural changes, since the predictions are no longer invariant to the choice of the repulsion exponent in the potential employed.

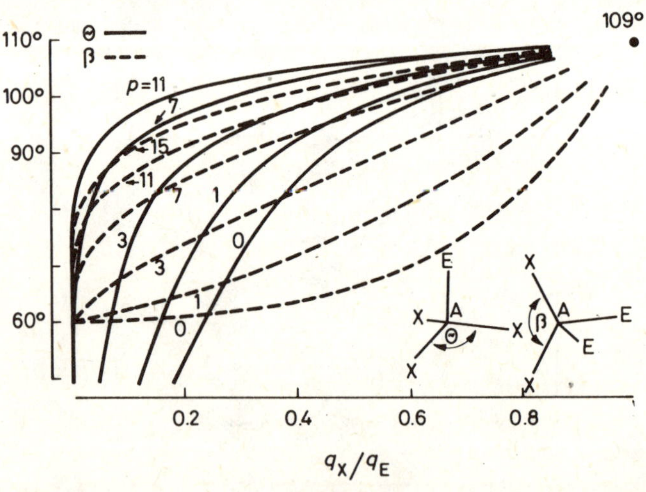

Fig. 92. Variation of the bond angles in the AX_3E (Θ) and AX_2E_2 (β) systems *vs.* the ratio of the two charges employing various values for the repulsion exponent in the expression of the potential (Hargittai and Baranyi, 1977)

Ab initio molecular orbital calculations have been carried out (Schmiedekamp, Cruickshank, Skaarup, Pulay, Hargittai and Boggs, 1979) in order to gain more insight into the structural variations discussed in this Section, and to correlate them with bonding properties. Most of the molecules appearing in Tables 79 and 81 were still too large for such calculations, whereas some other systems, *viz.*, e.g., H_2SO_2, H_2SO, H_2S, seemed to be of interest to include for comparison. The latter series of molecules, as some others as well, have already been subject to *ab initio* calculations (Van Wazer and Absar. 1972; Absar and Van Wazer, 1978). However, there has been no systematic comparison of the molecules relating wave-function properties obtained with a consistent system of basis sets, and the bond lengths and bond angles optimized to the lowest energy structure. As the results of the systematic *ab initio* calculations have become available, it seemed to be of considerable importance to carry out a reassessment of the validity and applicability of the VSEPR model. In this reassessment an enhanced emphasis was given to the total space requirement of the bonding and non-bonding electron pairs rather than to examine primarily the variations of the bond angles.

The calculated bond angles including the "bond angles" of the lone pairs of electrons E are compiled in Table 82. The centroid of charge was used for defining the position of the lone pair of electrons. The calculated bond angles were in good agreement with the available experimental results (*cf.* previous Tables). This, in itself, was an important achievement. The examination of the localized molecular orbitals, however, led to further progress in understanding the structural variations. The prediction, based on the VSEPR model, according to which the bond angles are expected to decrease with increasing number of lone pairs of electrons, was found to be valid only for the following pairs of BX_3E and BX_2E_2 configurations:

NH_3 and NH_2^-
NF_3 and NF_2^-
SH_3^+ and SH_2

Thus the validity of the above-mentioned prediction was restricted to pairs of systems with the same central atom. At the same time, the tendency observed in the experimental results, *viz.* that the bond angle of the CX_2E_2 molecule is larger than that of the BX_3E molecule, was observed in the results of the calculations as well. The only exception was the NF_3, OF_2 molecule pair with such a small difference that does not warrant a detailed analysis.

The central atom is unchanged in the series SO_2F_2, SOF_2, SF_2. The only change here is the subsequent "substitution" of the S=O bonds by lone pairs of electrons. The calculated geometries for this series are also in good agreement with the experimental data. Thus the F—S—F bond angle of sulphur difluoride is larger, and not smaller than that of thionyl fluoride. The results of the calculations for the partly hypothetical series SO_2H_2, SOH_2, SH_2 are similar to those obtained for the fluorine derivatives. This

TABLE 82

Calculated bond angles (*ab initio*, spd basis and in parentheses sp basis) for some simple molecules (Schmiedekamp, Cruickshank, Skaarup, Pulay, Hargittai and Boggs, 1979), E denotes the lone pair of electrons

Angle (°)	NH_4^+	NH_3E	$NH_2E_2^-$	NF_3E	$NF_2E_2^-$
∢H(F)—N—H(F)	109.5 (109.5)	105.4 (111.4)	96.7 (97.7)	(102.4)	99.4 (99.1)
∢H(F)—N—E		113.3 (107.5)	107.3 (107.4)	(115.9)	103.3 (102.9)
∢E—N—E			126.8 (125.8)		138.3 (139.7)

Angle (°)	SH_3E^+	SH_2E_2	$SHFE_2$	SF_2E_2
∢H(F)—S—H(F)	95.5 (99.6)	93.6 (95.6)	96.3 (95.6)	98.3 (98.1)
∢H—S—E	121.3 (118.1)	107.8 (108.2)	107.6 (108.5)	
∢F—S—E			105.1 (104.8)	104.4 (104.3)
∢E—S—E		126.9 (124.5)	130.0 (129.2)	135.2 (135.8)

Angle (°)	OH_2E_2	OF_2E_2
∢H(F)—O—H(F)	104.9 (108.1)	102.0 (101.7)
∢H(F)—O—E	108.2 (108.2)	104.3 (104.0)
∢E—O—E	118.5 (115.9)	133.8 (135.0)

Angle (°)	SO_2H_2	SOH_2E	SOF_2E
∢H(F)—S—H(F)	97.5 (98.1)	89.2 (92.8)	(91.7)
∢H(F)—S=O	108.3 (107.7)	109.9 (107.7)	(107.6)
∢O=S=O	123.2 (124.6)		
∢H(F)—S—E		110.9 (111.0)	(109.9)
∢O=S—E		122.5 (122.5)	(125.8)

Angle (°)	PH_3E	PF_3E
∢H(F)—P—H(F)	94.0 (95.4)	(96.9)
∢H(F)—P—E	122.4 (121.3)	(120.2)

means that the observed bond angle variations are not specific features connected with the fluorine ligands.

Another deviation from the predictions by the generally formulated VSEPR rules is noted by comparing the bond angles in the following pairs:

SF_2 and SH_2
NF_2^- and NH_2^-
PF_3 and PH_3

(the data are found in Table 82). Since the electronegativity of fluorine is larger than that of hydrogen, the bond angle F—A—F should be expected to be smaller than H—

A—H according to the electronegativity rule. The bond angles in the molecule pair OF_2 and OH_2 follow this rule, indeed, as is the case for NF_3 and NH_3 as well, but not for other pairs. Thus the emerging pattern is rather complicated and suggests, to recall a remark by Gillespie (1972) according to which "bond angles in hydrides are generally smaller than would be expected from the electronegativity of hydrogen", ascribing the phenomenon to the consequence of the lack of hydrogen core electrons. Our discussion will return to the bond configuration of hydrides a little later.

The bond angles shown in Table 82 convincingly support the postulates of the VSEPR model relating to the large repulsions exercised by the lone pairs of electrons. The "bond angle" made by the two lone pairs of electrons in AX_2E_2 molecules is larger than the "bond angle" between a lone pair and a bond, or the bond angle between the two bonds.

The data of Table 82 provide possibility for further interesting comparisons. According to the results for the molecule pair SOF_2 and SF_2, the repulsion between the S═O bond and the lone pair of electrons is smaller than the repulsion between the two lone pairs. This observation is based on a 10° angular opening as a consequence of "substitution" of the S═O bond by a lone pair of electrons. Not only the bond angles are larger in the fluorine derivatives of the SF_2, SH_2 and also NF_2^-, NH_2^- pairs, but also the angle made by the two lone pairs, by some 10°. Thus the variations in the angles assigned to the lone pairs of electrons are in complete agreement with the postulates of the VSEPR model.

It is particularly revealing to compare the so-called triple average angles. The triple average angle is the average value of the three angles made by a bond or a lone pair of electrons and is thus a measure of the space requirement of each electron pair. The triple average angles are shown in Table 83. According to these data there is a remarkable constancy of the angular requirements of a given bonding pair or lone pair orbital in various molecules and ions. This suggests that the angular geometries are primarily determined by the interactions among the orbitals. Note that the average angular space required by S—F, O—F, and N—F bonds is only slightly smaller than that by S—H, O—H, and N—H bonds. The requirement of an S═O double bond is nearly as large as that of a lone pair of sulphur and both are considerably larger than those of the single bonds. Note also that the lone pair requirements of nitrogen and oxygen are near that of sulphur. For any particular molecule the following relationship holds for the angles:

$$\angle \mathrm{E—A—E} > \angle \mathrm{E—A—X} > \angle \mathrm{X—A—X}$$

where A is the central atom, X and E are the ligands and the lone pairs of electrons. Thus for these molecules the calculations convincingly support the statement of Gillespie (1972) concerning the relative magnitudes of electron pair repulsions according to which the interactions decrease in the following order

$$\mathrm{E/E} > \mathrm{E/X} > \mathrm{X/X}.$$

It may then be concluded that there is no real controversy between the experimental bond angles or their variations presented in Figs 90 and 91 and the VSEPR model.

TABLE 83
Space requirement of bonds and lone pairs of electrons (E) as expressed by the triple average angles from *ab initio* calculations, after Schmiedekamp, Cruickshank, Skaarup, Pulay, Hargittai and Boggs (1979)

Orbital	Molecule	Triple average angle (°)	
		spd basis	sp basis
S—F	HSF	102.2	101.7
	SF_2	102.4	102.2
	SOF_2		102.9
S—H	HSF	103.8	104.2
	SH_2	103.1	104.0
	SOH_2	103.1	103.8
	SO_2H_2	104.7	104.5
	SH_3^+	104.1	105.8
S=O	SOH_2	113.6	112.6
	SO_2H_2	113.3	113.3
	SOF_2		113.7
S—E	SH_2	114.2	113.6
	HSF	114.2	114.2
	SOH_2	114.8	114.8
	SF_2	114.7	114.8
	SOF_2		114.9
	SH_3^+	121.3	118.1
O—H	OH_2	107.1	108.2
O—F	OF_2	103.5	103.2
O—E	OH_2	111.6	110.8
	OF_2	114.1	114.3
N—F	NF_3		106.9
	NF_2^-	102.0	101.6
N—H	NH_3	108.0	110.1
	NH_2^-	103.8	104.2
N—E	NF_3		115.9
	NH_3	113.5	107.5
	NF_2^-	115.0	115.2
	NH_2^-	113.8	113.5
P—F	PF_3		104.7
P—H	PH_3	103.5	104.0
P—E	PH_3	122.4	121.3
	PF_3		120.2

Thus, while for instance, the F—S—F bond angle of SF_2 is larger than the F—P—F bond angle of PF_3, the space requirement of the S—F bond is smaller than that of the P—F bond. On the other hand, the space requirement of the phosphorus lone pair is larger than that of the sulphur lone pair. Accordingly, in discussing and interpreting the geometrical variations, considerations should not be restricted to the bond angles, but the positions of the lone pairs should also be examined. This is obviously not possible without powerful *ab initio* calculations.

The calculations have also resolved the apparent controversy between the structural variations of hydrides as well as halides and the VSEPR model. Consider the space requirement of the bonds and lone pairs in the SH_2, SF_2, HSF series. According to the data of Table 82, the repulsion between a lone pair and the S—H bond in the HSF molecule is somewhat larger (∢E—S—H 107.6°) than the repulsion between a lone pair and the S—F bond (∢E—S—F 105.1°). Thus if the fluorine ligand of HSF is substituted by hydrogen, there will be two larger E—S—H angles and, accordingly, the H—S—H angle has to decrease in order to achieve the typical triple average angle of the S—H bond. On the other hand, if the hydrogen ligand of HSF is substituted by fluorine, there will be two smaller E—S—F angles and, accordingly, the F—S—F angle may open in order to achieve the typical triple average angle of the S—F bond. From all this follows that the larger F—S—F bond angle in SF_2 as compared with the H—S—H bond angle in SH_2 does not mean that the space requirement of the S—F bond is larger than that of the S—H bond. The conclusion is that the importance of electron pair repulsions is much better revealed by the examination of the space requirements of the bonds involved than by considering only the bond angles.

Let us return now to the X—S—X bond angle variations in the SX_2, SOX_2, SO_2X_2 series. It has already been seen that the space requirement of the sulphur–oxygen double bond is nearly the same as that of the sulphur lone pair. It was also found from the positions of the charge centroids that the sulphur–oxygen double bonds are less space-filling than the sulphur lone pairs. Accordingly, the sulphur–hydrogen and sulphur–fluorine bonds may have more space in the sulphones than in the analogous sulphoxides. On the other hand, the sulphur lone pair in a sulphoxide has considerably larger space requirement than in a sulphide. As compared with a sulphide molecule, the ligands in a sulphoxide get further away from the lone pair and in order to achieve the typical triple average angles, the X—S—X angle of the sulphoxide has to decrease. Looking in the other direction, the space requirement of the second oxygen atom of the sulphone molecule is smaller than that of the lone pair of the sulphoxide. Accordingly, the X—S—X bond angle increases as compared with the analogous bond angle of the sulphoxide. So far calculations are available for SO_2H_2 only but it is hoped that future calculations for SO_2F_2 will be consistent with this picture. The experimental data on the SF_2, SOF_2, SO_2F_2 series are in complete agreement with the model presented above. The notion on the large size of the sulphur lone pair of sulphoxide is supported by the bond angles of the molecule pairs PX_3, POX_3 presented in Table 80. The triple average angle of the phosphorus lone pair is 117.1° *vs.* the smaller triple average angle of the phosphorus–oxygen double bond, *viz.* 114.4°.

A final observation from the *ab initio* calculations concerns the experience with the polarization functions. The application of d orbitals increased the charge density in the vicinity of the bonds drawing away charge from the central atom and most of all from its lone pair of electrons.

4.12. GEOMETRICAL VARIATIONS IN THE REST OF THE MOLECULE

The changes of the bond angles and bond lengths in the rest of the sulphone molecule may reveal information on the influence of the sulphur bonding system on the whole molecule. The present discussion is restricted to the compilation and comparison of some selected features in related molecules. In addition to sulphones, data on sulphoxides and sulphides are also involved.

The bond angles at the two-coordinated oxygen atom have been determined in several sulphone molecules. They are all larger than the C—O—C angle in dimethyl ether ($111.5 \pm 1.5°$, Kimura and Kubo, 1959, *cf.* Table 30). Indeed, the bond angles at the two-coordinated oxygen atoms show a great variety, thus, for instance, this angle is 103° in F_2O (*cf.* Table 30) and 155° in $F_3SiOSiF_3$ (*cf.* Table 84). There seem to be some regularities in the variations of these bond angles. Consider the data collected in Table 84. The X—O—X bond angle is always larger than the analogous X—O—C angle, where X is B, Si, P or S. The largest differences in the X—O—X and X—O—C angles occur in the boron and silicon derivatives. This difference is gradually diminishing in the phosphorus and the sulphur derivatives. The absolute magnitudes of the bond angles also diminish in this direction. If it is true that the presence of electronegative ligands contracts the d orbitals of the X atom and thus makes them more easily available for bonding, and if the consequence of this is an increase in the oxygen bond angle, then the relatively small S—O—C and S—O—S bond angles may indicate relatively little participation by the d orbitals in these bonding systems. For the silicon and phosphorus derivatives a comparison is possible between molecules in which fluorine atoms or methyl groups are the ligands of the X atom. The oxygen bond angles are considerably larger in the fluorinated molecules than in the methyl or methoxy derivatives. Finally the controversy concerning the F_2POPF_2 structure will be mentioned. In addition to the results cited in Table 84 an earlier investigation (Arnold and Rankin, 1972/1973) reported considerably different results, including *r*(P—O) 153.3 ± 0.6 pm and ∢P—O—P $145.1 \pm 1.2°$. These parameters were obtained under the assumption of one conformer, while according to the reinvestigation (Yow, Rudolph and Bartell, 1975) the electron scattering data can be interpreted by an equilibrium of three rotational forms. Vilkov and Khaikin (1975) analysed the differences in detail and argued for the results cited in Table 84. Their conclusions are indirectly supported by the structural determinations of the sulphones. Thus it may be expected that the variations of the S—O bond lengths are paralleled by a lengthening of the P—O bond in F_2POPF_2 as compared with that of F_2POCH_3 (*cf.* Table 84).

The nitrogen bond configuration seems to be influenced by the presence of ligands containing silicon, phosphorus, sulphur or chlorine, in a decreasing order, as compared with ammonia or its methyl derivatives. This is well demonstrated by the mean values of the nitrogen bond angles (α_N) shown in Table 85. The mean values of the available mean bond angles ($\tilde{\alpha}_N$) are 119.5, 118.5, 115.8, and 106.9° in the silicon, phosphorus,

TABLE 84
Oxygen bond angles in boron, silicon, phosphorus and sulphur derivatives, and phosphorus–oxygen and sulphur–oxygen bond lengths

∢B—O—C	∢B—O—B	∢Si—O—C	∢Si—O—Si
		F_3SiOCH_3 (c) 131.4(32)°	$F_3SiOSiF_3$ (d) 155(2)°
		$\Delta = 24°$	
$B(OCH_3)_3$ (a) 121.4(5)°	$[(CH_3)_2B]_2O$ (b) 144.4(27)°	$(CH_3)_3SiOCH_3$ (e) 122.5(6)°	$[(CH_3)_3Si]_2O$ (f) 148(3)°
		$CH_3Si(OCH_3)_3$ (g) 123.7(10)°	
$\Delta = 23°$		$\Delta = 25°$	
∢P—O—C	**∢P—O—P**	**∢S—O—C**	**∢S—O—S**
F_2POCH_3 (h) 123.7(7)°	F_2POPF_2 (i) 135.2(9)°	FO_2SOCH_3 (l) 116.5(7)°	FO_2SOSO_2F (m) 123.6(12)°
$\Delta = 11°$		$\Delta = 7°$	
r(P—O)		r(S—O)	
156.0(20) pm	163.1(5) pm	155.8(7) pm	161.1(5) pm
$P(OCH_3)_3$ (j) 119.2(11)°			
Cl_2SPOCH_3 (k) 114.4(24)°			

(a) Gundersen (1976)
(b) Gundersen and Vahrenkamp (1976)
(c) Airey, Glidewell, Robiette and Sheldrick (1971a)
(d) Airey, Glidewell, Rankin, Robiette, Sheldrick and Cruickshank (1970)
(e) Csákvári, Wagner, Gömöry, Hargittai, Rozsondai and Mijlhoff (1976)
(f) Csákvári, Wagner, Gömöry, Mijlhoff, Rozsondai and Hargittai (1976)
(g) Gergő, Hargittai and Schultz (1976)
(h) Codding, Jones and Schwendeman (1974)
(i) Yow, Rudolph and Bartell (1975)
(j) Zaripov, Naumov and Tuzova (1974b)
(k) Bezzubov and Naumov (1976)
(l) Hargittai, Seip, Nair, Britt, Boggs and Cyvin (1977)
(m) Hencher and Bauer (1973)

sulphur and chlorine derivatives, respectively. The sulphones seem to exercise a similar influence to that of the sulphoxide and sulphide.

The X—N=C bond angles and X—N= bond lengths of isocyanates and isothiocyanates are collected in Table 86, with X = Si, P, S or Cl. The discrepancies in Si—N=C bond angles, as reported by different studies, are thought to be a consequence of different averaging procedures over intramolecular vibrations and have already been commented upon. The bond angles gradually decrease on going from the silicon to the chlorine derivatives. Thus the general trends in the bond angle variations are similar in the systems where atom X is linked to two-coordinated and three-coordinated nitrogen atoms (see above).

The variations of the mean nitrogen bond angles and the X—N bond lengths appear to be correlated. As the bond angles decrease from 180° (two-coordinated nitrogen) or 120° (three-coordinated nitrogen), the double-bond character of the X—N bonds

TABLE 85
Triple average angles of the nitrogen bonds and the X—N bond lengths in X = silicon, phosphorus, sulphur and chlorine derivatives

Compound		α_N(°)	r(Si—N), pm
$(H_3Si)_3N$	*(a)*	119.7	173.4(2)
$(H_3Si)_2NMe$	*(b)*	120.0	172.1(3)
$(H_3Si)_2NN(SiH_3)_2$	*(c)*	120.0	173.1(4)
$[(Me_3Si)_2N]_2Be$	*(d)*	120.0	172.2(7)
H_3SiNMe_2	*(e)*	117.0	171.5(4)
F_3SiNMe_2	*(f)*	120	165.4(15)
Cl_3SiNMe_2	*(f)*	119.8	165.7(12)
$ClSi(NMe_2)_3$	*(g)*	119.8	171.5(4)

Compound		α_N(°)	r(P—N), pm
$P(NMe_2)_3$	*(h)*	117.5	170.0(5)
$ClP(NMe_2)_2$	*(i)*	120.0	173.0(5)
Cl_2PNMe_2	*(j)*	120	169(3)
Cl_2OPNMe_2	*(j)*	116	167(4)
F_2PNH_2	*(k)*	120.0	165.0(4)
F_2PNMe_2	*(l)*	120	166
	(m)	116.1	168.4(8)

Compound		α_N(°)	r(S—N), pm
$S(NMe_2)_2$	*(n)*	117.4	168.8(6)
$SO(NMe_2)_2$	*(o)*	115.4	169.3(4)
$SO_2(NMe_2)_2$	*(p)*	116.1	165.1(3)
SO_2ClNMe_2	*(q)*	115.4	161.8(5)

Compound		α_N(°)	r(Cl—N), pm
Cl_3N	*(r)*	107.1	175.9(2)
Cl_2NMe	*(s)*	108.7	174(2)
$ClNH_2$	*(t)*	104.8	175.2(1)

(a) Beagley and Conrad (1970)
(b) Glidewell, Rankin, Robiette and Sheldrick (1969)
(c) Glidewell, Rankin, Robiette and Sheldrick (1970a)
(d) Clark and Haaland (1970)
(e) Glidewell, Rankin, Robiette and Sheldrick (1970b)
(f) Airey, Glidewell, Robiette, Sheldrick and Freeman (1971)
(g) Vilkov and Tarasenko (1969)
(h) Vilkov, Khaikin and Evdokimov (1972)
(i) Zaripov, Naumov and Tuzova (1974a)
(j) Vilkov and Khaikin (1966)
(k) Brittain, Smith, Lee, Cohn and Schwendeman (1971)
(l) Forti, Damiani and Favero (1973)
(m) Naumov, Gulyaeva and Pudovik (1972)
(n) Hargittai and Hargittai (1973a)
(o) Hargittai and Vilkov (1970)
(p) Hargittai, Vajda and Szőke (1973)
(q) Hargittai and Brunvoll (1976)
(r) Bürgi, Stedman and Bartell (1971)
(s) Stevenson and Schomaker (1940)
(t) Cazzoli, Lister and Favero (1972)

TABLE 86
X—N=C bond angles and X—N bond lengths in vapour-phase isocyanates and isothiocyanates

Compound		∢X—N=C(°)	r(X—N), pm
SiH_3NCO	*(a)*	180	169.9
	(b)	152.2(12)	170.3(4)
SiH_3NCS	*(c)*	180	167.25(5)
	(b)	163.8(26)	170.4(6)
$Si(NCO)_4$	*(d)*	146.4	168.8(3)
$ClSi(NCO)_3$	*(e)*	145(2)	168.4(3)
$Cl_2Si(NCO)_2$	*(e)*	136(1)	168.7(4)
Cl_3SiNCO	*(e)*	138.0(4)	164.6(8)
F_3SiNCO	*(f)*	160.7(12)	164.8(10)
F_2PNCO	*(g)*	130.6(8)	168.3(6)
F_2PNCS	*(g)*	140.5(7)	168.6(7)
Cl_2OPNCO	*(h)*	120(15)	168.4(10)
ClO_2SNCO	*(i)*	122.4	165.6
ClNCO	*(j)*	118.8(5)	170.5(5)

(a) Gerry, Thompson and Sugden (1966)
(b) Glidewell, Robiette and Sheldrick (1972)
(c) Dössel and Sutter (1977)
(d) Hjortaas (1967)
(e) Hilderbrandt and Bauer (1969)
(f) Airey, Glidewell, Robiette and Sheldrick (1971b)
(g) Rankin and Cyvin (1972)
(h) Naumov, Semashko and Shatrukov (1973)
(i) Brunvoll, Hargittai and Seip (1978b)
(j) Hocking, Williams and Gerry (1975)

decreases as well. The latter is demonstrated by the decrease of the relative shortening of the observed X—N bond lengths compared with the estimated single bond lengths, see Table 87. This comparison is rather crude, since there are not enough data to group them according to the electronegativities of the ligands attached to the X atom. Note that the silicon derivatives show a shortening of the X—N bonds with increasing ligand electronegativity in agreement with expectation.

The N=C bond is longer than the C=O bond in most isocyanates studied to date in the vapour phase. The relevant data are compiled in Table 88. Accordingly, one of the two bond length combinations may be preferred, as was found in the structure analysis of sulphonyl chloride isocyanate. However, most of the data of Table 88 originate from electron diffraction studies, and the parameters characterizing the two bonds in question have strong correlation hindering their accurate determination. It is suspected that double solutions for the relationship of the N=C and C=O bond lengths could have been obtained in more studies if tested. On the other hand, the N=C bond lengths reported for isocyanates are consistent with those determined in isothiocyanates. This is reassuring, since this bond length can be determined more reliably in isothiocyanates than in isocyanates. The lengths of the S=C bonds are around 156 pm (see Hargittai and Paul, 1977), thus no other bond of similar length hinders the determination of r(N=C) in the thio derivatives. Accordingly, the structure of methyl isocyanate (*cf.* Table 88) is somewhat suspect.

TABLE 87
Mean values of experimentally determined X—N bond lengths, estimated single bond lengths and their differences (X = Si, P, S or Cl)

Bond	Experi-mental*	Estimated**		Difference	
		r_1	r_2	Δ_1	Δ_2
r(X—N=), pm					
X = Si	169	182	179	13	10
P	168	176	177	8	9
S	166	172	173	6	7
Cl	170	169	172	−1	2
r(X—N<), pm					
X = Si	171	182	179	11	8
P	168	176	177	8	9
S	166	172	173	6	7
Cl	175	169	172	−6	−3

* The data referring to the two-coordinated and three-coordinated nitrogen are collected in Tables 86 and 85, respectively.

** $r_1 = r_X + r_N - 4\Delta\chi$

$r_2 = r_X + r_N - 7.3|\Delta\chi|^2$

Covalent radii from Pauling (1960) and Glidewell (1976), *cf.* Table 2, electronegativities from Pauling (1960)

The length of the C=C bonds in ethylene and some of its derivatives are given below in order to compare them with the C=C bonds of the vinyl derivatives appearing among the sulphur structures:

Compound	r_g(C=C), pm	
$H_2C{=}CH_2$	133.7(2)	*(a)*
$F_2C{=}CF_2$	131.1(7)	*(b)*
$(CF_3)_2C{=}CH_2$	137.4(13)	*(c)*
$H_2C{=}CHSCH_3$	134.2(3)	*(d)*
$H_2C{=}C(SCH_3)_2$	134.9(5)	*(d)*
$(CH_3)_2C{=}C(CH_3)_2$	135.3(4)	*(e)*
$H_2C{=}CHPOCl_2$	133.9	*(f)*
$(CH_2{=}CH)_2POCl$	134.6(9)	*(f)*
$H_2C{=}CHSO_2Cl$	135.7(18)	*(d)*
$(CH_2{=}CH)_2SO_2$	133.3(4)	*(d)*

(a) Kuchitsu (1966)
(b) Carlos, Karl and Bauer (1974)
(c) Hilderbrandt, Andreassen and Bauer (1970)
(d) See source in the discussion of the molecule
(e) Tokue, Fukuyama and Kuchitsu (1974); Eisma, Altona, Geise, Mijlhoff and Renes (1974); Carlos and Bauer (1974)
(f) Naumov and Shaidulin (1976)

TABLE 88
N=C and C=O bond lengths in vapour-phase isocyanates and isothiocyanates

Compound		r(N=C), pm	r(C=O), pm
HNCO	(a)	120.7(1)	117.1(1)
CH_3NCO	(b)	116.8(5)	120.2(5)
SiH_3NCO	(c)	121.6(9)	116.4(8)
	(c)	120.0(5)	118.0
	(d)	115.0	117.9
$(CH_3)_3SiNCO$	(e)	120.0(1)	118(1)
F_3SiNCO	(f)	[119.0]	116.8(25)
Cl_3SiNCO	(g)	121.9(7)	113.9(8)
$Cl_2Si(NCO)_2$	(g)	121.7(5)	114.6(5)
$ClSi(NCO)_3$	(g)	121.3(5)	114.4(5)
$Si(NCO)_4$	(h)	120.9(2)	116.5(2)
GeH_3NCO	(i)	119.0(7)	118.2(7)
F_2PNCO	(j)	125.6(6)	116.5(6)
$Cl_2(O)PNCO$	(k)	116.1(5)	122.1(15)
$Cl(O_2)SNCO$	(l)	122.1	115.9
	(l)	115.2	121.8
ClNCO	(m)	122.6(5)	116.2(5)
HNCS	(n)	121.6(2)	
CH_3NCS	(b)	119.2(6)	
SiH_3NCS	(c)	119.7(7)	
	(o)	122.08(5)	
$(CH_3)_3SiNCS$	(e)	118(1)	
F_2PNCS	(j)	122.1(6)	

(a) Jones, Shoolery, Shulman and Yost (1950)
(b) Anderson, Rankin and Robertson (1972)
(c) Glidewell, Robiette and Sheldrick (1972)
(d) Gerry, Thompson and Sugden (1966)
(e) Kimura, Katada and Bauer (1966)
(f) Airey, Glidewell, Robiette and Sheldrick (1971b)
(g) Hilderbrandt and Bauer (1969)
(h) Hjortaas (1967)
(i) Murdoch, Rankin and Beagley (1976)
(j) Rankin and Cyvin (1972)
(k) Naumov, Semashko and Shatrukov (1973)
(l) Brunvoll, Hargittai and Seip (1978b)
(m) Hocking, Williams and Gerry (1975)
(n) Dousmanis, Sanders, Townes and Zeiger (1953)
(o) Dössel and Sutter (1977)

The C=C bond of divinyl sulphone has the same length within experimental error, as that of ethylene. One of the longest bonds appears in ethene sulphonyl chloride. It is poorly determined, however. As noted by Yokozeki and Bauer (1975), the C=C bonds in substituted ethylenes are longer than in ethylene itself. The fluorinated derivatives are exceptions, while $(CF_3)_2C{=}CH_2$ appears to be again an exception from them.

The mean carbon–carbon bond lengths determined in benzene derivatives are the same within experimental error as the benzene C—C bond length, *viz.* $r_g = 139.9 \pm 0.1$ pm (Tamagawa, Iijima and Kimura, 1976).

In conclusion, the remarkably long carbon–carbon bond of thiirane-1,1-dioxide (Nakano, Saito and Morino, 1970) is mentioned. It is interesting to compare it with the sulphoxide (Saito, 1969b) and sulphide (Okiye, Hirose, Lister and Sheridan, 1974) analogues.

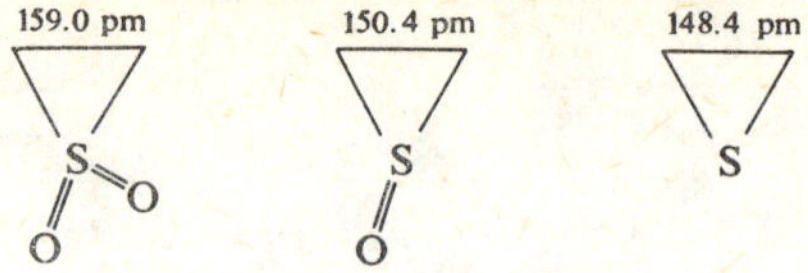

Hoffmann, Fujimoto, Swenson and Wan (1973) carried out semi-empirical molecular orbital calculations on this series of molecules and showed that the structural variations could be interpreted both with and without invoking sulphur d orbitals.

4.13. EMPIRICAL CORRELATIONS OF GEOMETRICAL AND VIBRATIONAL PARAMETERS OF THE S═O BOND

Empirical correlations between geometrical and vibrational parameters of a large series of relatively simple molecules with S═O bond were established by Gillespie and Robinson (1963) and these relationships have been extensively used (see, e.g., Banister, Moore and Padley, 1968; Szmant, 1971). As was, however, noted by Gillespie and Robinson (1963), at the time of establishing the relationships involving the S═O bond lengths and stretching frequencies, relatively few S═O bond lengths were available and they were considered in most cases to be less accurate than the corresponding frequencies. Note also that data on sulphones and sulphoxides and ionic crystals as well as covalent molecules have been used together in establishing the empirical correlations.

During the past years interest in empirical relationships has increased as the body of experimental data has grown. Also, critical selection of the data may enhance the usefulness of the empirical relationships. Empirical relationships were reported between geometrical and vibrational parameters of the P—Cl and S—Cl bonds by Bel'skii, Naumov and Nuretdinov (1974) and of the Si—O, Ge—O, and P—O bonds by Lazarev, Mirgorodskii, Ignat'ev and Muldagaliev (1975). As reliable and consistent geometrical parameters for a relatively large series of sulphones and some sulphoxides in the vapour phase became available, it was considered timely to revise the empirical relationships related to the S═O bond. Detailed studies have been applied to the sulphone molecules (Brunvoll and Hargittai, 1978; Hargittai and Brunvoll, 1979a), followed by an examination of the sulphoxide data (Hargittai, 1979d). First the empirical relationships established for the sulphone molecules are presented preceded by some comments on the data used, the force constant calculations, and the question of averaging the symmetric and antisymmetric S═O bond stretching frequencies.

Geometrical data. Only vapour-phase data have been used. The parameters available at the time of establishing the correlation relationships are cited in Table 89.

TABLE 89

Experimental bond lengths, bond angles and vibrational frequencies of the SO_2 group of sulphone molecules, used in establishing correlation relationships

Compound	Geometrical data				Vibrational frequencies*			
	r_g(S=O), pm		∢O=S=O (°)		ν_s(S=O)	ν_{as}(S=O)	$\delta(SO_2)$	
					(cm^{-1})			
SO_2F_2	139.8(2)	*(a)*	125.1(2)	*(b)*	1269	1502	544	*(c)*
SO_2Cl_2	140.5(4)	*(d)*	123.5(10)	*(d)*	1182	1414	560	*(c)*
SO_2FBr	140.7	*(e)*	123.7	*(e)*	1205	1437	564	*(f)*
SO_2FCl	140.8	*(g)*	123.7(10)	*(g)*	1228	1455	480	*(c)*
CH_3OSO_2F	141.0(3)	*(h)*	124.4(7)	*(h)*	1235	1465		*(c)*
CH_3SO_2F	141.1(3)	*(i)*	123.1(15)	*(i)*	1223	1415	531	*(j)*
$C_6H_5SO_2Cl$	141.8(12)	*(k)*	122.5(36)	*(k)*	1186	1377	572	*(l)*
CH_3OSO_2Cl	142.0(4)	*(m)*	122.2(15)	*(m)*	1192	1404	588	*(n)*
$(CH_3)_2NSO_2Cl$	142.2(4)	*(o)*	122.7(23)	*(o)*	1182	1395	572	*(ö)*
CH_3SO_2Cl	142.5(3)	*(p)*	120.8(8)	*(p)*	1173	1378	533	*(q)*
SO_2	143.2(2)	*(r)*	119.3(10)	*(s)*	1151	1362	518	*(c)*
$(CH_3)_2NSO_2N(CH_3)_2$	143.3(10)	*(t)*	120.2	*(u)*	1150	1353	526	*(ö)*
$(CH_3)_2SO_2$	143.6(3)	*(v)*	119.7(11)	*(v)*	1165	1343	510	*(w)*
$(C_6H_5)_2SO_2$	143.9(12)	*(x)*	119.3	*(u)*	1149	1305	580	*(y)*
$(CH_2)_2SO_2$	143.9(6)	*(z)*	121.4(5)	*(z)*	1159	1309	446	*(aa)*
$(CH_2{=}CH)_2SO_2$	144.0(4)	*(bb)*	119.2	*(u)*	1124	1320	550	*(cc)*
$(CH_2)_4SO_2$	145(1)	*(dd)*	114.6(30)	*(dd)*	1147	1301	567	*(ee)*

* Liquid-phase infrared spectroscopic data except for SO_2F_2, SO_2FBr and $(CH_2)_2SO_2$ liquid-phase Raman, $(CH_3)_2NSO_2Cl$ solution, SO_2 gas phase

(a) Hagen, Cross and Hedberg (1978)

(b) Calculated from the electron diffraction bond length *(a)* and from the mean value of the two microwave spectroscopic O...O distances, Lide, Mann and Fristrom (1957)

(c) After Gillespie and Robinson (1963)

(d) Hargittai (1969)

(e) Raley, Wollrab and Lovejoy (1973)

(f) Reed and Lovejoy (1968)

(g) Holt and Gerry (1971)

(h) Hargittai, Seip, Nair, Britt, Boggs and Cyvin (1977)

(i) Hargittai and Hargittai (1973b)

(j) Geiseler and Nagel (1973)

(k) Brunvoll and Hargittai (1976)

(l) Uno, Machida and Hanai (1968)

(m) Hargittai, Schultz and Kolonits (1977)

(n) Nagel, Stark, Fruwert and Geiseler (1976)

(o) Hargittai and Brunvoll (1976)

(ö) Török, Páldi, Dobos and Fogarasi (1970)

(p) Hargittai and Hargittai (1973)

(q) Cyvin, Dobos, Hargittai, Hargittai and Augdahl (1973)

(r) Clark and Beagley (1971)

(s) Morino, Kikuchi, Saito and Hirota (1964)

(t) Hargittai, Vajda and Szőke (1973)

(u) Calculated from r_g(S=O) and the assumed value of 248.4 pm for the O...O distance

(v) Hargittai and Hargittai (1974)

(w) Spoliti, Chackalackal and Stafford (1967)

(x) Rozsondai, Moore, Gregory and Hargittai (1979)

(y) Nagel, Steiger, Fruwert and Geiseler (1975)

(z) r_0 parameter, Nakano, Saito and Morino (1970)

(aa) Kuz'yants and Aleksanyan (1972)

(bb) Hargittai, Rozsondai, Nagel, Bulcke, Robinet and Labarre (1978)

(cc) Bulcke (1975)

(dd) Naumov, Semashko and Shaidulin (1973)

(ee) Katon and Feairheller (1965)

The bond angle of sulphonyl difluoride was calculated from the electron diffraction S=O bond length and microwave spectroscopic O...O distance. Although the physical meaning of the interatomic distances determined by these two techniques differs, the systems under consideration and in particular the SO_2 groups are rather rigid, so this difference originating from different averaging procedure over intramolecular vibrations is ignored. Assumed parameters have not been used in determining the empirical relationships. The data on the cyclic compounds were not considered in some of the calculations.

Vibrational frequencies. The symmetric and antisymmetric S=O bond stretching frequencies and SO_2 bending frequencies used in the calculations are cited in Table 89. The data originated mostly from infrared spectroscopic studies on liquid-phase samples similarly to those used by Gillespie and Robinson (1963). Naturally, vapour-phase data would have been preferred, but too little was available. It is difficult to predict the changes in frequencies accompanying the transition from the vapour phase to the pure liquid or a dissolved state. It has been noted (see, e.g., Banister, Moore and Padley, 1968) that the stretching frequencies are generally smaller and the bending frequencies larger in the liquid phase than in the vapour phase. Infrared gaseous and liquid data quoted below from a study by Spoliti, Chackalackal and Stafford (1967) show, however, that no definite trend can be really established at least for the stretching frequencies:

		Vapour	Liquid (condensed phase)	Difference
CH_3SO_2F	$\nu_{as}(S{=}O)$, cm^{-1}	1451	1443	8
	$\nu_s(S{=}O)$, cm^{-1}	1225	1233	−8
	$\delta(SO_2)$, cm^{-1}	534	543	−9
CH_3SO_2Cl		1404	1386	18
		1192	1185	7
		543	548	−5
$CH_3SO_2CH_3$		1355	1343	12
		1162	1165	−3
		496	510	−14

The changes are relatively small but probably real, since all the data originate from concurrent experiments. The above data on dimethyl sulphone may be compared with those from another investigation by Uno, Machida and Hanai (1971):

	Infrared (solid state)	Raman (liquid)
$\nu_{as}(S{=}O)$, cm^{-1}	1307	1289
$\nu_s(S{=}O)$, cm^{-1}	1143	1138
$\delta(SO_2)$, cm^{-1}	504	502

A final example for gas-liquid comparison with considerable difference is quoted from the infrared study of Bulcke (1975) on divinyl sulphone

	Vapour	Liquid	Difference
$\nu_{as}(S{=}O)$, cm^{-1}	1352	1310	42
$\nu_s(S{=}O)$, cm^{-1}	1150	1124	26

Thus it may be concluded that the next step in the further refinement of correlation relationships between the structural parameters of sulphone molecules necessitates a consistent set of vapour-phase vibrational frequencies.

Stretching force constants. Even when all the vibrational frequencies of a molecule are available, certain approximations may be needed in the determination of the molecular force field. The valence force constants produced are to some extent dependent on the approximations used. For several of the molecules discussed by Brunvoll and Hargittai (1978), not even a complete set of experimental frequencies was available. Accordingly, further simplifications were necessary. The SO_2 group was looked upon as a separate entity for which the well-known expressions of a bent model were applied (*cf.* Cyvin, 1968).

The force constants of column *(i)*, Table 90 were obtained in the following way. The *G*-matrix elements corresponding to kinematic coupling between the SO_2 group and the rest of the molecule were ignored as were coupling force constants between the SO_2 group and the rest of the molecule and the force constant for the SO_2 stretching–bending interaction. Furthermore the *G*-matrix element connected with kinematic coupling between symmetric S═O stretching and O═S═O bending was also ignored in obtaining the force constants of column *(ii)*, Table 90. The latter approximation caused an 8 to 17 N m^{-1} increase in f_r(S═O). Ignoring the *G*-matrix elements of the

TABLE 90
Calculated stretching force constants
f_r(S═O) in N m^{-1}

Molecule	*(i)**	*(ii)**
SO_2F_2	1208.1	1220.2
SO_2Cl_2	1054.0	1068.0
SO_2FBr	1092.7	1106.8
SO_2FCl	1132.9	1142.1
CH_3OSO_2F		1157.4
CH_3SO_2F	1094.2	1106.3
$C_6H_5SO_2Cl$	1028.1	1043.2
CH_3OSO_2Cl	1052.0	1068.3
$(CH_3)_2NSO_2Cl$	1038.0	1053.0
CH_3SO_2Cl	1017.5	1030.5
SO_2	985.6	998.0
$(CH_3)_2NSO_2N(CH_3)_2$	965.4	978.2
$(CH_3)_2SO_2$	985.5	996.7
$(C_6H_5)_2SO_2$	939.3	956.0
$(CH_2)_2SO_2$	960.8	969.0
$(CH_2{=}CH)_2SO_2$	922.8	937.6
$(CH_2)_4SO_2$	934.9	950.7

* The approximations used in calculating these force constants are described in the text

kinematic coupling between the SO_2 group and the rest of the molecule presumably causes similar differences.

A vibrational analysis considering the whole molecule for sulphuryl chloride showed some slight influence of the coupling force constants on the S=O stretching frequencies. When all the off-diagonal valence force constants except f_{rr} were put equal to zero, 1033 N m^{-1} was obtained for f_r (*vs.* 1054 N m^{-1} obtained from the triatomic model calculation). With a different set of force constants 1052 N m^{-1} has been obtained (Hargittai and Cyvin, 1969). Considering R′RSO_2 molecules, R′ and R will give different contributions, of course, both for the removed *G*-matrix elements and force constants. However, the results on SO_2Cl_2 should give some realistic estimates of these effects which are probably not larger than 10–20 N m^{-1}.

On averaging the S=O stretching frequencies. Gillespie and Robinson (1963) used the arithmetic mean value of the antisymmetric stretching and symmetric stretching frequencies in determining various correlation relationships. Szmant (1971) argued, however, that the two kinds of stretching frequencies are influenced to different

TABLE 91
S=O bond stretching frequencies calculated from the force constants and averaged from the experimental frequencies

Compound	ν(S=O), cm^{-1}			
	(1)	*(2)*	*(3)*	*(4)*
SO_2F_2	1387	1394	1390	1386
SO_2Cl_2	1295	1304	1303	1298
SO_2FBr	1319	1327	1326	1321
SO_2FCl	1343	1348	1346	1342
CH_3OSO_2F		1357	1355	1350
CH_3SO_2F	1320	1327	1322	1319
$C_6H_5SO_2Cl$	1279	1289	1285	1282
CH_3OSO_2Cl	1294	1304	1302	1298
$(CH_3)_2NSO_2Cl$	1286	1295	1293	1289
CH_3SO_2Cl	1273	1281	1280	1276
SO_2	1253	1261	1261	1257
$(CH_3)_2NSO_2N(CH_3)_2$	1240	1248	1246	1243
$(CH_3)_2SO_2$	1252	1260	1257	1254
$(C_6H_5)_2SO_2$	1223	1234	1229	1227
$(CH_2)_2SO_2$	1237	1242	1236	1234
$(CH_2{=}CH)_2SO_2$	1212	1221	1221	1217
$(CH_2)_4SO_2$	1220	1230	1226	1224

(1) and *(2)* were calculated from the force constants of columns *(i)* and *(ii)* of Table 90, respectively

(3) $\nu(S{=}O) = \left[\frac{1}{2}(\nu_{as}^2 + \nu_s^2)\right]^{1/2}$ *(4)* $\nu(S{=}O) = \frac{1}{2}(\nu_{as} + \nu_s)$

extent by changes in the sulphone ligands. The frequencies of columns *(1)* and *(2)* of Table 91 were calculated from the force constants of columns *(i)* and *(ii)* of Table 90, respectively, assuming the S=O bond to be a diatomic molecule. If SO_2 is considered to be a triatomic molecule and with some more assumptions, the following formula may be obtained

$$\nu = \left[\frac{1}{2}(\nu_{as}^2 + \nu_s^2)\right]^{1/2}$$

TABLE 92

The constants A_0, A_1, and A_2 for the expression $y = A_0 + A_1x + A_2x^2$ found by least-squares procedure; $2A$ is the O=S=O bond angle and R is the S=O bond length*

y	x	A_0	A_1	A_2	Standard deviation
ν_s	R	79,031.6	−1066.31	3.64878	9.8
ν_s	R	5404.21	−29.6262		12.5
lg ν_s	lg R	10.6292	−3.50857		11.9
ν_{as}	R	42,738.9	−539.623	1.75	14.0
ν_{as}	R	7426.24	−42.4164		14.5
lg ν_{as}	lg R	12.3924	−4.29574		14.2
ν_s	ν_{as}	3888.88	−4.48191	0.00182282	11.2
ν_s	ν_{as}	375.297	0.583985		13.0
lg ν_s	lg ν_{as}	0.947270	0.676814		13.1
ν_s/ν_{as}	$2A$	−10.8438	0.193229	−0.00079786	0.0079
ν_s/ν_{as}	$2A$	1.06171	−0.00172407		0.0083
lg (ν_s/ν_{as})	lg ($2A$)	0.442552	−0.245630		0.0083
f_r *(a)*	R	1369.14	−1854.11	631.889	0.132
f_r	R	95.1364	−59.5477		0.193
lg f_r	lg R	6.88854	−69.1779		0.131
f_r *(b)*	R	1200.76	−1615.04	547.123	0.158
f_r	R	97.1161	−60.8273		0.196
lg f_r	lg R	20.2852	−8.01743		0.176
ν *(c)*	R	74,537.8	−995.277	3.37636	8.4
ν	R	6464.12	−36.3905		11.3
lg ν	lg R	11.6585	−3.97072		10.4
ν *(d)*	R	64,356.1	−850.965	28.6569	9.8
ν	R	6551.91	−36.9387		14.9
lg ν	lg R	11.7432	−4.00854		10.9
ν *(e)*	R	61,159.2	−806.83	2.71303	9.8
ν	R	6432.34	−36.1405		11.3
lg ν	lg R	11.5957	−3.94097		10.7

* The linear relationship between y and x is meant where A_2 is not given

(a) and *(b)* See columns *(i)* and *(ii)* of Table 90, respectively
(c), *(d)*, and *(e)* See columns *(1)*, *(2)*, and *(4)* of Table 91, respectively

and the corresponding values are listed in column *(3)* of Table 91 while the arithmetic mean values are given in column *(4)* of the same Table. The frequencies of columns *(1)* and *(2)* of Table 91 differ some 5 to 11 cm^{-1}. The values of column *(3)* are smaller than or the same as those of column *(2)*. Of course, the arithmetic mean values are smaller than the frequencies of column *(3)* but most of them are larger than those of column *(1)*, with the exception of SO_2F_2, CH_3SO_2F, and $(CH_2)_2SO_2$, but in these cases the largest difference is only 5 cm^{-1}. Thus the arithmetic mean compares well with the calculations considering the SO_2 group as a separate entity. The $\nu = \left[\frac{1}{2}(\nu_{as}^2 + \nu_s^2)\right]^{1/2}$ expression is preferred to $\nu = \frac{1}{2}(\nu_{as} + \nu_s)$ but the differences are really small.

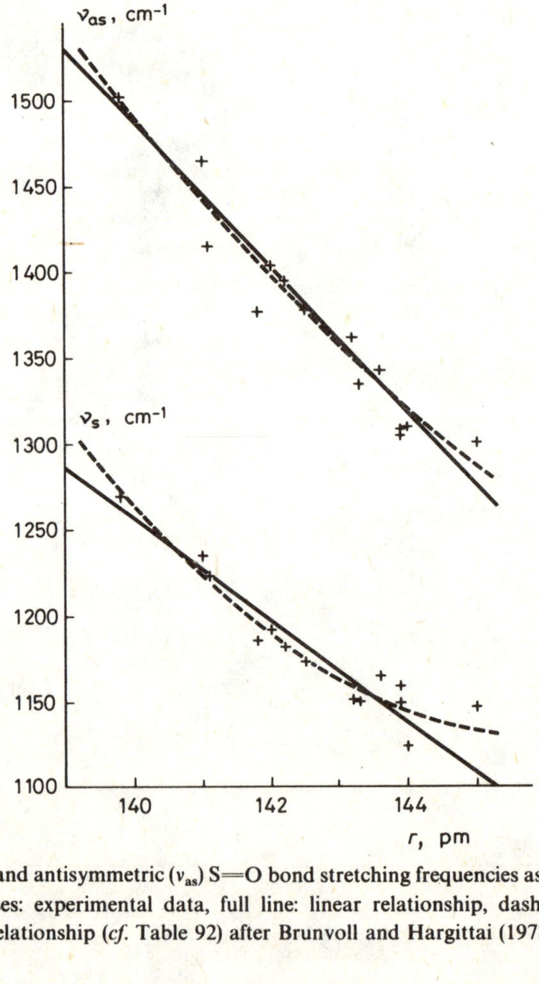

Fig. 93. Symmetric (ν_s) and antisymmetric (ν_{as}) S═O bond stretching frequencies as a function of the S═O bond length (r). Crosses: experimental data, full line: linear relationship, dashed line: second degree relationship (*cf.* Table 92) after Brunvoll and Hargittai (1978)

Correlation relationships. A standard least-squares procedure (Kuo, 1965) was used to establish the correlation relationships among the geometrical and vibrational parameters. The geometrical data were used with weights inversely proportional to the standard deviations parenthesized in Table 89. The vibrational data were given equal weights. The coefficients determined for 9 different relationships are listed in Table 92. Second degree, linear, and logarithmic expressions are given for each relationship. They are believed to be useful within the rather narrow intervals covered by the data. The linear relationships are preferred for extrapolation rather than the second degree expressions. Some of the relationships are illustrated by Figs. 93–95. In establishing the relationship between the average stretching frequency, $\tilde{\nu}=\left[\frac{1}{2}(\nu_{as}^2+\nu_s^2)\right]^{1/2}$, and the bond length,

$$\tilde{\nu}=-36.950r+6549.4 \quad (\sigma=9\ \mathrm{cm}^{-1})$$

some additional data have been used as compared with Table 89. This relationship is shown in Fig. 96 with the compounds indicated whose data were considered.

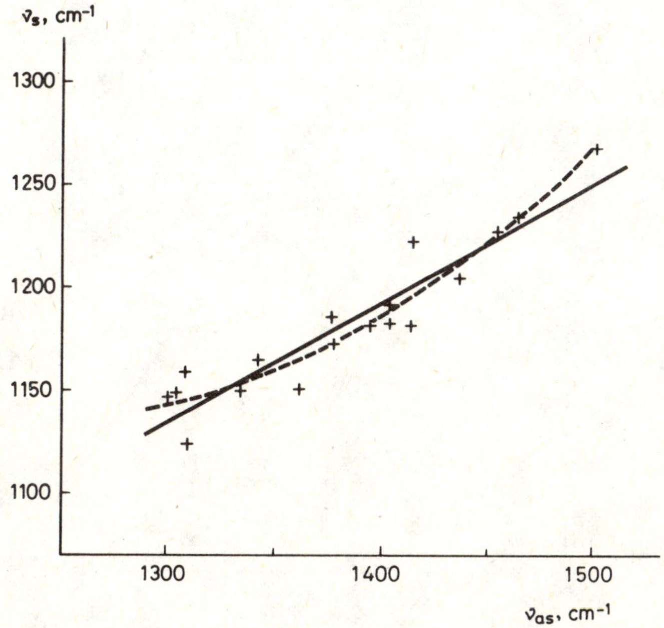

Fig. 94. Symmetric S=O stretching frequency (ν_s) as a function of the antisymmetric S=O stretching frequency (ν_{as}). Crosses: experimental data, full line: linear relationship, dashed line: second degree relationship (*cf.* Table 92) after Brunvoll and Hargittai (1978)

4.13.1. SOME APPLICATIONS

Experience with the correlation relationships has prompted an electron diffraction reinvestigation of the sulphuryl chloride molecular geometry (Hargittai and Hargittai, 1981). The estimated 140.8 pm from the correlation relationships are in complete agreement with the assumed S=O bond length (140.7 pm) used in the microwave spectroscopic investigation of sulphuryl chloride fluoride by Holt and Gerry (1971). The correlation relationships give 141.4 pm for *r*(S=O) of sulphuryl bromide fluoride which is somewhat larger than the assumed 140.7 pm in its microwave spectroscopic study by Raley, Wollrab and Lovejoy (1973). The bond angles O=S=O are 123.8 and 123.1° for SO_2ClF and SO_2BrF, respectively, from the estimated S=O bond lengths and the O. . .O distances accurately determined from the microwave spectra.

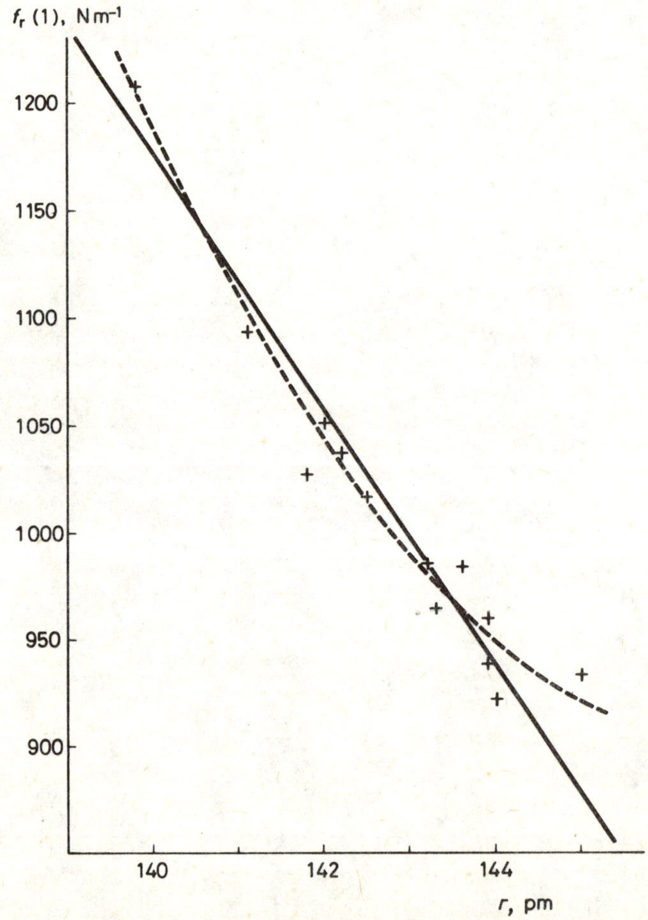

Fig. 95. The bond stretching force constant, f_r (1) of column *(i)* of Table 90 as a function of the S=O bond length (*r*). Crosses: calculated force constants, full line: linear relationship, dashed line: second degree relationship (*cf.* Table 92) after Brunvoll and Hargittai (1978)

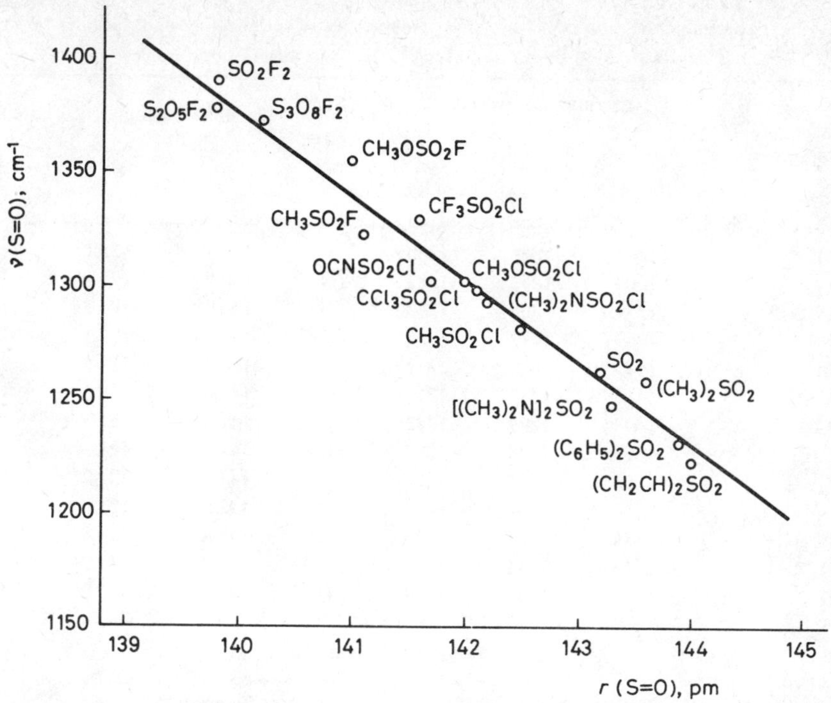

Fig. 96. Empirical relationship between the average stretching frequency, $\tilde{\nu}(S{=}O)=\left\{\frac{1}{2}\left[\nu_{as}^2(S{=}O)+\nu_s^2(S{=}O)\right]\right\}^{1/2}$, and the bond distance $r(S{=}O)$ for sulphones. Circles: experimental data, full line: $\tilde{\nu}=-36.950r+6549.4$ ($\sigma=9$ cm^{-1}) after Hargittai (1979a)

A striking example of the usefulness of the empirical relationships was the discovery of the error in the S=O bond length and O=S=O bond angle communicated for trichloromethane sulphonyl chloride (Alekseev, 1967). The results of the electron diffraction reinvestigation (Brunvoll, Hargittai and Seip, 1978a) are in excellent agreement with the estimated parameters (*cf.* the discussion of this molecule).

At the time of establishing the correlation relationships given in Table 92, the geometrical parameters of sulphonyl chloride isocyanate were not yet available. Thus it is of interest to compare the observed r_g(S=O) 141.7±0.3 pm with the estimated 142.1 pm based on the vibrational frequencies (ν_{as} 1412 and ν_s 1182 cm^{-1}) of Kanesaka and Kawai (1970). The agreement is quite satisfactory, the difference corresponding to realistic expectations. Since the data are within the interval covered by the experimental parameters used in establishing the correlation relationships, the second degree relationships may also be used giving 141.9 pm which is even slightly closer to the observed value. Judging the uncertainties, the experimental error of the bond length

TABLE 93

Experimental S=O bond stretching frequencies of some sulphone molecules and estimated S=O bond lengths and O=S=O bond angles

Compound	Experimental frequencies cm^{-1}			Reference	Estimated geometrical parameters		
	ν_{as}	ν_s	ν *(a)*		r, pm *(b)*	∢ (°) *(c)*	∢ (°) *(d)*
$FOSO_2F$	1501	1248	1380	*(e)*	139.9	125.1	125.2
$FOOSO_2F$	1500	1252	1382	*(f)*	139.8	125.1	125.3
CF_3OOSO_2F	1490	1250	1375	*(f)*	140.0	124.9	125.0
CF_3SO_2F	1463	1239	1356	*(e)*	140.6	124.1	124.2
FSO_2OH	1445	1230	1342	*(g)*	140.9	123.5	123.6
$C_6H_5SO_2F$	1412	1212	1316	*(h)*	141.6	122.5	122.5
$ClSO_2OH$	1408	1209	1312	*(g)*	141.7	122.3	122.4
CH_3SO_2Br	1391	1178	1289	*(i)*	142.4	121.4	121.5
$HOSO_2OH$	1368	1195	1284	*(g)*	142.5	121.2	121.3
$C_6H_5SO_2OCH_3$	1365	1187	1279	*(h)*	142.6	121.0	121.1
$C_6H_5SO_2NH_2$	1334	1157	1249	*(j)*	143.4	119.8	120.0
$CH_3SO_2NH_2$	1315	1145	1233	*(k)*	143.9	119.2	119.4
$CH_3SO_2CH{=}CH_2$	1312	1139	1229	*(l)*	144.0	119.0	119.2

(a) $\nu = \left[\frac{1}{2}(\nu_{as}^2 + \nu_s^2)\right]^{1/2}$

(b) S=O bond lengths, estimated from the empirical relationship $r = (6549.4 - \nu)/36.950$

(c) O=S=O bond angles, estimated from the empirical relationship ∢ $= -1.477\,r + 331.7$

(d) O=S=O bond angles, estimated from the trigonometrical relationship ∢ $= 2 \arcsin(248.4/2r)$

(e) Robinson (1961)
(f) Carter, Kirchmeier and Shreeve (1973)
(g) Birchall and Gillespie (1966)
(h) Ham, Hambly and Laby (1960)
(i) Geiseler and Nagel (1973)
(j) Uno, Machida and Hanai (1971)
(k) Hanai, Okuda, Uno and Machida (1975)
(l) Price and Gillis (1953)

determination and the unknown error in the frequency determination should be considered in addition to the standard deviation of the empirical relationship itself. The relationship illustrated by Fig. 96 yields 142.0 pm for the S=O bond length of sulphonyl chloride isocyanate.

The S=O bond length of tetrahydrothiophene-1,1-dioxide is estimated to be 144.1 pm from the vibrational frequencies (Katon and Feairheller, 1965). This corresponds then to a 118.9° O=S=O bond angle according to the *r*(S=O) *vs.* ∢O=S=O empirical relationship, or to a 119.2° bond angle supposing the constancy of the O...O distance. The estimated and observed bond lengths are consistent as are the bond angles if the relatively large experimental error of the observed ∢O=S=O value is considered. The estimated bond angle is more consistent with the data on the whole sulphone series.

Some S=O bond lengths and O=S=O bond angles are predicted for molecules whose structure has not yet been determined. The parameters given in Table 93 were estimated using the relationship

$$r = (6549.4 - \nu)/36.950$$

corresponding to the one shown in Fig. 96 and the two relationships shown in Fig. 89:

$$\measuredangle = -1.477r + 331.7$$

and

$$\measuredangle = 2 \arcsin (248.4/2r)$$

The empirical relationships can be used, of course, for estimating vibrational frequencies as well. While, e.g., the molecular geometry of ethene sulphonyl chloride has already been determined, its vibrational spectrum has not yet been reported. From r_g(S=O) 142.1 pm and using the relationships of Table 92, 1395 and 1187 cm^{-1} are estimated for the antisymmetric and symmetric bond stretching frequencies. The largest uncertainty in these estimates is the 0.6 pm experimental error of the bond length determination.

In conclusion the structure of sulphuryl fluoride is referred to again. The microwave spectroscopic and electron diffraction investigations yielded different S=O bond lengths and O=S=O bond angles (or rather, O...O distances). Combining the preferred electron diffraction S=O bond length and microwave spectroscopic O...O distance, the O=S=O bond angle is obtained to be 125.1°. From the expressions illustrated by Figs. 89 and 96 the following relationship may be given between the O=S=O bond angle and the average stretching frequency

$$\nu(\text{S=O}) = 25.017 \measuredangle \text{O=S=O} - 1748.7$$

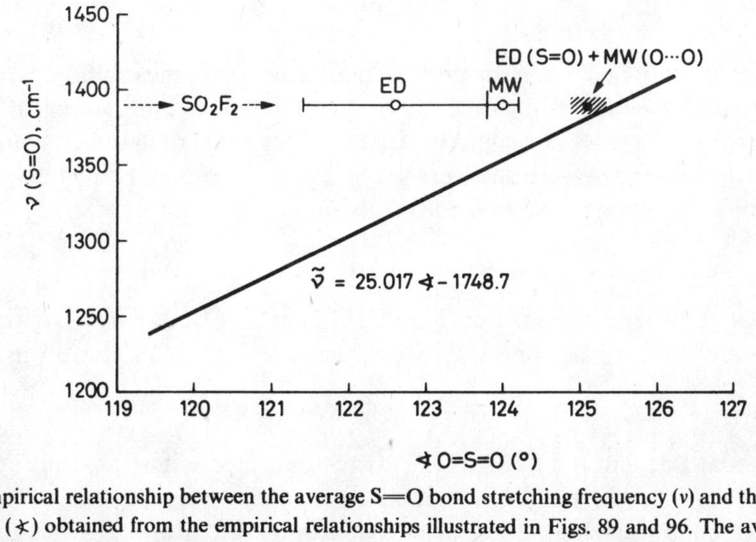

Fig. 97. Empirical relationship between the average S=O bond stretching frequency (ν) and the O=S=O bond angle ($\measuredangle$) obtained from the empirical relationships illustrated in Figs. 89 and 96. The average S=O bond stretching frequency of sulphonyl fluoride (1390 cm^{-1}) is indicated as are the O=S=O bond angles determined in this molecule by electron diffraction (ED) and microwave spectroscopy (MW) together with the communicated uncertainties. The bond angle O=S=O of SO_2F_2 as calculated from the electron diffraction S=O bond length, ED(S=O), and microwave spectroscopic O...O distance, MW(O...O), is also marked. For references see text and Tables

The straight line corresponding to this expression is shown in Fig. 97. The bond angles from the microwave spectroscopic (MW) and electron diffraction (ED) studies are also marked with the communicated uncertainties indicated. It is seen that the 125.1° bond angle obtained by combining the data is much more consistent with the empirical relationships than the bond angles reported from the individual studies.

4.13.2. EMPIRICAL RELATIONSHIPS FOR SULPHOXIDES

Experimental data on five sulphoxides are available for establishing empirical relationships between the parameters characterizing the sulphoxide S=O bond. The available data are the following:

	r_g(S=O), pm	ν(S=O), cm^{-1}
F_2SO	142.1(3)	1308 *(a)*
Cl_2SO	144.4(6)	1229 *(a)*
$[(CH_3)_2N]_2SO$	148.1(9)	1119 *(b)*
$(CH_3)_2SO$	148.6(7)	1069 *(c)*
$(C_6H_5)_2SO$	148.9(5)	1039 *(d)*

(a) Gillespie and Robinson, (1961)
(b) Török, Páldi, Dobos and Fogarasi, (1970)
(c) Gramstad (1963)
(d) Nagel, Steiger, Fruwert and Geiseler, (1975)

The sources of the bond lengths are given in the discussion of the sulphoxide molecules. The relationships were established by a least-squares procedure, in which the bond lengths were used again with weights inversely proportional to the experimental errors while the vibrational frequencies were given equal weights. One of the correlation relationships found for the sulphoxides is the following

$$\lg \nu(\mathrm{S{=}O}) = -4.62 \lg r(\mathrm{S{=}O}) + 13.064 \qquad \text{(I)}$$

with a relatively large standard deviation of 17 cm^{-1}. Combining 12 data pairs on sulphones and 5 data pairs on sulphoxides, the following relationship was obtained

$$\lg \nu(\mathrm{S{=}O}) = -4.30 \lg r(\mathrm{S{=}O}) + 12.370 \qquad \text{(II)}$$

with a standard deviation of 12 cm^{-1}. This relationship is illustrated in Fig. 98. The S=O bond lengths of some sulphoxides not yet studied are predicted in Table 94 on the basis of bond stretching frequencies found in the literature.

During the past years several papers dealt with empirical relationships involving parameters characterizing the S=O and Se=O bonds. Thus, e.g., according to Paetzold (1968)

$$f(\mathrm{Se{=}O}) = 0.593 f(\mathrm{S{=}O}) + 137 \qquad \text{(III)}$$

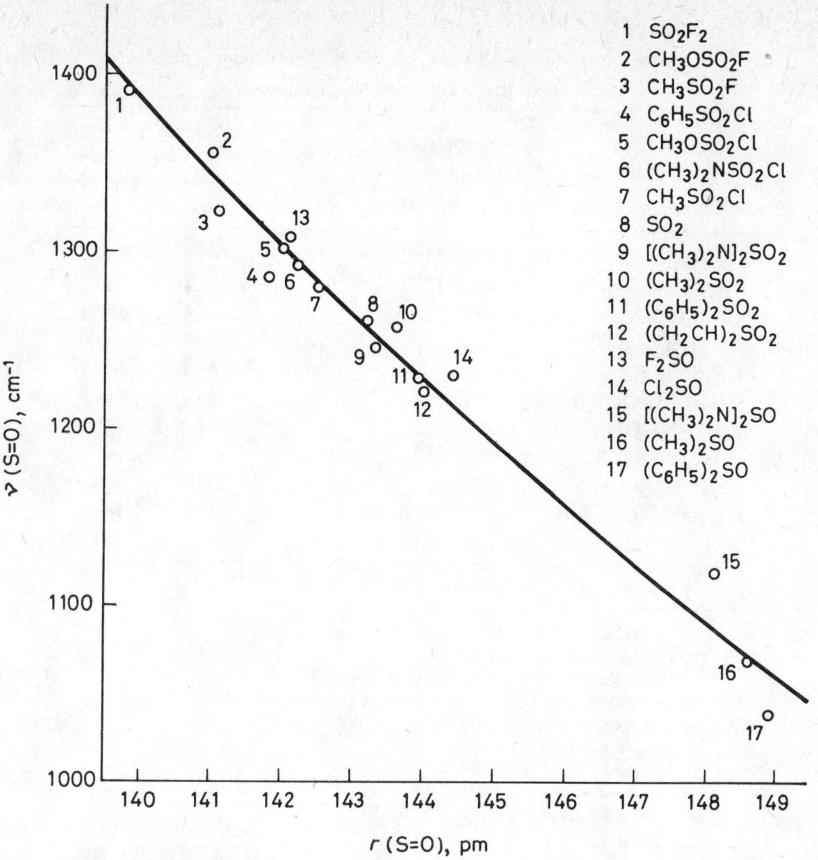

Fig. 98. Empirical relationship between S=O bond stretching frequencies and S=O bond lengths from data on sulphones *and* sulphoxides (Hargittai, 1979d)

where f is the bond stretching force constant in $N\,m^{-1}$, and according to Steudel (1970)

$$\nu(Se{=}O) = [0.475\,\nu^2(S{=}O) + 175{,}000]^{1/2} \tag{IV}$$

The above two relationships are shown together in Fig. 99. A further relationship was established by Paetzold (1970) between the length and the force constant of the Se=O bond

$$\lg f(Se{=}O) = -4.551\,\lg r(Se{=}O) + 12.887 \tag{V}$$

The bond stretching force constants f(S=O) and f(Se=O) can be estimated to a good approximation by employing the model of a diatomic molecule. Utilizing the expressions (I, III, IV, and V), the following Se=O bond lengths of the analogous

TABLE 94
Estimation of the S═O bond length in some simple sulphoxides from the bond stretching frequency

Compound	Experimental frequency cm^{-1}	Reference	Estimated r(S═O) pm (*a*)	Estimated r(S═O) pm (*b*)
$SOBr_2$	1228–1193	(*c*)	144.7	144.5
FS(O)Cl	1293–1276	(*d*)	142.8	142.5
$CF_3S(O)F$	1267–1257	(*d*)	143.4	143.1
$(CF_3)_2SO$	1242	(*e*)	143.9	143.6
$CF_3S(O)Cl$	1260–1205	(*d*)	144.1	143.9
$CH_3OS(O)OCH_3$	1209–1193	(*f*)	144.9	144.7
$(CH_3)_2NS(O)F$	1230	(*g*)	144.2	143.9
$(CH_3)_2NS(O)Cl$	1185	(*h*)	145.3	145.2
$CH_3OS(O)F$	1250	(*i*)	143.7	143.4
$CH_3OS(O)Cl$	1223–1190	(*j*)	144.8	144.6
$(CH_2CH)_2SO$	1059	(*k*)	148.9	149.0

(*a*) From the empirical relationship (I)
(*b*) From the empirical relationship (II)
(*c*) Steudel and Lautenbach (1969)
(*d*) Kirchmeier and Shreeve (1975)
(*e*) Carter, Shiow-Chyn Wang and Shreeve (1973)
(*f*) Simon, Kriegsmann and Dutz (1956); Huong and Raducanu (1974)
(*g*) Brown, Crosbie, Darragh, Ross and Sharp (1970)
(*h*) Paetzold and Rönsch (1970)
(*i*) Seel, Boudier and Gombler (1969)
(*j*) Huong and Raducanu (1974)
(*k*) Nagel and Stock (1978)

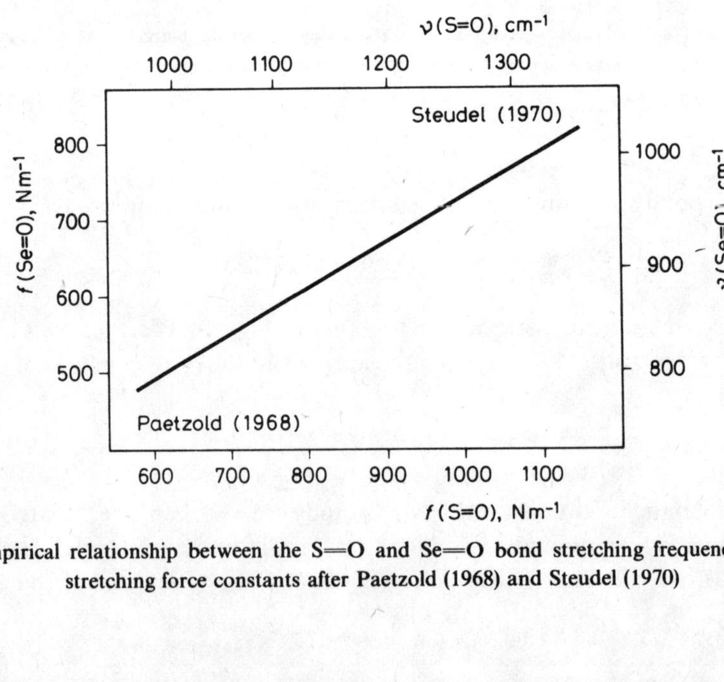

Fig. 99. Empirical relationship between the S═O and Se═O bond stretching frequencies and bond stretching force constants after Paetzold (1968) and Steudel (1970)

selenoxides may be predicted:

	r(Se=O), pm
F_2SeO	157.2
Cl_2SeO	161.3
$[(CH_3)_2N]_2SeO$	167.7
$(CH_3)_2SeO$	168.5
$(C_6H_5)_2SeO$	169.0

The observed Se=O bond lengths of F_2SeO and Cl_2SeO are 157.6(3) and 161.4(5) pm, respectively (*cf.* Table 45).

4.14. EMPIRICAL GROUP ELECTRONEGATIVITIES IN THE SULPHONE SERIES

The sulphur bond configuration variations observed in the XSO_2Y sulphone series have already been qualitatively correlated with the electronegativities of the X and Y ligands in an earlier section. It is more difficult to establish quantitative correlation relationships since well-defined electronegativity values are available for the elements only (see, e.g., Pauling, 1960; Allred and Rochow, 1958) and not for groups. As is well known, while the electronegativity of an element is considered to be a property of this element, the group electronegativity depends on the bond that the group forms. Thus, strictly speaking there is a specific group electronegativity scale for the sulphone series. At the same time the group electronegativities established for a well-studied compound series may be of considerable use in making various estimations of properties for compound series not yet studied so extensively. The estimation of group electronegativities in the sulphone series, or rather the establishment of a relative group electronegativity scale characteristic for the sulphones is attractive because a relatively large body of information is available on the S=O bond lengths and S=O bond stretching frequencies which show characteristic changes with the ligand electronegativity variations (Hargittai, 1979a).

An initial set of electronegativities was compiled for a few ligands, *viz.* F, Cl, CF_3, CCl_3, and CH_3. The choice of the first two was obvious, while group electronegativities for the three CX_3 ligands have been established by Hinze, Whitehead and Jaffé (1963) χ_{CF_3} 3.29, χ_{CCl_3} 2.79, and χ_{CH_3} 2.30. These values were consistent with the experience reported on the relative electronegativities of these five ligands in sulphones (Lagowski, 1959) and sulphoxides (Steudel, 1971, see also Steudel, 1970). The geometries of six sulphone molecules with various combinations of these ligands have been recently determined. These initial data are given in Table 95.

The following relationships were obtained by least-squares procedure, and are illustrated in Figures 100 and 101:

Linear

$$\Sigma\chi = -0.9128r + 135.54 \quad (\text{standard deviation}, \sigma = 0.10)$$
$$\Sigma\chi = 0.02546\nu - 27.3 \quad (\sigma = 0.08)$$

TABLE 95
Initial data used in establishing the correlation relations between S=O bond lengths or S=O bond stretching frequencies and group electronegativities

XSO_2Y	r(S=O), pm	ν(S=O), cm^{-1} *(a)*	$\chi_X + \chi_Y$
FSO_2F	139.8 *(b)*	1390 *(h)*	8.0
FSO_2CF_3		1356 *(i)*	7.3
FSO_2Cl	140.8 *(c)*	1346 *(h)*	7.0
FSO_2CH_3		1322 *(j)*	6.3
$ClSO_2CF_3$	141.6 *(d)*		6.3
$ClSO_2Cl$		1303 *(h)*	6.0
$ClSO_2CCl_3$	142.1 *(e)*	1298 *(i)*	5.8
$ClSO_2CH_3$	142.5 *(f)*	1280 *(k)*	5.3
$CH_3SO_2CH_3$	143.6 *(g)*	1257 *(l)*	4.6

(a) $\nu(S{=}O) = \left[\frac{1}{2}(\nu_{as}^2 + \nu_s^2)\right]^{1/2}$
(b) Hagen, Cross and Hedberg (1978)
(c) Holt and Gerry (1971)
(d) Brunvoll, Hargittai and Kolonits (1978)
(e) Brunvoll, Hargittai and Seip (1978a)
(f) Hargittai and Hargittai (1973)
(g) Hargittai and Hargittai (1974)
(h) Gillespie and Robinson (1963)
(i) Robinson (1961)
(j) Geiseler and Nagel (1973)
(k) Cyvin, Dobos, Hargittai, Hargittai and Augdahl (1973)
(l) Spoliti, Chackalackal and Stafford (1967)

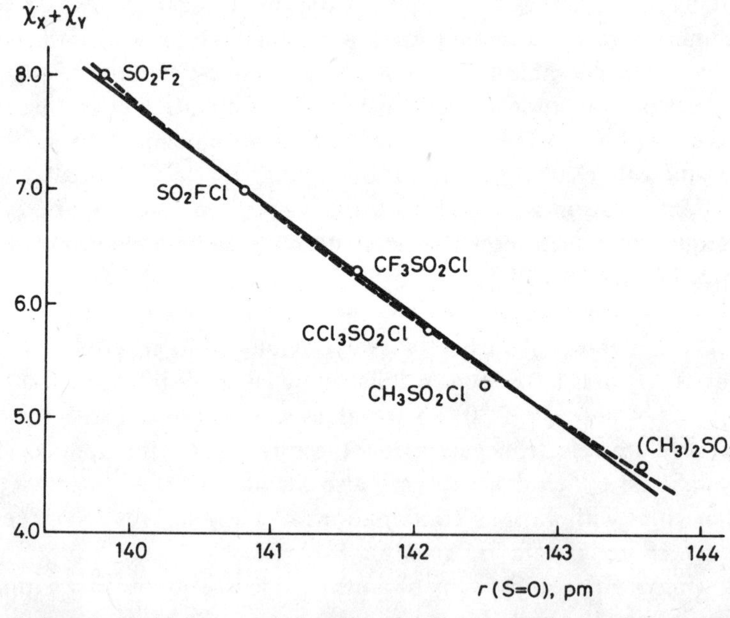

Fig. 100. Correlation relationship between the bond length, r(S=O) and the sum of the electronegativities of the ligands X and Y in the XSO_2Y sulphone molecules. The circles represent the initial data given in Table 95, the full line and dashed line are determined by the linear and second degree relationships described in the text. Within the interval covered by the initial data the second degree relationship is somewhat superior to the linear one. However, the linear relationship is recommended for extrapolation. After Hargittai (1979a)

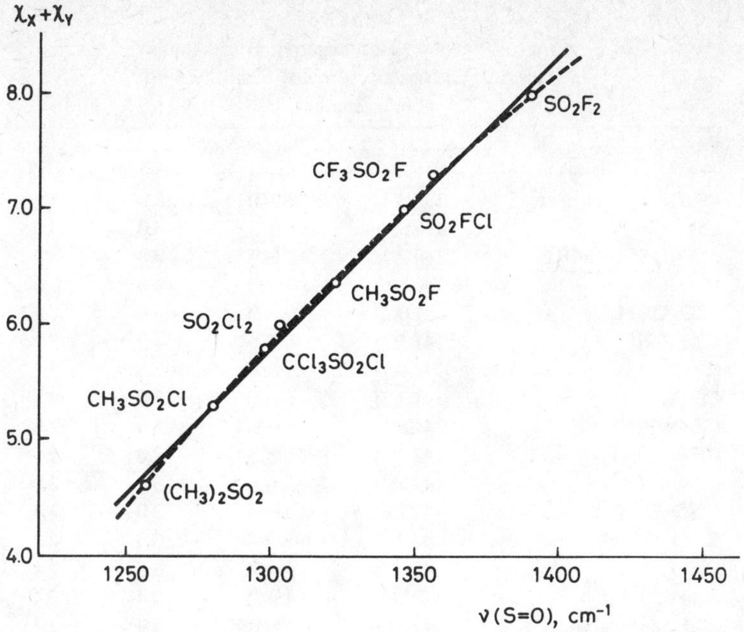

Fig. 101. Correlation relationship between the average stretching frequency, $\nu(\text{S=O}) = \left\{\frac{1}{2}\left[\nu_{as}^2(\text{S=O}) + \nu_s^2(\text{S=O})\right]\right\}^{1/2}$, and the sum of the electronegativities of the ligands X and Y in the XSO_2Y sulphone molecules. The circles represent the initial data given in Table 95, the full line and dashed line are determined by the linear and second degree relationships. The latter is somewhat superior to the linear one within the interval covered by the initial data. However, the linear relationship is recommended for extrapolation. After Hargittai (1979a)

Second degree

$$\Sigma\chi = 0.043669r^2 - 13.2865r + 1012.00 \quad (\sigma = 0.07)$$
$$\Sigma\chi = -0.00003247\nu^2 + 0.11139\nu - 84.1 \quad (\sigma = 0.06)$$

These relationships were then used to estimate other group electronegativities on the basis of available S=O bond lengths and bond stretching frequencies. The results are collected in Tables 96 and 97.

In addition to the usefulness of the group electronegativities, their tentative character should also be emphasized. Only some of the many possible sources of uncertainty are mentioned here. As was stated above, in a strict sense only electronegativities of groups linked to identical molecular moieties are comparable. Thus even within the sulphone series the group electronegativities χ_X in the sulphonyl chloride series (XSO_2Cl) should be distinguished from the χ_X values in the sulphonyl fluoride series (XSO_2F), etc. The experimental errors of the bond lengths and stretching

TABLE 96
Empirical group electronegativities of ligands in the XSO_2Y series as estimated from the S=O bond lengths*

XSO_2Y	r(S=O), pm	$\Sigma\chi$	χ_X	χ_Y
FSO_2F	139.8(2)	8.0(2)	4.0	4.0
FSO_2OSO_2F	139.8(2)	8.0(2)	4.0	4.0
$FSO_2OSO_2OSO_2F$	140.2(3)	7.6(3)	3.8	3.8
FSO_2Cl	140.8(6)	7.0(6)	4.0	3.0
FSO_2OCH_3	141.0(2)	6.8(2)	4.0	2.8
FSO_2CH_3	141.1(3)	6.7(3)	4.0	2.7
$ClSO_2CF_3$	141.6(7)	6.2(7)	3.0	3.2
$ClSO_2NCO$	141.7(3)	6.1(3)	3.0	3.1
$ClSO_2C_6H_5$	141.8(12)	6.0(12)	3.0	(3.0)
$ClSO_2OCH_3$	142.0(3)	5.9(3)	3.0	2.9
$ClSO_2CCl_3$	142.1(3)	5.8(3)	3.0	2.8
$ClSO_2CHCH_2$	142.1(6)	5.8(6)	3.0	(2.8)
$ClSO_2N(CH_3)_2$	142.2(4)	5.7(4)	3.0	2.7
$ClSO_2CH_3$	142.5(3)	5.4(3)	3.0	2.4
SO_2	143.2(2)	4.9(2)	2.4	2.4
$(CH_3)_2NSO_2N(CH_3)_2$	143.3(10)	4.8(10)	2.4	2.4
$CH_3SO_2CH_3$	143.6(3)	4.6(3)	2.3	2.3
$C_6H_5SO_2C_6H_5$	143.9(5)	4.3(5)	2.2	2.2
$CH_2CHSO_2CHCH_2$	144.0(3)	4.3(3)	2.1	2.1

* $\Sigma\chi = 0.043669r^2 - 13.2865r + 1012.0$; the experimental errors of the bond lengths and the corresponding uncertainties in the $\Sigma\chi$ values are parenthesized; Also parenthesized are some improbable values of χ_Y; The references to the experimental r(S=O) values have been given earlier

TABLE 97
Empirical group electronegativities of ligands in the XSO_2Y series as estimated from the S=O bond stretching frequencies

XSO_2Y	ν(S=O), cm^{-1}	$\Sigma\chi$ (*)	χ_X	χ_Y
FSO_2F	1390 *(a)*	8.0	4.0	4.0
FSO_2OOF	1382 *(b)*	7.8	4.0	3.8
FSO_2OF	1380 *(c)*	7.8	4.0	3.8
FSO_2OSO_2F	1378 *(c)*	7.7	4.0	3.7
FSO_2OOCF_3	1375 *(b)*	7.7	4.0	3.7
$FO_2SOSO_2OSO_2F$	1372 *(c)*	7.6	3.8	3.8
FSO_2CF_3	1356 *(c)*	7.2	4.0	3.2
FSO_2OCH_3	1355 *(a)*	7.2	4.0	3.2
FSO_2Cl	1346 *(a)*	7.0	4.0	3.0
FSO_2OH	1342 *(d)*	6.9	4.0	2.9
$ClSO_2CF_3$	1329 *(e)*	6.6	3.0	3.6
FSO_2Br	1326 *(f)*	6.5	4.0	2.5
$FSO_2C_6H_5$	1316 *(g)*	6.3	4.0	2.3
$ClSO_2OH$	1312 *(d)*	6.1	3.0	3.1

TABLE 97 (cont.)

XSO_2Y	$\nu(S{=}O)$, cm^{-1}	$\Sigma\chi$	χ_X	χ_Y
$ClSO_2Cl$	1303 *(a)*	5.9	3.0	3.0
$ClSO_2NCO$	1302 *(h)*	5.9	3.0	2.9
$ClSO_2OCH_3$	1302 *(i)*	5.9	3.0	2.9
$ClSO_2CCl_3$	1298 *(c)*	5.8	3.0	2.8
$ClSO_2N(CH_3)_2$	1293 *(j)*	5.6	3.0	2.6
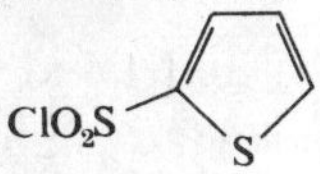	1291 *(k)*	5.4	3.0	2.4
$ClSO_2C_6H_5$	1285 *(l)*	5.4	3.0	2.4
$HOSO_2OH$	1284 *(d)*	5.4	2.7	2.7
$ClSO_2CH_2Cl$	1280 *(m)*	5.3	3.0	2.3
$ClSO_2CH_3$	1280 *(n)*	5.3	3.0	2.3
$CH_3OSO_2C_6H_5$	1279 *(g)*	5.2	2.9	2.4
$ClSO_2CH_2Br$	1278 *(m)*	5.2	3.0	2.3
SO_2	1261 *(a)*	4.7	2.4	2.4
$CH_3SO_2CH_3$	1257 *(o)*	4.6	2.3	2.3
		(**)		
$C_6H_5SO_2NH_2$	1249 *(l)*	4.5	2.4	2.1
$(CH_3)_2NSO_2N(CH_3)_2$	1246 *(j)*	4.4	2.2	2.2
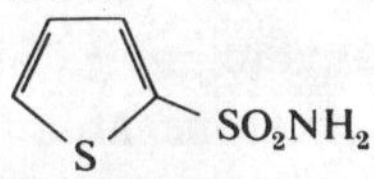	1243 *(p)*	4.3	2.4	1.9
$CH_3SO_2NH_2$	1233 *(q)*	4.1	2.3	1.8
$C_6H_5SO_2CHCH_2$	1230 *(r)*	4.0	2.0	2.0
$C_6H_5SO_2C_6H_5$	1229 *(s)*	4.0	2.0	2.0
$CH_3SO_2CHCH_2$	1229 *(t)*	4.0	2.0	2.0
$CH_2CHSO_2CHCH_2$	1221 *(u)*	3.8	1.9	1.9

* $\Sigma\chi = -0.00003247\,\nu^2 + 0.11139\nu - 84.1$

** $\Sigma\chi = 0.02546\nu - 27.3$

$$\nu = \left[\frac{1}{2}(\nu_{as}^2 + \nu_s^2)\right]^{1/2}$$

(a) Gillespie and Robinson (1963)
(b) Carter, Kirchmeier and Shreeve (1973)
(c) Robinson (1961)
(d) Birchall and Gillespie (1966)
(e) Holly (1978)
(f) Reed and Lovejoy (1968)
(g) Ham, Hambly and Laby (1960)
(h) Kanesaka and Kawai (1970)
(i) Nagel, Stark, Fruwert and Geiseler (1976)
(j) Török, Páldi, Dobos and Fogarasi (1970)
(k) Peron, Saumagne and Lebas (1970)
(l) Uno, Machida and Hanai (1971)
(m) Remizov and Samarina (1975)
(n) Cyvin, Dobos, Hargittai, Hargittai and Augdahl (1973)
(o) Spoliti, Chackalackal and Stafford (1967)
(p) Arcoria, Maccarone, Musumarra and Tomaselli (1974)
(q) Hanai, Okuda, Uno and Machida (1975)
(r) Remizov and Mannafov (1976b)
(s) Nagel, Steiger, Fruwert and Geiseler (1975)
(t) Price and Gillis (1953)
(u) Hargittai, Rozsondai, Nagel, Bulcke, Robinet and Labarre (1978)

frequencies should not be ignored. $\Delta r(S{=}O) = 0.5$ pm or $\Delta\nu(S{=}O) = 20\ cm^{-1}$ changes the estimated electronegativities by about 0.5. The estimated experimental errors of the bond lengths are indicated in Table 96 and their respective influence in the estimated group electronegativities is also given. It is seen that the relatively large errors of some of the bond lengths render the corresponding estimated group electronegativities meaningless. Unfortunately, no comparable quantities of the experimental error are available for the frequencies. Thus the estimated χ values should be treated especially carefully. Furthermore, the estimation procedure is somewhat influenced by the choice of the intial data (*cf.* Table 95). It is true, however, that the emerging pattern of group electronegativities seems to be consistent with expectation and previous experience and thus the initial choice of data is probably fortunate. Finally, it should be remembered that all the bond lengths in our tables refer to free (vapour-phase) molecules while the stretching frequencies originate from spectra of liquid-phase samples.

To summarize, the variations in the S=O bond lengths and bond stretching frequencies parallel the changes in the ligand electronegativities in the sulphone series. Accordingly, group electronegativities can be estimated from the geometrical and vibrational parameters. The present results support earlier notions as regards the relative electronegativities of the CF_3, CCl_3, CH_3 groups. Furthermore, they indicate the position of some other groups, notably the isocyanate and methoxy groups, on the electronegativity scale. The electronegativities of some of the important ligands follow the relationship:

$$F > OF > CF_3 > NCO > Cl > OCH_3 > CH_3, \text{ etc.}$$

It is also noted that the numerical χ values for some groups, e.g. OH, NH_2, C_6H_5, are very uncertain.

The empirical relationships discussed above can also be used to predict S=O bond lengths and ν(S=O) values on the basis of the ligand electronegativities. As an illustration consider the S=O bond lengths for some individual compounds* using the following relationship:

$$r(S{=}O) = (135.54 - \Sigma\chi)/0.9128$$

Lentz and Seppelt (1978) suggested that the ligands $OSeF_5$ and $OTeF_5$ are at least as electronegative as fluorine. Accordingly, the lengths of the S=O bonds in the SO_2 groups of the following molecules can be predicted to take the values listed below, assuming 4.0 for the electronegativities of the ligands $OSeF_5$ and $OTeF_5$:

Compound	Predicted r(S=O), pm
FSO_2OSeF_5	139.7
FSO_2OTeF_5	139.7
$ClSO_2OSeF_5$	140.8
$ClSO_2OTeF_5$	140.8
$F_5SeOSO_2OSeF_5$	139.7
$F_5TeOSO_2OTeF_5$	139.7

Using the above relationship, the S=O bond length for some more sulphone molecules are predicted below. Predicted O=S=O bond angles are also listed as estimated from Figure 89.

* For frequency predictions it would, of course, be of more value to use empirical relationships ν_{as}(S=O) *vs.* $\Sigma\chi$ and ν_s(S=O) *vs.* $\Sigma\chi$. Such relationships can be easily established on the basis of the available data.

		r(S=O) (pm)	∢O=S=O (°)
FSO_2NCO	(χ_{NCO} 3.0)	140.8	123.8
$(OCN)_2SO_2$	(χ_{NCO} 3.0)	141.9	122.1
FSO_2OF	(χ_{OF} 3.8)	139.9	125.1
FSO_2CF_3	(χ_{CF_3} 3.3)	140.5	124.3

The molecular geometry of none of these compounds has been determined yet. It is difficult of course to give a realistic uncertainty estimate for these predicted values. A change of 0.3 in $\Sigma\chi$, for example, corresponds to $\Delta r = 0.3$ pm and then to Δ∢ $= 0.5°$. The uncertainty in the electronegativity values should thus also be considered in addition to the approximate character of the empirical relationships themselves. Because of the inherent uncertainties of these estimations, the exact physical meaning of the intramolecular distances has not been considered.

4.15. SULPHATE MOLECULES

The geometries of only four inorganic sulphate molecules have been studied so far and electron diffraction was the only technique employed. The sulphates have low volatility and high-temperature experimental conditions are needed for their electron diffraction investigation. The visual electron diffraction data on caesium sulphate, Cs_2SO_4, were interpreted in terms of the classical covalent structure (Spiridonov, Khodchenkov and Akishin, 1965). According to this model, the geometrical arrangement is that of a distorted tetrahedron with the sulphur atom at the centre and the oxygen atoms at the apexes. There are two different types of sulphur–oxygen bonds in such a structure, roughly corresponding to single and double bonds as illustrated in the classical structural formula,

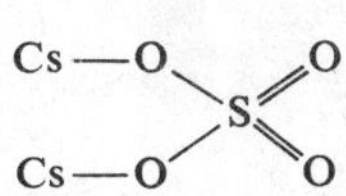

This structural pattern, so familiar from elementary textbooks, differed considerably from that established for this type of compound in liquid state. In the latter a regular or nearly regular tetrahedral arrangement of the oxygen atoms around the sulphur atom was found, see, e.g., the liquid-state structures of Li_2SO_4, Na_2SO_4, K_2SO_4 (Ukshe, 1966). Another result which contradicted the visual electron diffraction study was reported by Büchler, Stauffer and Klemperer (1967). In this work caesium sulphate was shown by molecular beam deflection measurements to have no permanent electric dipole moment. Consequently, the structure of caesium sulphate has been reinvestigated by Spiridonov and Lutoshkin (1970) using the sector-microphotometer electron diffraction technique. In addition, they also examined potassium sulphate, K_2SO_4. Yet another electron diffraction analysis of caesium sulphate by Ugarov, Ezhov and Rambidi (1975) appeared later along with a similar study on thallium sulphate, Tl_2SO_4 (Ugarov, Ezhov and Rambidi, 1973; 1975; Petrov, Ugarov and Rambidi, 1980). The choice of the thallium derivative was especially advantageous from the point of view of determining the position of the metal atom since thallium is quite a heavy element and thus makes a relatively large contribution to the total electron scattering. The accurate structural determination of these systems is usually hindered by the large-amplitude intramolecular motion as a consequence of which the contribution of the light atoms is quickly "washed out" with the increasing scattering angle.

The most recent structural studies on the sulphate molecules established that the SO_4 groups have a regular tetrahedral configuration within the experimental error. The metal atoms are located on axes perpendicular to the edges of the SO_4 tetrahedron as indicated in Fig. 102. These structures are often called "bicyclic" as they consist of two four-membered rings with their planes perpendicular to each other. The molecular parameters including some mean amplitudes of vibration are given in Table 98. The sulphur–oxygen bond lengths are considerably smaller than that of a single bond and these vapour-phase linkages should rather be considered as double bonds.

According to the recent results, it can be concluded that the sulphate structures in the vapour phase and those of the corresponding acids are different. The molecular geometry of sulphuric acid has been studied in all three phases. Two different sulphur–oxygen bonds were well established. Their lengths are given in Table 98(b) together with the bond angles determined in gas and crystal. Two models could fit the data in the liquid as far as the bond angles were concerned. The two different bond lengths are less different in crystal than in gas and the bond angles are also nearer to each other as well as to the ideal tetrahedral in crystal than in gas. The structural differences in gas and crystal have been attributed to the consequences of intermolecular hydrogen bond formation in the crystal.

In spite of the substantially different bond lengths and bond angles, the three different types of non-bonded O. . . O distances are very similar and are so in both gas and crystal:

	Gas	Crystal
=O. . .O=	250 pm	254 pm
=O. . .O—	243 pm	240 pm
—O. . .O—	243 pm	242 pm

Thus the four oxygen atoms seem to arrange a nearly regular tetrahedron.

One of the interesting structural peculiarities of the salts of oxoacids is the large vibrational amplitude associated with the metal–oxygen bonds (*cf.* Table 98). Both the lengths and the amplitudes indicate that the interaction between the metal atom and the remainder of the molecule is rather weak.

As in the molten salts (Ukshe, 1966), also in crystalline sulphates (*cf.* Banister, Moore and Padley, 1968), the molecular structures are similar to those of the vapour-phase species.

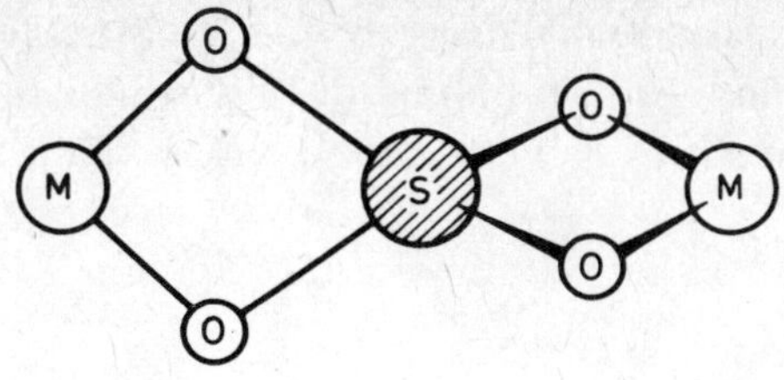

Fig. 102. Bicyclic model of M_2SO_4 sulphate molecules

TABLE 98

(a) Geometrical and vibrational parameters of M_2SO_4 sulphate molecules from high-temperature electron diffraction

Bond lengths (pm) and bond angles (°)					
Compound		*r* (S—O)	∢O—S—O	*r* (M—O)	∢O—M—O
K_2SO_4	*(a)*	147(1)	109(10)	245(3)	59(5)
Cs_2SO_4	*(b)*	148(1)	109(10)	260(3)	55(5)
Tl_2SO_4	*(c)*	146.3(9)	[109.5]	245(2)	58(2)

Mean amplitudes of vibration (pm)						
Compound		*l* (S—O)	*l* (M—O)	*l* (M...S)	*l* (M...O)	*l* (M...M)
K_2SO_4	*(a)*		15			
Cs_2SO_4	*(b)*	6(2)	34(5)	15(3)		
Tl_2SO_4	*(c)*	3(2)	17(2)	12(2)	31(5)	30(5)

(a) 1300–1500 °C; Spiridonov and Lutoshkin (1970)
(b) 1100 °C; Ugarov, Ezhov and Rambidi (1975)
(c) 700 °C; Petrov, Ugarov and Rambidi (1980)

(b) Geometrical parameters of sulphuric acid in gas, liquid and crystal

Parameter	Gas *(a)*	Liquid *(b)*	Crystal *(c)*
r(S=O), pm	142.2(10)	141.7(14)	142.6(15)
r(S—O), pm	157.4(10)	155.0(19)	153.5(15)
∢O=S=O (°)	123.3(10)		118.6(1)
∢O=S—O (°)	107.5		108.3
∢O—S—O (°)	101.3(10)		104(1)

(a) Microwave spectroscopy, Kuczkowski, Suenram and Lovas (1981)
(b) x-ray diffraction, Weidner, Geisenfelder and Zimmermann (1971); Two models could fit the data for the bond angles
(c) x-ray diffraction, Pascard-Billy (1965)

As a general characterization of the structures of sulphates as well as M_2XO_4 salts of other oxoacids Szpiridonov (1972) proposed the following structural formula:

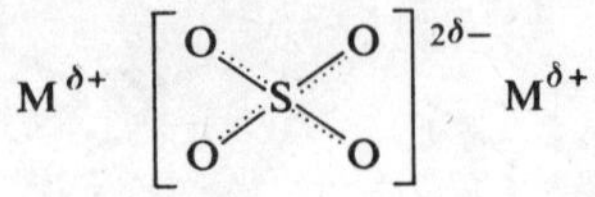

As in the classical structural formula, the central sulphur atom is hexavalent, having four bonds with a bond order of 1.5. The merits of such an approach can be summarized following the discussions by Szpiridonov (1972) and Rambidi (1975a; 1975b):

(i) it reflects the autonomous character of the SO_4 group;
(ii) it indicates that the SO_4 group remains unchanged in transition from one phase into another;
(iii) it shows the equality of the sulphur–oxygen bonds;
(iv) it gives emphasis to the rather loose character of the metal–oxygen linkages.

Chapter 5

MOLECULES WITH FIVE- AND SIX-COORDINATED SULPHUR

5.1. THIONYL TETRAFLUORIDE

Thionyl tetrafluoride, OSF_4, has analogous structure to that of sulphur tetrafluoride with the S═O bond in the place of the sulphur lone pair of electrons. There have been unexpected turns in the history of the elucidation of the structure of this interesting molecule, the only one with five-coordinated sulphur whose structure has been so far determined.

The C_{2v} symmetry and trigonal bipyramidal configuration with equatorial S═O bond of thionyl tetrafluoride have already been established by a liquid-phase vibrational spectroscopic investigation (Goggin, Roberts and Woodward, 1961). These findings could be consistent with both the structures presented below:

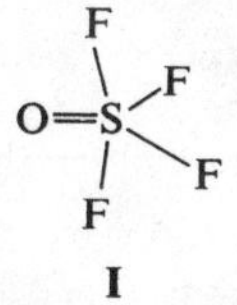

I

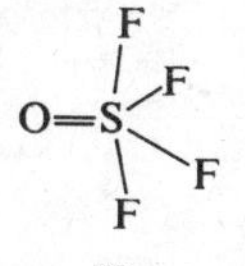

II

In fact structure **I** was reported as a result of the first electron diffraction investigation of thionyl tetrafluoride (Kimura and Bauer, 1963). It was also established that the axial sulphur–fluorine bonds are longer than the equatorial. There were some doubts however about the choice of structure **I** over **II** (*cf.* Bartell, 1966; Cruickshank, 1968) since the distortion of this structure from a more regular model was in a direction contrary to what would be expected from the VSEPR model. If the electron pair repulsions are considered, the axial S—F bonds would be expected to move *away from* the S═O bond rather than to turn in its direction.* Eventually it has been noticed that the electron diffraction radial distribution (*cf.* Fig. 103) could also be interpreted in a different way, and structure **II** was also found to be consistent with the experimental

* Similarly, in the originally reported structure the O═S—F_{eq} angle was smaller than 120° and, accordingly, the F_{eq}—S—F_{eq} angle was larger, again contradicting the VSEPR model.

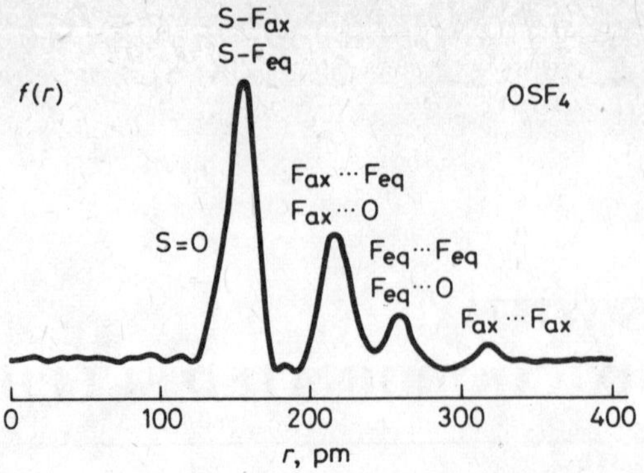

Fig. 103. Radial distribution of thionyl tetrafluoride and the assignment of internuclear distances after the electron diffraction study by Gundersen and Hedberg (1969)

TABLE 99
Four sets of geometrical parameters of thionyl tetrafluoride from the electron diffraction investigation by Gundersen and Hedberg (1969)

Parameter	*A*	*B*	*C*	*D*
r(S=O), pm	140.2(3)	140.3(3)	140.6(4)	140.5(4)
$r(\text{S—F})_{eq}$, pm	154.8(4)	155.2(4)	153.5(4)	153.3(4)
$r(\text{S—F})_{ax}$, pm	157.8(4)	157.5(4)	159.3(4)	159.6(4)
∢F_{eq}—S—F_{eq} (°)	117.9(22)	110.2(18)	118.7(33)	114.9(34)
∢F_{ax}—S—F_{ax} (°)	178.8	178.7	163.9	164.5
∢F_{eq}—S=O (°)	121.0(11)	124.9(9)	120.6(17)	122.5(17)
∢F_{ax}—S=O (°)	90.6(4)	90.6(4)	98.0(3)	97.8(3)
∢F_{eq}—S—F_{ax} (°)	89.7(2)	89.6(2)	85.9(2)	85.8(2)

data. The bond lengths and angles, actually two different sets of them, were then refined for structure **II** using the original measurements (Hencher, Cruickshank and Bauer, 1968). In the meantime an independent study has been initiated by Gundersen and Hedberg (1969) who found altogether four sets of parameters each one of which corresponded to a minimum in the least-squares refinement and had the overall configuration of structure **II**. These four sets of parameters differed in the relative magnitudes of the $F_{eq} \ldots F_{eq}$ and $F_{eq} \ldots O$, and the $F_{ax} \ldots F_{eq}$ and $F_{ax} \ldots O$ non-bond distances. Two of the four models corresponded to the two parameter sets previously communicated by Hencher, Cruickshank and Bauer (1968). Some characteristic parameters of the four models are shown in Table 99. Gundersen and Hedberg (1969) found no basis on which any of the four structures could be easily eliminated. They reasoned that the distortion of the F_{ax}—S=O angle from the ideal 90° is expected to be smaller than that of the F_{eq}—S=O angle from the ideal 120°. The repulsion between the axial and equatorial S—F bonds could be expected to grow more quickly with the decrease of the F_{ax}—S=O angle than the repulsion between the equatorial S—F bonds accompanying the decrease of the F_{eq}—S—F_{eq} angle. This renders models *A* and *B* to be preferred over models *C* and *D*. Finally Gundersen and Hedberg (1969) slightly favoured *B* over *A* using considerations on electron pair repulsions.

Gundersen and Hedberg's (1969) tentative choice of course originated from somewhat overextended inferences of the VSEPR model. It can be shown that the models *C* and especially *D* may be favoured over *A* and *B* from similar qualitative reasoning also based on the VSEPR model, but using the so-called average angles and considerations of atom–atom non-bonded interactions (Hargittai, 1979c).

Following the *ab initio* calculations of Schmiedekamp, Cruickshank, Skaarup, Pulay, Hargittai and Boggs (1979) on SO_2X_2, SOX_2E, SX_2E_2 molecules, it was observed that the triple average angle of the S=O and that of the S—E "bond" (E is the lone pair of electrons of sulphur), differ very little in the various molecules, the former being only smaller by some 1° than the latter.

Consider now the quadruple average angle of S—E in the SF_4 molecule*, *viz.* 111.3°

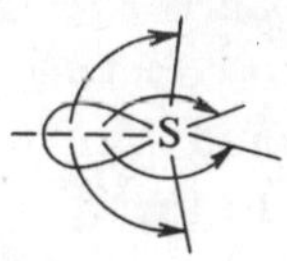

$\tilde{\alpha}_E$ 111.3°

which is smaller than the triple average angle of the sulphur lone pair in the sulphone–sulphoxide–sulphide series, *viz.* 114.2–114.8°. The decrease is apparently due to the sulphur being five-coordinated in SF_4 (or rather SF_4E) *vs.* the four-coordination in the series mentioned. Since the multiple character of the bonds has also to be considered, it is difficult to compare the electron pair repulsions in molecule pairs like, e.g. SOF_2 and SF_4. However, the atom–atom non-bonded interactions are definitely expected to be stronger in the latter because of the greater coordination number. If the triple average angles in the sulphone–sulphoxide–sulphide series are very similar, then the quadruple average angles in SF_4 and OSF_4 are expected to be even less different. The length of the S=O bond in OSF_4 also indicates that the space requirement of this bond is probably not smaller than that in the sulphone, or especially in the sulphoxide, as compared with the space requirement of the respective lone pairs. The four models of OSF_4 correspond to the following quadruple average angles:

	A	*B*	*C*	*D*
$\tilde{\alpha}_{S=O}$	105.8°	107.8°	109.3°	110.2°

Thus according to the above reasoning, model *D* is the most likely model.

Consider now the non-bond distances in these models and in particular those which are the shortest:

* Fortunately, the quadruple average angle of S—E in SF_4 (Tolles and Gwinn, 1962), as well as that of S=O in OSF_4, can be calculated from the bond angles by virtue of symmetry. The quadruple average angle for trigonal bipyramidal configurations has been introduced by Hargittai (1979c).

	A	B	C	D
$r(F_{ax}\ldots O)$, pm	212.2	212.1	226.7	226.5
$r(F_{ax}\ldots F_{eq})$, pm	220.5	220.4	213.3	213.1

The 1,3 non-bonded radii of fluorine and oxygen were postulated to be 108 and 113 pm, respectively (*cf.* Table 2). Thus the shortest F...O and F...F distances are expected to be 221 and 216 pm, respectively. On this basis models *C* and *D* seem to be also more plausible than models *A* and *B*.

The four models of thionyl tetrafluoride have been tested by *ab initio* calculations (Oberhammer and Boggs, 1979b) showing model *D* to be the most stable. Earlier, Christe, Schack and Curtis (1977) favoured model *D* (or *C*) following a vibrational spectroscopic investigation of thionyl tetrafluoride. Their conclusions were based on the large difference between the equatorial and axial SF_2 stretching force constants. Further refinement of the thionyl tetrafluoride force field is expected (Christe, 1979) as the assignment of its spectrum had to be modified following a vibrational spectroscopic reinvestigation of sulphur tetrafluoride (Christe, Willner and Sawodny, 1979).

A limited amount of microwave spectroscopic information on thionyl tetrafluoride is also available (Murty, 1965; Murty and Mohanty, 1971). The following three rotational constants and corresponding to them principal moments of inertia have been determined:

A	4095.97 MHz	I_a	123.38×10^4 amu pm^2
B	3358.98 MHz	I_b	150.46×10^4 amu pm^2
C	3240.10 MHz	I_c	155.98×10^4 amu pm^2

Thus two of the five geometrical parameters describing the thionyl tetrafluoride structure have to be assumed in order to calculate the other three from the microwave spectroscopic data. Murty and Mohanty (1971) assumed the axial and equatorial S—F bond lengths to be 162.8 and 153.5 pm, respectively and obtained the following parameters: r(S═O) 138.6(10) pm, ∢F_{ax}—S—F_{ax} 145(4) and ∢F_{eq}—S—F_{eq} 116(1)°. They have produced other series of parameters also consistent with the experimental data. Considering those parameter sets that contain an S═O bond length similar to the well-determined electron diffraction r(S═O) value, the results of Murty and Mohanty (1971) are nearest to model *D* of the four parameter sets found by Gundersen and Hedberg (1969).

The $F_{eq}\ldots F_{eq}$ and $F_{ax}\ldots F_{ax}$ distances, however, can be directly deduced from the microwave spectroscopic data, using the expressions

$$I_a + I_c - I_b = m_F r_0^2(F_{eq}\ldots F_{eq})$$

and

$$I_b + I_c - I_a = m_F r_0^2(F_{ax}\ldots F_{ax})$$

The resulting $r_0(F_{eq}\ldots F_{eq})$ 260.5 and $r_0(F_{ax}\ldots F_{ax})$ 310.4 pm can then be used together with the electron diffraction bond lengths to calculate the bond angles. Using the bond lengths of models *B* and *D* of Table 99, the following angles (°) are obtained:

	B	D
$\measuredangle F_{ax}$—S—F_{ax}	160.4	153.0
$\measuredangle F_{eq}$—S—F_{eq}	114.1	116.3
$\measuredangle F_{ax}$—S—F_{eq}	84.7	82.9
$\measuredangle F_{ax}$—S=O	99.8	103.5
$\measuredangle F_{eq}$—S=O	122.9	121.85

These two sets of bond angles correspond to the quadruple average angles 111.35 (*B*) and 112.7° (*D*). This quadruple average angle is rather uncertain as it depends on the particular choice of the bond lengths. It is qualitatively consistent with the quadruple average angle of sulphur tetrafluoride, although the quadruple average angle made by the S=O bond could really be expected to be somewhat smaller. Finally a disturbing circumstance is mentioned. It was seen above that the $F_{ax}\dots F_{ax}$ distance was obtained directly from the microwave spectroscopic data to be 310.4 pm. On the other hand, the electron diffraction models *A* through *D* all have 315–316 pm for this distance. Moreover, its contribution is indeed well resolved on the experimental radial distribution at around 316 pm (see Fig. 103). We have no explanation for the 6 pm large difference. Apparently the authors of the electron diffraction and microwave spectroscopic papers were not aware of each other's work and thus could not comment on this discrepancy. Oberhammer and Boggs' (1979b) calculated geometry corresponds to a 312 pm $F_{ax}\dots F_{ax}$ distance. The new theoretical calculations (Oberhammer and Boggs, 1979b) and considerations (Hargittai, 1979c) prompted a new analysis of the electron diffraction data of thionyl tetrafluoride (Hedberg, 1979) utilizing the results of a recent microwave spectroscopic study (*cf.* Graybeal, 1976).

The knowledge of an accurate force field and reliable mean amplitudes of vibration for thionyl tetrafluoride could have greatly facilitated the electron diffraction structure analysis. Mean vibrational amplitudes from two independent electron diffraction studies are cited in Table 100 together with those calculated by Cyvin (1970). Earlier

TABLE 100
Mean amplitudes of vibration (pm) of thionyl tetrafluoride

Atomic pair	Electron diffraction			Calculated
	(a)	*(b)*	*(c)*	*(d)*
S=O	3.0(3)	3.7(5)	3.5(9)	3.4
S—F_{eq}	4.5(3)	5.4(3)	4.3(6)	4.0
S—F_{ax}	5.7(3)		4.4(6)	4.2
$F_{eq}\dots F_{eq}$	5.3(6)	4.7(12)	6.9(68)	6.9
$F_{ax}\dots F_{ax}$	6.2(5)	6.1(12)	6.1(12)	5.0
$F_{eq}\dots F_{ax}$	11.2(3)	9.6(7)	6.0(5)	8.5
$F_{eq}\dots$O	8.7(9)	7.6(19)	8.4(43)	6.7
$F_{ax}\dots$O	5.7(3)	4.3(6)	6.1(9)	5.8

(a) Hencher, Cruickshank and Bauer (1968)
(b) and *(c)* for models *B* and *D*, respectively, *cf.* Table 99, Gundersen and Hedberg (1969)
(d) $T = 298$ K, Cyvin (1970)

calculations are also known (Sathianandan, Ramaswamy, Sundaram and Cleveland, 1964).

Methylene sulphur tetrafluoride

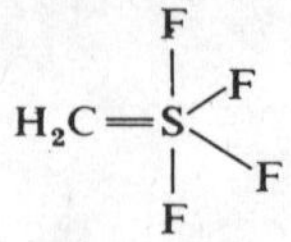

has been recently synthesized (Kleemann and Seppelt, 1978) and structural studies on this molecule have been initiated. According to the preliminary experimental results and *ab initio* calculations (Oberhammer and Boggs, 1979b) the equatorial F—S—F bond angle of methylene sulphur tetrafluoride is considerably smaller than that of thionyl tetrafluoride. At the same time the quadruple average angle of the S=C bond is 113° showing thus a mere 2° increase as compared with the quadruple average angle of the sulphur lone pair in sulphur tetrafluoride. It is noteworthy that the S=C bond of H_2CSF_4 is very short pointing to a relatively large space requirement of this bond in the valence shell of the central sulphur atom.

With respect to the structural variations of trigonal bipyramidal molecules, the structures of SF_4 and $(CF_3)_2SF_2$ are considered here again. Oberhammer, Kumar, Knerr and Shreeve (1981) stated: "The decrease of the equatorial bond angle from 101.5(5)° in SF_4 to 97.3(8)° in $(CF_3)_2SF_2$ is incompatible with the VSEPR model."

This is, however, only an *apparent* failure of the VSEPR model originating from the fact that only the variations of the bond angles have been looked at rather than the variations of all angles made by bonding and non-bonding pairs. This happens often and naturally so since the bond angles are usually the only parameters directly available from the experiment.

It is, however, the basic postulate of the VSEPR model that in determining the bond configuration around a central atom, the mutual repulsions of *all* electron pairs in its valence shell must be taken into account. That obviously includes the lone pairs of electrons. Similarly, when examining the compatibility of a structure with the VSEPR model, *all angles characterizing the electron pair configuration in the valence shell of the central atom should be considered* (Hargittai, 1982).

Both in SF_4 and in $(CF_3)_2SF_2$ there are two kinds of interactions that have to be taken into account, *viz.* E/b and b/b where b is a bonding pair. The only difference in the two systems is that the two ligands in the equatorial positions are less electronegative in $(CF_3)_2SF_2$ than the corresponding ones in SF_4.

Consider first the quadruple average angles (Hargittai, 1979c) of the lone pairs:

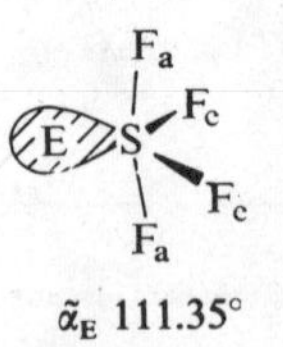

$\tilde{\alpha}_E$ 111.35°

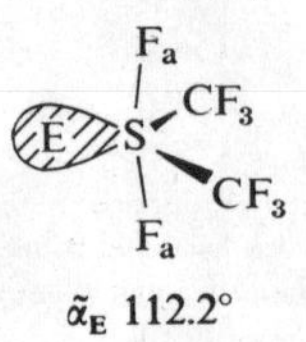

$\tilde{\alpha}_E$ 112.2°

Although the difference in the two quadruple average angles is small, its direction is in complete agreement with the prediction of the VSEPR model postulating the E/b repulsions to be greater when involving bonds to less electronegative ligands. Here the E/S—F_a repulsions in the two molecules are assumed to be equal.

Consider now the individual bond angles:

ESF_4		$E(CF_3)_2SF_2$		Comment
E—S—F_a	93.4$_5$°	E—S—F_a	93.0$_5$°	Equal within experimental error
E—S—F_e	129.2$_5$°	E—S—C	131.3$_5$°	$\Delta_1 = 2.1°$. Sign in agreement with the prediction of VSEPR model. Two such interactions
F_e—S—F_e	101.5°	C—S—C	97.3°	$\Delta_2 = -4.2°$. Sign in disagreement with the prediction of VSEPR model. One such interaction
F_a—S—F_e	87.8°	F_a—S—C	88.0°	Equal within experimental error

It is obvious that $2 \times |\Delta_1|$ and $|\Delta_2|$ are equal. It is also obvious that there are two competing effects acting in the equatorial plane of $(CF_3)_2SF_2$ as compared with SF_4. One of them is that the E/S—CF_3 repulsion is greater than the E/S—F_e one, thus tending to increase thc E—S—C angles. The other effect is that the S—CF_3/S—CF_3 repulsion is greater than the S—F_e/S—F_e one, thus tending to increase the C—S—C angle.

All angles being in the equatorial plane, either the E—S—C or the C—S—C angle can increase and can do so only at the expense of the other. The relative magnitude of the two competing effects decides which of them prevails. The experimentally observed decrease of the C—S—C angle in $(CF_3)_2SF_2$ as compared with the F_e—S—F_e angle in SF_4 unambiguously indicates the E/S—CF_3 repulsions prevailing, in complete agreement with the VSEPR model (Hargittai, 1982).

5.2. SULPHUR HEXAFLUORIDE AND ITS DERIVATIVES

Sulphur hexafluoride, SF_6. Its regular octahedral (O_h symmetry) geometry is characterized by a 156.1(2) pm sulphur–fluorine bond distance as determined in a recent electron diffraction reinvestigation (Bartell and Doun, 1978). The observed mean amplitudes of vibration are in excellent agreement with the calculated ones,

	Observed by Bartell and Doun (1978)	Calculated by Cyvin, Brunvoll and Müller (1968)
l(S—F), pm	4.4(1)	4.2
l(F...F) short, pm	6.1(2)	6.0
l(F...F) long, pm	5.6(6)	5.3

The spectroscopic calculations cited were preceded by a detailed examination of various vibrational effects and of the molecular force field (Meisingseth and Cyvin, 1962; Bye and Cyvin, 1963; Cyvin and Meisingseth, 1963).

TABLE 101
Bond lengths (pm) in SF_6 and derivatives from electron diffraction

Bond length	SF_6 *(a)*	SF_5Cl *(b)*	SF_5OSF_5 *(c)*	SF_5NF_2 *(d)*
$r(S—F)_{mean}$	156.1(2)	157.0(1)	156.0(4)	154.9(6)
$r(S—F)_{eq}$		156.6(3)	155.8(8)	154.6(6)
$r(S—F)_{ax}$		158.8(9)	157.2(34)	156.0(21)
$r(S—Cl)$		204.7(3)		
$r(S—O)$			158.6(11)	
$r(S—N)$				169.8(15)

(a) and *(b)* r_g, *(c)* and *(d)* probably r_a
(a) Bartell and Doun (1978)
(b) Marsden and Bartell (1976)
(c) Oberhammer and Seppelt (1978)
(d) Parameters from Haase, Oberhammer, Zeil, Glemser and Mews (1971), uncertainties from Landolt–Börnstein (1976)

Sulphur chloride pentafluoride, SF_5Cl. Its molecular structure has been determined by an electron diffraction investigation (Marsden and Bartell, 1976) which utilized the available microwave spectroscopic data (Kewley, Murty and Sugden, 1962). The bond lengths are shown in Table 101 together with those characterizing the sulphur bond configuration of F_5SOSF_5 and F_5SNF_2. The data on the latter seem to be rather uncertain (Haase, Oberhammer, Zeil, Glemser and Mews, 1971).

Bis(pentafluorosulphur)oxide, F_5SOSF_5. Its electron diffraction investigation by Oberhammer and Seppelt (1978) was paralleled by similar studies on the selenium and tellurium analogues, $F_5SeOSeF_5$ and $F_5TeOTeF_5$.

The equatorial sulphur–fluorine bonds turn somewhat away from the substituent both in sulphur chloride pentafluoride and bis(pentafluorosulphur) oxide, as shown by the respective bond angles:

SF_5Cl	$\measuredangle Cl—S—F_{eq}$	90.7(2)°
F_5SOSF_5	$\measuredangle O{=}S—F_{eq}$	92.1(9)°

The early electron diffraction studies of the molecular structures of SF_5OF (Crawford, Dudley and Hedberg, 1959) and F_5SOOSF_5 (Harvey and Bauer, 1954) are only mentioned here.

Selenium hexafluoride, SeF_6, has also been investigated by electron diffraction (Ewing and Sutton, 1963) yielding 168.9 pm for the Se—F distance. *Tellurium hexafluoride*, TeF_6, has been examined twice by electron diffraction (Seip and Stølevik, 1966; Gundersen, Hedberg and Strand, 1978). Experimental data were collected at three different temperatures. The Te—F bond was found to be 181.5(2) pm (r_a, 20°C).

Twelve isotopic species of *tellurium chloride pentafluoride*, TeF_5Cl, were subjected to microwave spectroscopic analysis by Legon (1973). This study yielded rich information mainly over the Te—Cl bond since the fluor atom has no isotopes. Supposing that the molecule has C_{4v} symmetry and assuming the equatorial and axial tellurium–fluorine bonds to be equal, the following parameters were determined:

$$r(\text{Te—F}) \quad 183.05(12)\ \text{pm} \quad \text{and} \quad \measuredangle \text{F}_{eq}\text{—Te—F}_{eq}\ 88°\ 15' \pm 7'$$

Some of the geometrical parameters of *bis(pentafluoroselenium)oxide* and *bis(pentafluorotellurium)oxide* are as follows (Oberhammer and Seppelt, 1978):

	$F_5SeOSeF_5$	$F_5TeOTeF_5$	
r(Se—F_{eq})	168.3(9) pm	182.0(10) pm	r(Te—F_{eq})
r(Se—F_{ax})	166.5(31) pm	179.9(47) pm	r(Te—F_{ax})
∡O—Se—F_{eq}	91.1(8)°	90.2(9)°	∡O—Te—F_{eq}

The lengths of the axial and equatorial bonds show interesting variations in the monosubstituted derivatives of sulphur hexafluoride and also in their selenium and tellurium analogues. The sulphur–fluorine bonds are longer in sulphur chloride pentafluoride than in sulphur hexafluoride (*cf.* Table 101). According to the VSEPR model the substitution of one of the fluorines by chlorine increases the repulsions in the valence shell of the central sulphur atom as the less electronegative chlorine ligand draws away less electron density from the vicinity of the central atom. As a result of increased repulsions, the equatorial S—F bonds lengthen. This is indeed observed but, what is more interesting, the axial S—F bond gets even longer. This was ascribed by Bartell and Doun (1978) to a "secondary effect" of relaxation caused by the repulsion of the equatorial bonds. This is illustrated by Fig. 104. The opening of the Cl—S—F_{eq} bond angle as compared with the ideal 90° means enhanced repulsion for the axial bonds. Brooks, Eshaque, Lau and Passmore (1976) calculated the changes in force

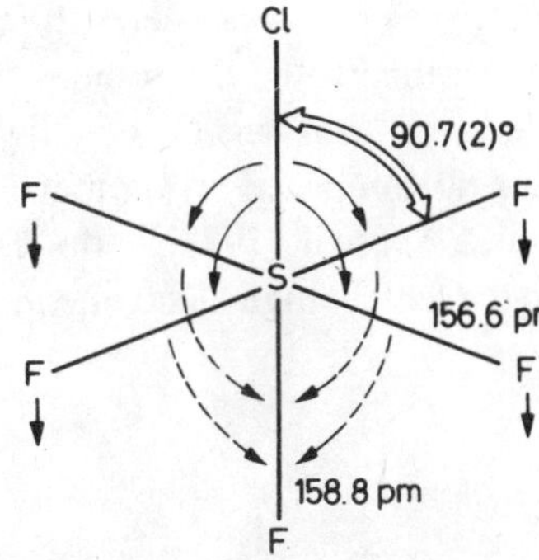

Fig. 104. Bond configuration and geometrical parameters of sulphur chloride pentafluoride from the electron diffraction investigation by Marsden and Bartell (1976). The primary effects are indicated by full line arrows, while dashed line arrows show the "secondary effects" of relaxation

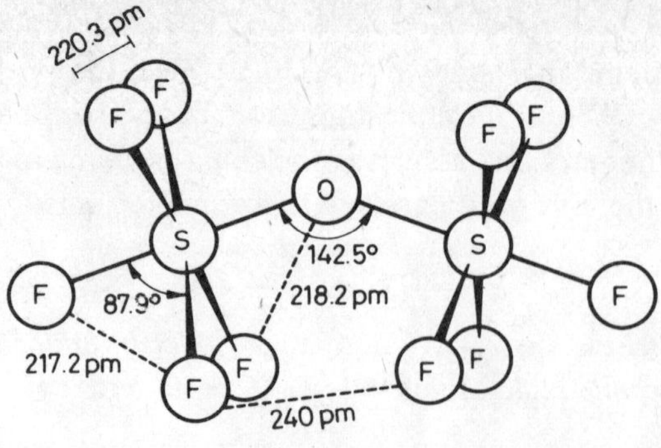

Fig. 105. Molecular model, bond angles and non-bond distances of F_5SOSF_5 after Oberhammer and Seppelt (1978)

constants for the $SF_6 \rightarrow SF_5Cl$ transition showing the diminishing bond strength. However, they gave larger force constants for the axial bonds than for the equatorial S—F bonds. According to their calculations, the Se—F and Te—F bonds show analogous variations. The force field of SF_5Cl is discussed by Marsden and Bartell (1976) and Brooks, Eshaque, Lau and Passmore (1976).

The Se—F and Te—F bonds in the F_5XOXF_5 series apparently follow a tendency similar to the one observed by Brooks, Eshaque, Lau and Passmore (1976). In the structure of the sulphur hexafluoride derivatives, however, the fluorine–fluorine non-bonded repulsions may be of great importance. For illustrating this point, some non-bond distances are indicated in Fig. 105. The sums of 1,3 non-bonded radii are noted here for comparison $2r(1,3)_F = 216$ pm, $r(1,3)_F + r(1,3)_O = 221$ pm (*cf.* Table 2). Since the observed distances are similar or even smaller, the atom–atom non-bonded interactions cannot be ignored in discussing the geometrical peculiarities of the sulphur hexafluoride derivatives. These interactions probably play a diminishing role in the structure of $F_5TeOTeF_5$. In this case it is to be expected that the electronegativity of the $OTeF_5$ ligand as compared with fluorine is decisive since $F_5TeOTeF_5$ can be considered to be a tellurium hexafluoride derivative. Since the tellurium bond configurations have been determined to be the same within experimental error in the two molecules, it may be deduced that the $OTeF_5$ ligand has an electronegativity similar to that of fluorine. This notion based on geometrical evidence is consistent with the observations of Lentz and Seppelt (1978), who on the basis of chemical and spectroscopic data suggested extremely high electronegativity for the $OTeF_5$ ligand (as well as for $OSeF_5$).

BIBLIOGRAPHY

Abbar, C. (1963): *Compt. Rend. Acad. Sci. Paris* **257,** 2022.

Abbar, C. (1965): *Compt. Rend. Acad. Sci. Paris* **261,** 365.

Abbar, C., Journel, G. and Moise, A. (1965): *Compt. Rend. Acad. Sci. Paris* **261,** 5047.

Abrahams, S. C. (1957): *Acta Crystallogr.* **10,** 417.

Absar, I. and Van Wazer, J. R. (1978): *Angew. Chem. Int. Ed. Engl.* **17,** 80.

Acha, L., Cromie, E. R. and Rankin, D. W. H. (1981): *J. Mol. Struct.* **73,** 111.

Adams, W. J. and Bartell, L. S. (1977): *J. Mol. Struct.* **37,** 261.

Adams, W. J., Geise, H. J. and Bartell, L. S. (1970): *J. Amer. Chem. Soc.* **92,** 5013.

Airey, W., Glidewell, C., Rankin, D. W. H., Robiette, A. G., Sheldrick, G. M. and Cruickshank, D. W. J. (1970): *Trans. Faraday Soc.* **66,** 551.

Airey, W., Glidewell, C., Robiette, A. G. and Sheldrick, G. M. (1971a): *J. Mol. Struct.* **8,** 413.

Airey, W., Glidewell, C., Robiette, A. G. and Sheldrick, G. M. (1971b): *J. Mol. Struct.* **8,** 435.

Airey, W., Glidewell, C., Robiette, A. G., Sheldrick, G. M. and Freeman, J. M. (1971): *J. Mol. Struct.* **8,** 423.

Akishin, P. A., Rambidi, N. G. and Ezhov, Yu. S. (1960): *Zh. Neorg. Khim.* **5,** 747.

Akishin, P. A. and Spiridonov, V. P. (1959): *Dokl. Akad. Nauk SSSR* **129,** 1317.

Akishin, P. A. and Spiridonov, V. P. (1960): *Dokl. Akad. Nauk SSSR* **131,** 557.

Akishin, P. A., Spiridonov, V. P. and Mishulina, R. A. (1960): *Vestnik Mosk. Univ. Khim.* **17,** 23.

Alekseev, N. V. (1967): *Zh. Strukt. Khim.* **8,** 532.

Alekseev, N. V. and Velichko, F. K. (1967): *Zh. Strukt. Khim.* **8,** 8.

Alléaume, M. and Decap, J. (1965a): *Acta Crystallogr.* **18,** 731.

Alléaume, M. and Decap, J. (1965b): *Acta Crystallogr.* **19,** 934.

Alléaume, M. and Decap, J. (1968): *Acta Crystallogr.* **B24,** 214.

Allred, A. L. and Rochow, E. G. (1958): *J. Inorg. Nucl. Chem.* **5,** 264.

Almenningen, A., Bastiansen, O., Ewing, V., Hedberg, K. and Traetteberg, M. (1963): *Acta Chem. Scand.* **17,** 2455.

Almenningen, A., Bastiansen, O. and Svendås, P. (1958): *Acta Chem. Scand.* **12,** 1671.

Almenningen, A., Fernholt, L. and Kveseth, K. (1977): *Acta Chem. Scand.* **A31,** 297.

Almenningen, A., Fernholt, L. and Seip, H. M. (1968): *Acta Chem. Scand.* **22,** 51.

Almenningen, A., Fernholt, L., Seip., H. M. and Henriksen, L. (1974): *Acta Chem. Scand.* **A28,** 1037.

Almenningen, A., Hargittai, I., Kloster-Jensen, E. and Stølevik, R. (1970): *Acta Chem. Scand.* **24,** 3463.

Almenningen, A., Hedberg, K. and Seip, R. (1963): *Acta Chem. Scand.* **17,** 2264.

Almenningen, A., Seip, H. M. and Vassbotn, P. (1973): *Acta Chem. Scand.* **27,** 21.

Almenningen, A., Seip, H. M. and Willadsen, T. (1969): *Acta Chem. Scand.* **23,** 2748.

Almond, V., Charles, S. W., Macdonald, J. N. and Owen, N. L. (1977): *J. C. S. Chem. Commun.* 483.

Amano, T., Hirota, E. and Morino, Y. (1967): *J. Phys. Soc. Japan* **22,** 399.

Anderson, D. W. W., Rankin, D. W. H. and Robertson, A. (1972): *J. Mol. Struct.* **14,** 385.

Arbuzov, B. A., Naumov, V. A. and Anonimova, I. V. (1970): *Dokl. Akad. Nauk SSSR* **192,** 327.

Arbuzov, B. A., Naumov, V. A., Zaripov, N. M. and Pronicheva, L. D. (1970): *Dokl. Akad. Nauk SSSR* **195,** 1333.

Arbuzov, B. A., Remizov, A. B. and Nuretdinova, O. N. (1973): *Zh. Prikl. Spektr.* **19,** 733.

Arcoria, A., Maccarone, E., Musumarra, G. and Tomaselli, G. A. (1974): *Spectrochim. Acta* **30A,** 611.

Arnold, D. E. L. and Rankin, D. W. H. (1972/1973): *J. Fluorine Chem.* **2,** 405.

Arnold, W. and Rogowski, F. (1965): *Z. Naturforsch.* **20b,** 806.

Auberg, E., Samdal, S. and Seip, H. M. (1979): *J. Mol. Struct.* **57,** 95.

Averbukh, B. S., Mayants, L. S. and Shaltuper, G. B. (1969): *J. Mol. Spectrosc.* **30,** 310.

Azrak, R. G. and Wilson, E. B. (1970): *J. Chem. Phys.* **52,** 5299.

Bak, B., Christensen, D., Rastrup-Andersen, J. and Tannenbaum. E. (1956): *J. Chem. Phys.* **25,** 892.

Bak, B., Christensen, D., Hansen-Nygaard, L. and Rastrup-Andersen, J. (1961): *J. Mol. Spectrosc.* **7,** 58.

Bak, B., Nielsen, O. J. and Svanholt, H. (1978): *J. Mol. Spectrosc.* **69,** 401.

Bak, B., Nielsen, O., Svanholt, H., Almenningen, A., Bastiansen, O., Fernholt, L., Gundersen, G., Nielsen, C. J., Cyvin, B. N. and Cyvin, S. J. (1978): *Acta Chem. Scand.* **A32,** 1005.

Bak, B., Nielsen, O., Svanholt, H. and Christiansen, J. J. (1979): *J. Mol. Spectrosc.* **75,** 134.

Bak, B. and Skaarup, S. (1971): *J. Mol. Struct.* **10,** 385.

Bak, B., Svanholt, H. and Larsen, C. (1977): *J. Mol. Struct.* **36,** 55.

Banister, A. J., Moore, L. F. and Padley, J. S. (1968): *Inorganic Sulphur Chemistry* (Ed.: Nickless, G.). Elsevier, Amsterdam.

Bartell, L. S. (1960): *J. Chem. Phys.* **32,** 827.

Bartell, L. S. (1962): *Tetrahedron* **17,** 177.

Bartell, L. S. (1966): *Trans. Amer. Crystallogr. Assoc.* **2,** 134.

Bartell, L. S. (1968): *J. Chem. Educ.* **45,** 754.

Bartell, L. S. and Andersen, B. (1973): *J. C. S. Chem. Commun.* 786.

Bartell, L. S. and Brockway, L. O. (1960): *J. Chem. Phys.* **32,** 512.

Bartell, L. S. and Doun, S. K. (1978): *J. Mol. Struct.* **43,** 245.

Bartell, L. S. and Hirst, L. C. (1959): *J. Chem. Phys.* **31,** 449.

Bartlett, N., Wechsberg, M., Jones, G. R. and Burbank, R. D. (1972): *Inorg. Chem.* **11,** 1124.

Barzdain, P. P. and Alekseev, N. V. (1968): *Zh. Strukt. Khim.* **9.** 520.

Bastiansen, O., Fernholt, L., Seip, H. M., Kambara, H. and Kuchitsu, K. (1973): *J. Mol. Struct.* **18,** 163.

Bastiansen, O., Fritsch, F. N. and Hedberg, K. (1964): *Acta Crystallogr.* **17,** 538.

Bastiansen, O. and Hassel, O. (1946): *Tidsskr. Kjemi, Bergv., Metallurgi* **6,** 96.

Bastiansen, O., Seip, H. M. and Boggs, J. E. (1971): in *Perspectives in Structural Chemistry,* Vol. 4 (Eds.: Dunitz, J. D. and Ibers, J. A.). Wiley, New York.

Bastiansen, O. and Viervoll, H. (1948): *Acta Chem. Scand.* **2,** 702.

Beagley, B., Brown, D. P. and Freeman, J. M. (1974): *J. Mol. Struct.* **20,** 315.

Beagley, B., Chantrell, S. J., Kirby, R. G. and Schmidling, D. G. (1975): *J. Mol. Struct.* **25,** 319.

Beagley, B., Clark, A. H. and Hewitt, T. G. (1968): *J. Chem. Soc.* (A) 658.

Beagley, B. and Conrad, A. R. (1970): *Trans. Faraday Soc.* **66,** 2740.

Beagley, B., Eckersley, G. H., Brown, D. P. and Tomlinson, D. (1969): *Trans. Faraday Soc.* **65,** 1.

Beagley, B. and Hewitt, T. G. (1968): *Trans. Faraday Soc.* **64,** 2561.

Beagley, B. and McAloon, K. T. (1971): *Trans. Faraday Soc.* **67,** 3216.

Beagley, B. and Medwid, A. R. (1977): *J. Mol. Struct.* **38,** 229.

Beagley, B., Moutran, R., Narula, S. P. and Ulbrecht, V. (1979): *J. Mol. Struct.* **56,** 207.

Beagley, B., Ulbrecht, V., Katsumata, S., Lloyd, D. R., Connor, J. A. and Hudson, G. A. (1977): *J. C. S. Faraday Trans.* II, **73,** 1278.

Beall, R., Herdklotz, J. K. and Sass, R. L. (1970): *Acta Crystallogr.* **B26,** 1633.

Beecher, J. F. (1966): *J. Mol. Spectrosc.* **4,** 414.

Bel'skii, V. E., Naumov, V. A. and Nuretdinov, I. A. (1974): *Dokl. Akad. Nauk SSSR* **215,** 355.

Beppu, T., Hirota, E. and Morino, Y. (1970): *J. Mol. Spectrosc.* **36**, 386.

Bevan, J. W., Legon, A. C. and Millen, D. J. (1974): *J. C. S. Chem. Commun.* 659.

Bevan, J. W., Legon, A. C. and Millen, D. J. (1977): *Proc. Roy. Soc. (London)* **A354**, 491.

Bezzubov, V. M. and Naumov, V. A. (1976): *Zh. Strukt. Khim.* **17**, 98.

Bigotto, A., Galasso, V., Pappalardo, G. C. and Scarlata, G. (1978): *Spectrochim. Acta* **34A**, 435.

Birchall, T. and Gillespie, R. J. (1966): *Spectrochim. Acta* **22**, 681.

Bjørseth, A. (1974): *Acta Chem. Scand.* **A28**, 113.

Bjørseth, A., Kloster-Jensen, E., Marstokk, K. M. and Møllendal, H. (1970): *J. Mol. Struct.* **6**, 181.

Bjørseth, A. and Marstokk, K. M. (1972): *J. Mol. Struct.* **11**, 15.

Blackman, R. D., Brown, R. D., Burden, F. R. and Kent, J. E. (1967): *Chem. Phys. Lett.* **1**, 379.

Bleijerveld, R. H. T. and Mijlhoff, F. C. (1975): *Rec. Trav. Chim.* **94**, 190.

Bliefert, C. and Wanczek, K.-P. (1970): *Z. Naturforsch.* **25a**, 1770.

Block, E., Corey, E. R., Penn, R. E., Renken, T. L. and Sherwin, P. F. (1976): *J. Amer. Chem. Soc.* **98**, 5715.

Blukis, U., Kasai, P. H. and Myers, R. J. (1963): *J. Chem. Phys.* **38**, 2753.

Bock, H. (1977): *Angew. Chem. Int. Ed. Engl.* **16**, 613.

Bock, H. and Tabatabai, A. (1976): as cited by Bock (1977).

Bogatyreva, I. K., Kotov, A. V. and Shishkina, M. V. (1973): *Izv. Akad, Nauk SSSR Khim.* 2248.

Boggia, L. M., Filgueira, R. R., Marañon, J. and Sorarrain, O. M. (1978): *Spectrosc. Lett.* **11**, 143.

Boggs, J. E. (1975): private communication.

Bonham, R. A. and Bartell, L. S. (1959): *J. Amer. Chem. Soc.* **81**, 3491.

Bonham, R. A. and Momany, F. A. (1963): *J. Phys. Chem.* **67**, 2474.

Bouchy, A. and Roussy, G. (1973): *Compt. Rend. Acad. Sci. Paris Ser.* **C277**, 143.

Bowater, I. C., Brown, R. D. and Burden, F. R. (1967): *J. Mol. Spectrosc.* **23**, 272.

Bowater, I. C., Brown, R. D. and Burden, F. R. (1968a): *J. Mol. Spectrosc.* **28**, 454.

Bowater, I. C., Brown, R. D. and Burden, F. R. (1968b): *J. Mol. Spectrosc.* **28**, 461.

Bowen, H. J. M. (1953): *Nature* **172**, 171.

Bowen, H. J. M. (1954a): *Trans. Faraday Soc.* **50**, 444.

Bowen, H. J. M. (1954b): *Trans. Faraday Soc.* **50**, 452.

Bowers, M. T., Beaudet, R. A., Goldwhite, H. and Tang, R. (1969): *J. Amer. Chem. Soc.* **91**, 17.

Boyd, A. S. F., Laurenson, G. S. and Rankin, D. W. H. (1981): *J. Mol. Struct.* **71**, 217.

Bradford, W. F., Fitzwater, S. and Bartell, L. S. (1977): *J. Mol. Struct.* **38**, 185.

Brand, J. C. D., Callomon, J. H., Moule, D. C., Tyrre, J. and Goodwin, T. H. (1965): *Trans. Faraday Soc.* **61**, 2365.

Brassington, N. J., Edwards, H. G. M., Farwell, D. W., Long, D. A. and Mansour, H. R. (1978): *J. Raman Spectrosc.* **7**, 154.

Brauer, D. J. and Stucky, G. D. (1969): *J. Amer. Chem. Soc.* **91**, 5642.

Brendhaugen, K., Nilssen, E. W. and Seip, H. M. (1973): *Acta Chem. Scand.* **27**, 2965.

Brittain, A. H., Smith, J. E., Lee, P. L., Cohn, K. and Schwendeman, R. H. (1971): *J. Amer. Chem. Soc.* **93**, 6772.

Brooks, W. V. F. and Cyvin, S. J. (1962): *Acta Chem. Scand.* **16**, 820.

Brooks, W. V. F., Cyvin, S. J. and Kvande, P. C. (1965): *J. Phys. Chem.* **69**, 1489.

Brooks, W. V. F., Eshaque, M., Lau, C. and Passmore, J. (1976): *Can. J. Chem.* **54**, 817.

Brown, D. H., Crosbie, K. D., Darragh, J. I., Ross, D. S. and Sharp, D. W. A. (1970): *J. Chem. Soc.* (A) 914.

Brown, R. D. and Crofts, J. G. (1973): *Chem. Phys.* **1**, 217.

Bruce, R. B., Gillespie, R. J. and Slim, D. R. (1978): *Can. J. Chem.* **56**, 2927.

Brunvoll, J. (1962): *Thesis*, Trondheim.

Brunvoll, J. (1967): *Acta Chem. Scand.* **21**, 820.

Brunvoll, J., Exner, O. and Hargittai, I. (1981): *J. Mol. Struct.* **73**, 99.

Brunvoll, J. and Hargittai, I. (1976): *J. Mol. Struct.* **30**, 361.

Brunvoll, J. and Hargittai, I. (1977): *Acta Chim. (Budapest)* **94**, 333.

Brunvoll, J. and Hargittai, I. (1978): *Acta Chim. Acad. Sci. Hung.* **96**, 337.

Brunvoll, J., Hargittai, I. and Kolonits, M. (1978): *Z. Naturforsch.* **33a**, 1236.

Brunvoll, J., Hargittai, I. and Rozsondai, B. (1982): *J. Mol. Struct.* **84**, 153.

Brunvoll, J., Hargittai, I. and Seip, R. (1978a): *Z. Naturforsch.* **33a**, 222.

Brunvoll, J., Hargittai, I. and Seip, R. (1978b): *J. C. S. Dalton Trans.* 299.

Brunvoll, J., Hargittai, I., Székely, T. and Pappalardo, G. C. (1980): *J. Mol. Struct.* **66**, 173.

Bryan, P. S. and Kuczkowski, R. L. (1972): *Inorg. Chem.* **11**, 553.

Bulcke, P. (1975): *Diplomarbeit*, KMU (Leipzig).

Burrus, C. A. and Gordy, W. (1953): *Phys. Rev.* **92**, 274.

Büchler, A., Stauffer, J. L. and Klemperer, W. (1967): *J. Chem. Phys.* **46**, 605.

Bürger, H., Burczyk, K., Blaschette, A. and Safari, H. (1971): *Spectrochim. Acta* **27A**, 1073.

Bürger, H., Pawelke, G., Haas, A., Willner, H. and Downs, A. J. (1978): *Spectrochim. Acta* **34A**, 287.

Bürgi, H. B., Stedman, D. and Bartell, L. S. (1971): *J. Mol. Struct.* **10**, 31.

Bye, B. H. and Cyvin, S. J. (1963): *Acta Chem. Scand.* **17**, 1804.

Calleri, M., Chiari, G., Chiesi Villa, A., Gaetani Manfredotti, A., Guastini, C. and Viterbo, D. (1976): *Acta Crystallogr.* **B32**, 1032.

Calleri, M., Chiari, G., Chiesi Villa, A. and Guastini, C. (1976): *Cryst. Struct. Commun.* **5**, 113.

Caminati, W., Cervellati, R. and Mirri, A. M. (1974): *J. Mol. Spectrosc.* **51**, 288.

Caminati, W., Mirri, A. M. and Maccagnani, G. (1977): *J. Mol. Spectrosc.* **66**, 368.

Careless, A. J., Croto, H. W. and Landsberg, B. M. (1973): *Chem. Phys.* **1**, 371.

Carlos, J. L. and Bauer, S. H. (1974): *J. C. S. Faraday Trans.* II. **70**, 171.

Carlos, J. L., Karl, R. R. and Bauer, S. H. (1974): *J. C. S. Faraday Trans.* II. **70**, 177.

Caron, A., Palenik, G. J., Goldish, E. and Donohue, J. (1964): *Acta Crystallogr.* **17**, 102.

Carpenter, J. H. (1974): *J. Mol. Spectrosc.* **50**, 182.

Carpenter, J. H., Rimmer, D. F., Smith, G. and Whiffen, D. H. (1975): *J. C. S. Faraday Trans.* II. **71**, 1752.

Carter, H. A., Kirchmeier, R. L. and Shreeve, J. M. (1973): *Inorg. Chem.* **12**, 2237.

Carter, H. A., Shiow-Chyn Wang, C. and Shreeve, J. M. (1973): *Spectrochim. Acta* **29A**, 1479.

Cassoux, P., Kuczkowski, R. L. and Creswell, R. A. (1977): *Inorg. Chem.* **16**, 2959.

Cazzoli, G., Favero, P. G. and Dal Borgo, A. (1974): *J. Mol. Spectrosc.* **50**, 82.

Cazzoli, G., Lister, D. G. and Favero, P. G. (1972): *J. Mol. Spectrosc.* **42**, 286.

Centofanti, L. F. and Kuczkowski, R. L. (1968): *Inorg. Chem.* **7**, 2582.

Chachaty, C., Pappalardo, G. C. and Scarlata, G. (1976): *J. C. S. Perkin Trans.* II. 1234.

Chadwick, B. M., Grzybowski, J. M. and Long, D. A. (1978): *J. Mol. Struct.* **48**, 139.

Chadwick, D. and Millen, D. J. (1971): *Trans. Faraday Soc.* **67**, 1539.

Chawdhury, S. A. (1976): *Acta Crystallogr.* **B32**, 1069.

Chiang, J. F. (1971): *J. Amer. Chem. Soc.* **93**, 5044.

Chiang, J. F. and Lu, K. C. (1977): *J. Phys. Chem.* **81**, 1682.

Chiurdoglu, G., Reisse, J. and Rogier, M.V.S. (1961): *Chem. Ind.* 1874.

Christe, K. O. (1979): private communication.

Christe, K. O. and Curtis, E. C. (1972): *Spectrochim. Acta* **28A**, 1889.

Christe, K. O., Schack, C. J. and Curtis, E. C. (1977): *Spectrochim. Acta* **33A**, 323.

Christe, K. O., Willner, H. and Sawodny, W. (1979): *Spectrochim. Acta* **35A**, 1347.

Christen, D. (1980): *J. Mol. Struct.* **66**, 211.

Christen, D., Oberhammer, H., Zeil, W., Haas, A. and Darmadi, A. (1980): *J. Mol. Struct.* **66**, 203.

Churchill, M. R., DeBoer, B. G., Kalra, K. L., Reich-Rohrwig, P. and Wojcicki, A. (1972): *J. C. S. Chem. Commun.* 981.

Clark, A. H. and Beagley, B. (1971): *Trans. Faraday Soc.* **67**, 2216.

Clark, A. H. and Haaland, A. (1970): *Acta Chem. Scand.* **24**, 3024.

Clark, A. H. and Hewitt, T. G. (1971): *J. Mol. Struct.* **9**, 33.

Clever, H. L. and Westrum, E. F. (1970): *J. Phys. Chem.* **74**, 1309.

Codding, E. G., Jones, C. E. and Schwendeman, R. H. (1974): *Inorg. Chem.* **13**, 178.

Cohen, E. R. and Taylor, B. N. (1973): *J. Phys. Chem. Ref. Data* **2,** 663.

Colapietro, M., Domenicano, A. and Vaciago, A. (1976): *Acta Crystallogr.* **B32,** 2581.

Cook, R. L. and Kirchhoff, W. H. (1967): *J. Chem. Phys.* **47,** 4521.

Corkill, M. J., Cox, A. P. and Ewart, I. C. (1976): *J. C. S. Chem. Commun.* 546.

Corosine, M., Crasnier, M.-C., Labarre, M.-C., Labarre, J.-F. and Leibovici, C. (1973): *Chem. Phys. Lett.* **20,** 111.

Cotton, F. A. and Stokely, P. F. (1970): *J. Amer. Chem. Soc.* **92,** 294.

Cradock, S., Laurenson, G. S. and Rankin, D. W. H. (1981): *J. C. S. Dalton Trans.* 187.

Crawford, R. A., Dudley, F. B. and Hedberg, K. (1959): *J. Amer. Chem. Soc.* **81,** 5287.

Cruickshank, D. W. J. (1968): *Inorganic Sulphur Chemistry* (Ed.: Nickless, G.). Elsevier, Amsterdam.

Csákvári, B., Wagner, Zs., Gömöry, P., Hargittai, I., Rozsondai, B. and Mijlhoff, F. C. (1976): *Acta Chim. (Budapest)* **90,** 149.

Csákvári, B., Wagner, Zs., Gömöry, P., Mijlhoff, F. C., Rozsondai, B. and Hargittai, I. (1976): *J. Organometal. Chem.* **107,** 287.

Cunningham, G. L., Boyd, A. W., Myers, R. J., Gwinn, W. D. and Le Van, W. I. (1951): *J. Chem. Phys.* **19,** 676.

Curl, R. F. (1959): *J. Chem. Phys.* **30,** 1529.

Cyvin, B. N. (1975): private communication.

Cyvin, B. N. and Cyvin, S. J. (1969): *Acta Chem. Scand.* **23,** 3139.

Cyvin, B. N. and Cyvin, S. J. (1972): *Acta Chem. Scand.* **26,** 1284.

Cyvin, S. J. (1961): *Tidsskr. Kjemi, Bergv., Metallurgi* **21,** 236; (1962): **22,** 44; 73.

Cyvin, S. J. (1968): *Molecular Vibrations and Mean Square Amplitudes.* Universitetsforlaget, Oslo and Elsevier, Amsterdam.

Cyvin, S. J. (1969a): *Molecular Vibrations and Mean Square Amplitudes.* Part. I of Suppl. to Chapts 12 and 15. Det Kgl Norske Videns. Selsk. Shrift. Nr. 1, 1969.

Cyvin, S. J. (1969b): *Acta Chem. Scand.* **23,** 576.

Cyvin, S. J. (1970): *Z. anorg. allg. Chem.* **378,** 117.

Cyvin, S. J. (1973): private communication.

Cyvin, S. J., Brunvoll, J. and Müller, A. (1968): *Acta Chem. Scand.* **22,** 2739.

Cyvin, S. J., Brunvoll, J. and Robiette, A. G. (1971): *Chem. Phys. Lett.* **11,** 263.

Cyvin, S. J., Brunvoll, J. and Stølevik, R. (1969): *Acta Chem. Scand.* **23,** 333.

Cyvin, S. J., Cyvin, B. N., Brockner, W. and Demiray, A. F. (1978): *Z. Naturforsch.* **33c,** 714.

Cyvin, S. J., Dobos, S., Hargittai, I., Hargittai, M. and Augdahl, E. (1973): *J. Mol. Struct.* **18,** 203.

Cyvin, S. J. and Hargittai, I. (1969): *Acta Chim. Acad. Sci. Hung.* **61,** 159.

Cyvin, S. J. and Meisingseth, E. (1963): *Acta Chem. Scand.* **17,** 1805.

D'Antonio, P., George, C., Lowrey, A. H. and Karle, J. (1971): *J. Chem. Phys.* **55,** 1071.

Davis, M. I. and Hanson, H. P. (1965): *J. Phys. Chem.* **69,** 4091.

Davis, R. W. and Gerry, M. C. L. (1977): *J. Mol. Spectrosc.* **65,** 455.

De Alti, G., Decleva, P. and Sgamellotti, A. (1979): *J. Mol. Struct.* **53,** 129.

Delker, G. L., Wang, Y., Stucky, G. D., Lambert, R. L., Haas, C. K. and Seyfert, D. (1976): *J. Amer. Chem. Soc.* **98,** 1779.

Den Engelsen, D. (1967): *J. Mol. Spectrosc.* **22,** 426.

Den Engelsen, D. (1969): *J. Mol. Spectrosc.* **30,** 474.

Derissen, J. L. and Bijen, J. M. J. M. (1973): *J. Mol. Struct.* **16,** 289.

Derissen, J. L., Kocken, J. W. M. and van Weelden, R. H. (1971): *Acta Crystallogr.* **B27,** 1692.

Deroche, J.-C., Bürger, H., Schulz, P. and Willner, H. (1981): *J. Mol. Spectrosc.* **89,** 269.

De Rooij, J., Mijlhoff, F. C. and Renes, G. (1975): *J. Mol. Struct.* **25,** 169.

Dobyns, V. and Pierce, L. (1963): *J. Amer. Chem. Soc.* **85,** 3553.

Donohue, J. (1978): *J. Cryst. Mol. Struct.* **8,** 141.

Donohue, J. and Schomaker, V. (1948): *J. Chem. Phys.* **16,** 92.

Dorney, A. J., Hoy, A. R. and Mills, I. M. (1973): *J. Mol. Spectrosc.* **45,** 253.

Dousmanis, G. C., Sanders, T. M., Townes, C. H. and Zeiger, H. J. (1953): *J. Chem. Phys.* **21,** 1416.
Dowd, P., Dyke, T. and Klemperer, W. (1970): *J. Amer. Chem. Soc.* **92,** 6327.
Dössel, K.-F. and Robiette, A. G. (1977): *Z. Naturforsch.* **32a,** 462.
Dössel, K.-F. and Sutter, D. H. (1977): *Z. Naturforsch.* **32a,** 473.
Dössel, K.-F. and Sutter, D. H. (1978): *Z. Naturforsch.* **33a,** 1190.
Dössel, K.-F. and Sutter, D. H. (1979): *Z. Naturforsch.* **34a,** 482.
Dräger, M. and Gattow, G. (1971): *Chem. Ber.* **104,** 1429.
Dreizler, H. and Dendl, G. (1964): *Z. Naturforsch.* **19a,** 512.
Dreizler, H., Rudolph, H. D. and Schleser, H. (1970): *Z. Naturforsch.* **25a,** 1643.
Dubrulle, A. (1972): *Thèse,* L'Université des Sciences et Techniques de Lille.
Dubrulle, A. and Boucher, D (1974): *Compt. Rend. Acad. Sci. Paris* **278,** Ser. B, 211.
Duckett, J. A., Robiette, A. G. and Mills, I. M. (1976): *J. Mol. Spectrosc.* **62,** 34.
Duncan, J. L., McKean, D. C., Mallinson, P. D. and McCulloch, R. D. (1973): *J. Mol. Spectrosc.* **46,** 232.
Dupont, L. and Dideberg, O. (1972): *Acta Crystallogr.* **B28,** 2340.
Durig, J. R. and Bucy, W. E. (1977): *J. Mol. Spectrosc.* **64,** 474.
Durig, J. R., Compton, D. A. C. and Jalilian, M.-R. (1979): *J. Phys. Chem.* **83,** 511.
Durig, J. R., Craven, S. M., Mulligan, J. H., Hawley, C. W. and Bragin, J. (1973): *J. Chem. Phys.* **58,** 1281.
Durig, J. R. and Griffin, M. G. (1977): *J. Chem. Phys.* **67,** 2220.
Durig, J. R., Hudgens, B. A. and Odom, J. D. (1974): *Inorg. Chem.* **13,** 2306.
Durig, J. R., Kalasinsky, K. S. and Kalasinsky, V. F. (1978a): *J. Phys. Chem.* **82,** 438.
Durig, J. R., Kalasinsky, K. S. and Kalasinsky, V. F. (1978b): *J. Chem. Phys.* **69,** 918.
Durig, J. R., Li, Y. S. and Sullivan, J. F. (1979): *J. Chem. Phys.* **71,** 1041.
Durig, J. R., Sullivan, J. F., Cox, A. W. and Streusand, B. J. (1978): *Spectrochim. Acta* **34A,** 719.
Duxbury, G., Horani, M. and Rostas, J. (1972): *Proc. Roy. Soc. (London)* **A331,** 109.
Dyke, T. R., Howard, B. J. and Klemperer, W. (1972): *J. Chem. Phys.* **56,** 2442.
Edwards, T. H., Moncur, N. K. and Snyder, L. E. (1967): *J. Chem. Phys.* **46,** 2139.
Eisma, S. W., Altona, C., Geise, H. J., Mijlhoff, F. C. and Renes, G. H. (1974): *J. Mol. Struct.* **20,** 251.
Eliel, E. L. and Thill, B. P. (1963): *Chem. Ind.* 88.
Emerson, K. (1966): *Acta Crystallogr.* **21,** 970.
Emken, W. C. and Hedberg, K. (1973): *J. Chem. Phys.* **58,** 2195.
Endo, Y., Saito, S., Hirota, E. and Chikaraishi, T. (1979): *J. Mol. Spectrosc.* **77,** 222.
Ewing, V. C. and Sutton, L. E. (1963): *Trans. Faraday Soc.* **59,** 1241.
Exner, O. (1979): private communication.
Exner, O., Dembech, P. and Vivarelli, P. (1971): *J. Chem. Soc.* (A) 620.
Eysel, H. H. (1970): *J. Mol. Struct.* **5,** 275.
Ezhov, Yu. S., Tolmachev, S. M. and Rambidi, N. G. (1970): *Zh. Strukt. Khim.* **11,** 527.
Ezhov, Yu. S., Tolmachev, S. M., Spiridonov, V. P. and Rambidi, N. G. (1968): *Teplofiz. vys. temp.* **6,** 68.
Feder, W., Dreizler, H., Rudolph, H. D. and Typke, V. (1969): *Z. Naturforsch.* **24a,** 266.
Ferguson, R. C. (1954): *J. Amer. Chem. Soc.* **76,** 850.
Fernholt, L. (1978): private communication.
Fernholt, L., Haaland, A., Hargittai, M., Seip, R. and Weidlein, J. (1981): *Acta Chem. Scand.* **A35,** 529.
Fernholt, L. and Kveseth, K. (1978): *Acta Chem. Scand.* **A32,** 63.
Fernholt, L. and Kveseth, K. (1979): *Acta Chem. Scand.* **A33,** 335.
Feuerhahn, M. and Vahl, G. (1979): *Chem. Phys. Lett.* **65,** 322.
Flood, E. and Boggs, J. E. (1976): *J. Mol. Struct.* **34,** 147.
Flood, E., Pulay, P. and Boggs, J. E. (1977): *J. Amer. Chem. Soc.* **99,** 5570.
Forti, P., Damiani, D. and Favero, P. G. (1973): *J. Amer. Chem. Soc.* **95,** 756.
Fraser, A., Laurenson, G. S. and Rankin, D. W. H. (1981): unpublished work as cited by Cradock, Laurenson and Rankin (1981).
Frenzel, C. A., Blick, K. E., Bennett, C. R. and Niedenzu, K. (1970): *J. Chem. Phys.* **53,** 198.

Fukuyama, T., Oyanagi, K. and Kuchitsu, K. (1976): *Bull. Chem. Soc. Japan* **49,** 638.
Galasso, V., De Alti, G. and Bigotto, A. (1971): *Tetrahedron,* **27,** 6151.
Galasso, V., Pappalardo, G. C. and Scarlata, G. (1977): *Z. Naturforsch.* **32a,** 517
Gallaher, K. L. and Bauer, S. H. (1975): *J. C. S. Faraday Trans.* II. **71,** 1173.
Gallo, A. R. and Innes, K. K. (1974): *J. Mol. Spectrosc.* **54,** 472.
Gebhardt, O. (1971): *Dissertation,* University of Trondheim, Trondheim.
Gebhardt, O. and Cyvin, S. J. (1972): *J. Mol. Struct.* **12,** 205.
Geise, H. J., Adams, W. J. and Bartell, L. S. (1969): *Tetrahedron* **25,** 3045.
Geise, H. J., Altona, C. and Romers, C. (1967): *Tetrahedron* **23,** 439.
Geise, H. J. and Van Laere, E. (1975): *Bull. Soc. Chim. Belg.* **84,** 775.
Geiseler, G. and Hanschmann, G. (1971): *J. Mol. Struct.* **8,** 293.
Geiseler, G. and Nagel, B. (1973): *J. Mol. Struct.* **16,** 79.
Georgiou, K., Kroto, H. W. and Landsberg, B. M. (1979): *J. Mol. Spectrosc.* **77,** 365.
Gergő, É., Hargittai, I. and Schultz, Gy. (1976): *J. Organometal. Chem.* **112,** 29.
Gerry, M. C. L., Thompson, J. C. and Sugden, T. M. (1966): *Nature* **211,** 846.
Gillespie, R. J. (1972): *Molecular Geometry.* Van Nostrand Reinhold Co., London.
Gillespie, R. J. and Robinson, E. A. (1961): *Can. J. Chem.* **39,** 2171.
Gillespie, R. J. and Robinson, E. A. (1963): *Can. J. Chem.* **41,** 2074; and references therein.
Gleisberg, F., Haberl, A. and Zeil, W. (1975): *Z. Naturforsch.* **30a,** 549.
Gleiter, R. and Hoffmann, R. (1968): *Tetrahedron* **24,** 5899.
Glemser, O., Heussner, W. D. and Haas, A. (1963): *Naturwissenschaften* **50,** 402.
Glemser, O. and Schröder, H. (1956): *Z. anorg. allg. Chem.* **284,** 97.
Glidewell, C. (1973): *Inorg. Chim. Acta Rev.* **7,** 69.
Glidewell, C. (1975): *Inorg. Chim. Acta* **12,** 219.
Glidewell, C. (1976): *Inorg. Chim. Acta* **20,** 113.
Glidewell, C. (1977): *Inorg. Chim. Acta* **25,** 77.
Glidewell, C. (1979): *Inorg. Chim. Acta* **36,** 135.
Glidewell, C., Rankin, D. W. H., Robiette, A. G. and Sheldrick, G. M. (1969): *J. Mol. Struct.* **4,** 215.
Glidewell, C., Rankin, D. W. H., Robiette, A. G. and Sheldrick, G. M. (1970a): *J. Chem. Soc.* (A) 318.
Glidewell, C., Rankin, D. W. H., Robiette, A. G. and Sheldrick, G. M. (1970b): *J. Mol. Struct.* **6,** 231.
Glidewell, C., Rankin, D. W. H., Robiette, A. G., Sheldrick, G. M., Beagley, B. and Cradock, S. (1970): *J. Chem. Soc.* (A) 315.
Glidewell, C., Robiette, A. G. and Sheldrick, G. M. (1972): *Chem. Phys. Lett.* **16,** 526.
Goggin, P. L., Roberts, H. L. and Woodward, L. A. (1961): *Trans. Faraday Soc.* **78,** 1877.
Gogoi, B. N. and Hargraves, A. (1970): *Acta Crystallogr.* **B26,** 2132.
Gordy, W. and Cook, R. L. (1970): *Microwave Molecular Spectra.* Interscience Publ., New York.
Goubeau, J. and Wittmeier, H. W. (1952): *Z. anorg. allg. Chem.* **270,** 16.
Gramstad, T. (1963): *Spectrochim. Acta* **19,** 829.
Graybeal, J. D. (1976): *Sixth Austin Symposium on Gas Phase Molecular Structure.* Austin, Texas.
Green, J. H. S. (1962): *Spectrochim. Acta* **18,** 39.
Green, J. H. S. (1968): *Spectrochim. Acta* **24A,** 1627.
Gregory, D., Hargittai, I. and Kolonits, M. (1976): *J. Mol. Struct.* **31,** 261.
Gropen, O. and Seip, H. M. (1971): *Chem. Phys. Lett.* **11,** 445.
Grunwell, J. R. and Danison, W. C. (1971): *Tetrahedron* **27,** 5315.
Gruttadauria, S., Maccarone, E., Pappalardo, G. C. and Tomaselli, G. A. (1976): *Gazz. Chim. Ital.* **106,** 657.
Guarnieri, A. (1968): *Z. Naturforsch.* **23a,** 1867.
Guarnieri, A. (1970): *Z. Naturforsch.* **25a,** 18.
Guarnieri, A., Charpentier, L. and Kück, B. (1973): *Z. Naturforsch.* **28a,** 1721.
Guin, H. W. (1969): *Dissertation,* University of Texas, Austin.
Gundersen, G. (1976): *J. Mol. Struct.* **33,** 79.

Gundersen, G. (1978): *Kém. Közlem.* **49,** 261.

Gundersen, G. and Hedberg, K. (1969): *J. Chem. Phys.* **51,** 2500.

Gundersen, G., Hedberg, K. and Strand, T. G. (1978): *J. Chem. Phys.* **68,** 3548.

Gundersen, G. and Vahrenkamp, H. (1976): *J. Mol. Struct.* **33,** 97.

Haaland, A., Stokkeland, O. and Weidlein, J. (1975): *J. Organometal. Chem.* **94,** 353.

Haas, A. and Willner, H. (1978): *Spectrochim. Acta* **34A,** 541.

Haas, A. and Willner, H. (1979): *Z. anorg. allg. Chem.* **454,** 17.

Haase, J. and Krebs, A. (1972): *Z. Naturforsch.* **27a,** 624.

Haase, J., Oberhammer, H., Zeil, W., Glemser, O. and Mews, R. (1970): *Z. Naturforsch.* **25a,** 153.

Haase, J., Oberhammer, H., Zeil, W., Glemser, O. and Mews, R. (1971): *Z. Naturforsch.* **26a,** 1333.

Hadži, D., Kidrič, J., Keževic, Z. V. and Barlič, B. (1976): *Spectrochim. Acta* **32A,** 693.

Hagen, G., Hargittai, I. and Schultz, Gy. (1975): *Acta Chim. (Budapest)* **86,** 219.

Hagen, K., Cross, V. R. and Hedberg, K. (1978): *J. Mol. Struct.* **44,** 187.

Hagen, K. and Hedberg, K. (1973a): *J. Amer. Chem. Soc.* **95,** 1003.

Hagen, K. and Hedberg, K. (1973b): *J. Amer. Chem. Soc.* **95,** 4796.

Ham, N. S., Hambly, A. N. and Laby, R. H. (1960): *Aust. J. Chem.* **13,** 443.

Ham, N. S. and Wills, J. B. (1960): *Spectrochim. Acta* **16,** 279.

Hamada, K. and Morishita, H. (1978): *J. Mol. Struct.* **44,** 119.

Hamilton, W. C. (1964): *Statistics in Physical Science.* The Ronald Press Co., New York.

Hamilton, W. C. (1965): *Acta Crystallogr.* **18,** 502.

Hamm, R. (1979): *Z. Naturforsch.* **34a,** 325.

Hanai, K., Okuda, T., Uno, T. and Machida, K. (1975): *Spectrochim. Acta* **31A,** 1217.

Hargittai, I. (1968a): *Magy. Kém. Foly.* **74,** 362.

Hargittai, I. (1968b): *Magy. Kém. Foly.* **74,** 596.

Hargittai, I. (1969): *Acta Chim. Acad. Sci. Hung.* **60,** 231.

Hargittai, I. (1973): *Természet Világa* **104,** 78.

Hargittai, I. (1974a): The electron diffraction interatomic distance (in Hungarian). *A kémia újabb eredményei* (Ed.: Csákvári, B.), Vol. 21. Akadémiai Kiadó, Budapest.

Hargittai, I. (1974b): *Second European Crystallographic Meeting, Collected Abstracts* pp. 441–443, Keszthely.

Hargittai, I. (1978): Sulphone Molecular Structures. *Lecture Notes in Chemistry*, Vol. 6. Springer-Verlag, Berlin–Heidelberg–New York.

Hargittai, I. (1979a): *Z. Naturforsch.* **34a,** 755.

Hargittai, I. (1979b): *J. Mol. Struct.* **54,** 287.

Hargittai, I. (1979c): *J. Mol. Struct.* **56,** 301.

Hargittai, I. (1979d): *Kém. Közlem.* **52,** 473.

Hargittai, I. (1981): Gas Electron Diffraction — A Tool of Structural Chemistry in Perspectives. *Topics in Current Chemistry* **96,** 43. Springer-Verlag, Berlin–Heidelberg–New York.

Hargittai, I. (1982): *Inorg. Chem.* **21,** 4334.

Hargittai, I. and Baranyi, A. (1977): *Acta Chim. (Budapest)* **93,** 279.

Hargittai, I. and Brunvoll, J. (1976): *Acta Chem. Scand.* **A30,** 634.

Hargittai, I. and Brunvoll, J. (1978): *J. Mol. Struct.* **44,** 107.

Hargittai, I. and Brunvoll, J. (1979a): *Kém. Közlem.* **51,** 293.

Hargittai, I. and Brunvoll, J. (1979b): unpublished calculations.

Hargittai, I. and Cyvin, S. J. (1969): *Acta Chim. Acad. Sci. Hung.* **61,** 51.

Hargittai, I. and Hargittai, M. (1973a): *Acta Chim. (Budapest)* **75,** 129.

Hargittai, I. and Hargittai, M. (1973b): *J. Mol. Struct.* **15,** 399.

Hargittai, I. and Mijlhoff, F. C. (1973): *J. Mol. Struct.* **16,** 69.

Hargittai, I. and Orville-Thomas, W. J. (Eds.) (1981): *Diffraction Studies on Non-Crystalline Substances.* Akadémiai Kiadó, Budapest and Elsevier, Amsterdam.

Hargittai, I. and Paul, I. C. (1977): Structural Chemistry of the Cyanates and Their Thio Derivatives. *The Chemistry of Cyanates and Their Thio Derivatives* (Ed.: Patai, S.). Wiley, New York.
Hargittai, I., Rozsondai, B., Nagel, B., Bulcke, P., Robinet, G. and Labarre, J.-F. (1978): *J. C. S. Dalton Trans.* 861.
Hargittai, I., Schultz, Gy. and Kolonits, M. (1977): *J. C. S. Dalton Trans.* 1299.
Hargittai, I., Seip, R., Nair, K. P. R., Britt, Ch. O., Boggs, J. E. and Cyvin, B. N. (1977): *J. Mol. Struct.* **39**, 1.
Hargittai, I., Vajda, E. and Szőke, A. (1973): *J. Mol. Struct.* **18**, 381.
Hargittai, I. and Vilkov, L. V. (1970): *Acta Chim. Acad. Sci. Hung.* **63**, 143.
Hargittai, M. and Hargittai, I. (1973): *J. Chem. Phys.* **59**, 2513.
Hargittai, M. and Hargittai, I. (1974): *J. Mol. Struct.* **20**, 283.
Hargittai, M. and Hargittai, I. (1977): *The Molecular Geometries of Coordination Compounds in the Vapour Phase.* Akadémiai Kiadó, Budapest and Elsevier, Amsterdam.
Hargittai, M. and Hargittai, I. (1981): *J. Mol. Struct.* **73**, 253.
Harlow, R. L., Sammes, M. P. and Simonsen, S. H. (1974): *Acta Crystallogr.* **B30**, 2903.
Harlow, R. L., Simonsen, S. H., Pfluger, C. E. and Sames, M. P. (1974): *Acta Crystallogr.* **B30**, 2264.
Harris, D. O., Harrington, H. W., Luntz, A. C. and Gwinn, W. D. (1966): *J. Chem. Phys.* **44**, 3467.
Harris, S. J., Janda, K. C., Novick, S. E. and Klemperer, W. (1975): *J. Chem. Phys.* **63**, 881.
Harris, S. J., Novick, S. E., Klemperer, W. and Falconer, W. E. (1974): *J. Chem. Phys.* **61**, 193.
Harris, W. C., Holtzclaw, J. R. and Kalasinsky, V. F. (1977): *J. Chem. Phys.* **67**, 3330.
Harshbarger, W. R. and Bauer, S. H. (1970): *Acta Crystallogr.* **B26**, 1010.
Harvey, R. B. and Bauer, S. H. (1954): *J. Amer. Chem. Soc.* **76**, 859.
Hassel, O. and Viervoll, H. (1947): *Acta Chem. Scand.* **1**, 149.
Hastie, J. W., Hauge, R. H. and Margrave, J. L. (1969): *J. Amer. Chem. Soc.* **91**, 2536.
Hayashi, M., Imaishi, H. and Kuwada, K. (1974): *Bull. Chem. Soc. Japan* **47**, 2382.
Hayashi, M., Shiro, Y., Murakami, M. and Murata, H. (1965): *Bull. Chem. Soc. Japan* **38**, 1740.
Hayashi, M., Shiro, Y. and Murata, H. (1966): *Bull. Chem. Soc. Japan* **39**, 112.
Hayashi, M., Shiro, Y., Oshima, T. and Murata, H. (1965): *Bull. Chem. Soc. Japan* **38**, 1734.
Hedberg, K. (1955): *J. Amer. Chem. Soc.* **77**, 6491.
Hedberg, K. (1979): private communication.
Hencher, J. L. and Bauer, S. H. (1973): *Can. J. Chem.* **51**, 2047.
Hencher, J. L., Cruickshank, D. W. J. and Bauer, S. H. (1968): *J. Chem. Phys.* **48**, 518.
Hencher, J. L., Shen, Q. and Tuck, D. G. (1976): *J. Amer. Chem. Soc.* **98**, 899.
Herman, Z. S. and Weiss, K. (1975): *Inorg. Chem.* **14**, 1592.
Herzog, K., Steger, E., Rosmus, P., Scheithauer, S. and Mayer, R. (1969): *J. Mol. Struct.* **3**, 339.
Higgins, R. J., Combs, L. L., Malloy, T. B. and Cook, R. L. (1975): *J. Mol. Struct.* **28**, 121.
Hilderbrandt, R. L., Andreassen, A. L. and Bauer, S. H. (1970): *J. Phys. Chem.* **74**, 1586.
Hilderbrandt, R. L. and Bauer, S. H. (1969): *J. Mol. Struct.* **3**, 325.
Hill, R. A. and Edwards, T. H. (1965): *J. Chem. Phys.* **42**, 1391.
Hinchliffe, A. (1979): *J. Mol. Struct.* **55**, 127.
Hinze, J., Whitehead, M. A. and Jaffé, H. H. (1963): *J. Amer. Chem. Soc.* **85**, 148.
Hirose, C. (1974): *Bull. Chem. Soc. Japan* **47**, 1311.
Hirota, E. (1958): *Bull. Chem. Soc. Japan* **31**, 130.
Hirota, E., Sugisaki, R., Nielsen, C. J. and Sørensen, G. O. (1974): *J. Mol. Spectrosc.* **49**, 251.
Hjortaas, K. E. (1967): *Acta Chem. Scand.* **21**, 1381.
Hocking, W. H., Williams, M. L. and Gerry, M. C. L. (1975): *J. Mol. Spectrosc.* **58**, 250.
Hocking, W. H. and Winnewisser, G. (1975): *J. C. S. Chem. Commun.* **63.**
Hocking, W. H. and Winnewisser, G. (1976a): *Z. Naturforsch.* **31a**, 422.
Hocking, W. H. and Winnewisser, G. (1976b): *Z. Naturforsch.* **31a**, 438.
Hocking, W. H. and Winnewisser, G. (1976c): *Z. Naturforsch.* **31a**, 995.
Hocking, W. H. and Winnewisser, G. (1977): *Z. Naturforsch.* **32a**, 1108.

Hoffmann, R., Fujimoto, H., Swenson, J. R. and Wan, Ch.-Ch. (1973): *J. Amer. Chem. Soc.* **95,** 7644.
Holly, S. (1977/1978): unpublished data.
Holt, C. W. and Gerry, M. C. L. (1971): *Chem. Phys. Lett.* **9,** 621.
Horrocks, W. D. and Cotton, F. A. (1961): *Spectrochim. Acta* **17a,** 134.
Hosoya, S. (1966a): *Acta Crystallogr.* **20,** 429.
Hosoya, S. (1966b): *Acta Crystallogr.* **21,** 21.
Howell, J. M., Absar, I. and Van Wazer, J. R. (1973): *J. Chem. Phys.* **59,** 5895.
Huong, P. V. and Raducanu, E. (1974): *J. Mol. Struct.* **23,** 81.
Iijima, T. (1972): *Bull. Chem. Soc. Japan* **45,** 1291.
Iijima, T. (1973): *Bull. Chem. Soc. Japan* **46,** 2311.
Iijima, T., Tsuchiya, S. and Kimura, M. (1977): *Bull. Chem. Soc. Japan* **50,** 2564.
Ikeda, T., Kewley, R. and Curl, R. F. (1972): *J. Mol. Spectrosc.* **44,** 459.
Imaishi, H. and Hayashi, M. (1974): *J. Sci. Hiroshima Univ.* **A38,** 21.
Isvén, M. K., Ralowski, W. M. and Ljunggren, S. O. (1977): *Chem. Phys. Lett.* **47,** 294.
Iwasaki, M. and Hedberg, K. (1962): *J. Chem. Phys.* **36,** 589.
Jache, A. W., Moser, P. W. and Gordy, W. (1956): *J. Chem. Phys.* **25,** 209.
Jacob, E. J. and Lide, D. R. (1971): *J. Chem. Phys.* **54,** 4591.
Jacob, E. J. and Samdal, S. (1977): *J. Amer. Chem. Soc.* **99,** 5656.
Jacob, E. J., Danielson, D. D. and Samdal, S. (1980): *J. Mol. Struct.* **62,** 143.
Jacob, E. J., Thompson, H. B. and Bartell, L. S. (1967): *J. Chem. Phys.* **47,** 3736.
Jandacek, R. J. (1975): *Dissertation*, University of Texas, Austin.
Jandal, P., Seip, H. M. and Torgrimsen, T. (1976): *J. Mol. Struct.* **32,** 369.
Jen, M. and Lide, D. R. (1962): *J. Chem. Phys.* **36,** 2525.
Jenkins, D. R., Kewley, R. and Sugden, T. M. (1962): *Trans. Faraday Soc.* **58,** 1284.
Johansen, R., Nilssen, E. W., Seip, H. M. and Siebert, W. (1973): *Acta Chem. Scand.* **27,** 3015.
Johansen, R., Seip, H. M. and Siebert, W. (1975): *Acta Chem. Scand.* **A29,** 644.
Johnson, D. R. and Powell, F. X. (1969): *Science* **164,** 950.
Johnson, D. R., Powell, F. X. and Kirchhoff, W. H. (1971): *J. Mol. Spectrosc.* **39,** 136.
Jones, L. H. (1967): *Inorg. Chem.* **6,** 429.
Jones, L. H., Shoolery, J. N., Shulman, R. G. and Yost, D. M. (1950): *J. Chem. Phys.* **18,** 990.
Jordan, T., Smith, H. W., Lohr, L. L. and Lipscomb, W. N. (1963): *J. Amer. Chem. Soc.* **85,** 846.
Kalasinsky, V. F., Block, E., Powers, D. E. and Harris, W. C. (1979): *Appl. Spectrosc.* **33,** 361.
Kaldor, A. and Maki, A. G. (1973): *J. Mol. Struct.* **15,** 123.
Kaldor, A., Maki, A. G., Dorney, A. J. and Mills, I. M. (1973): *J. Mol. Spectrosc.* **45,** 247.
Kálmán, A., Duffin, B. and Kucsman, Á. (1971): *Acta Crystallogr.* **B27,** 586.
Kanesaka, I. and Kawai, K. (1970): *Bull. Chem. Soc. Japan* **43,** 3298.
Kao, J. (1977): *J. Mol. Struct.* **41,** 321.
Kao, J. (1978): *J. Amer. Chem. Soc.* **100,** 4685.
Kao, J. and Allinger, N. L. (1977): *Inorg. Chem.* **16,** 35.
Kao, J. and Radom, L. (1979): *J. Amer. Chem. Soc.* **101,** 311.
Kapur, A., Steer, R. P. and Mezey, P. G. (1978): *J. Chem. Phys.* **69,** 968.
Kapur, A., Steer, R. P. and Mezey, P. G. (1979a): *J. Chem. Phys.* **70,** 745.
Kapur, A., Steer, R. P. and Mezey, P. G. (1979b): *J. Chem. Phys.* **71,** 588.
Karakida, K., Fukuyama, T. and Kuchitsu, K. (1974): *Bull. Chem. Soc. Japan* **47,** 299.
Karakida, K. and Kuchitsu, K. (1975): *Bull. Chem. Soc. Japan* **48,** 1691.
Karakida, K. and Kuchitsu, K. (1976): *Inorg. Chim. Acta* **16,** 29.
Karl, R. R. and Bauer, S. H. (1972): *Acta Crystallogr.* **B28,** 2619.
Karl, R. R. and Bauer, S. H. (1975): *Inorg. Chem.* **14,** 1859.
Karle, I. L. (1973): *Z. Kristallogr.* **138,** 184.
Karle, J. (1966): *J. Chem. Phys.* **45,** 4149.

Kasai, P. H. and Myers, R. J. (1959): *J. Chem. Phys.* **30,** 1096.
Kato, C., Konaka, S., Iijima, T. and Kimura, M. (1969): *Bull. Chem. Soc. Japan* **42,** 2148.
Katon, J. E. and Feairheller, W. R. (1965): *Spectrochim. Acta* **21,** 199.
Kewley, R. (1973): *Can. J. Chem.* **51,** 529.
Kewley, R. (1978): *Can. J. Chem.* **56,** 772.
Kewley, R., Hemphill, D. C. and Curl, R. F. (1972): *J. Mol. Spectrosc.* **44,** 443.
Kewley, R., Murty, K. S. R. and Sugden, T. M. (1962): *Trans. Faraday Soc.* **58,** 1284.
Khaikin, L. S., Vilkov, L. V., Vasil'ev, A. F., Mel'nikov, N. N., Tulyakova, T. F. and Anashkin, M. G. (1972): *Dokl. Akad. Nauk SSSR* **203,** 1090.
Kiers, C. Th. and Vos, A. (1972): *Rec. Trav. Chim.* **91,** 126.
Kimura, K. and Bauer, S. H. (1963): *J. Chem. Phys.* **39,** 3172.
Kimura, K., Katada, K. and Bauer, S. H. (1966): *J. Amer. Chem. Soc.* **88,** 416.
Kimura, K. and Kubo, M. (1959): *J. Chem. Phys.* **30,** 151.
Kirchhoff, W. H. (1969): *J. Amer. Chem. Soc.* **91,** 2437.
Kirchhoff, W. H., Johnson, D. R. and Powell, F. X. (1973): *J. Mol. Spectrosc.* **48,** 157.
Kirchhoff, W. H. and Wilson, E. B. (1962): *J. Amer. Chem. Soc.* **84,** 334.
Kirchmeier, R. L. and Shreeve, J. M. (1975): *Inorg. Chem.* **14,** 2431.
Kiss, Á. I. (1979): private communication.
Kitchin, R. W., Avirah, T. K., Malloy, T. B. and Cook, R. L. (1975): *J. Mol. Struct.* **24,** 337.
Klaboe, P. and Smith, Z. (1978): *Spectrochim. Acta* **34A,** 489.
Kleemann, G. and Seppelt, K. (1978): *Angew. Chem. Int. Ed. Engl.* **17,** 516.
Klug, H. P. (1968): *Acta Crystallogr.* **B24,** 792.
Kniseley, R. N., Hirshmann, R. P. and Fassel, V. A. (1967): *Spectrochim. Acta* **23,** 109.
Knudsen, R. E., George, C. F. and Karle, J. (1966): *J. Chem. Phys.* **44,** 2334.
Kohrmann, H. J. and Zeil, W. (1975): *Z. Naturforsch.* **30a,** 183.
Kojima, T. (1960): *J. Phys. Soc. Japan* **15,** 1284.
Kojima, T. and Nishikawa, T. (1957): *J. Phys. Soc. Japan* **12,** 680.
Komiya, S., Huffman, J. C. and Kochi, J. K. (1977): *Inorg. Chem.* **16,** 2138.
Konaka, S. and Kimura, M. (1973): *Bull. Chem. Soc. Japan* **46,** 413.
Krebs, V. B. and Beyer, H. (1969): *Z. anorg. allg. Chem.* **369,** 199.
Krishna Pillai, M. G. and Cleveland, F. F. (1960): *J. Mol. Spectrosc.* **5,** 212.
Kronfeld, L. R. and Sass, R. L. (1968): *Acta Crystallogr.* **B24,** 981.
Kroto, H. W. (1975): *Molecular Rotation Spectra.* John Wiley, London.
Kroto, H. W. and Landsberg, B. M. (1976): *J. Mol. Spectrosc.* **62,** 346.
Kuchitsu, K. (1966): *J. Chem. Phys.* **44,** 906.
Kuchitsu, K. (1972): Gas Electron Diffraction. *MTP International Review of Science. Physical Chemistry Series One*, Vol. 2. *Molecular Structure and Properties* (Ed.: Allen, G.). Medical and Technical Publ. Co., Oxford.
Kuchitsu, K. (1976): *Kém. Közlem.* **45,** 311.
Kuchitsu, K., Guillory, J. P. and Bartell, L. S. (1968): *J. Chem. Phys.* **49,** 2488.
Kuchitsu, K., Shibata, T., Yokozeki, A. and Matsumura, C. (1971): *Inorg. Chem.* **10,** 2584.
Kuczkowski, R. L. (1963): *J. Amer. Chem. Soc.* **85,** 3047.
Kuczkowski, R. L. (1964): *J. Amer. Chem. Soc.* **86,** 3617.
Kuczkowski, R. L. (1968): *J. Amer. Chem. Soc.* **90,** 1705.
Kuczkowski, R. L. and Lide, D. R. (1967): *J. Chem. Phys.* **46,** 357.
Kuczkowski, R. L., Suenram, R. D. and Lovas, F. J. (1981): *J. Amer. Chem. Soc.* **103,** 2561.
Kuczkowski, R. L. and Wilson, E. B. (1963): *J. Amer. Chem. Soc.* **85,** 2028.
Kuo, S. S. (1965): *Numerical Methods and Computers.* Addison-Wesley, Reading, Mass.
Kuyper, J., Isselmann, P. H., Mijlhoff, F. C., Spelbos, A. and Renes, G. (1975): *J. Mol. Struct.* **29,** 247.
Kuz'yants, G. M. and Aleksanyan, V. T. (1972): *Zh. Strukt. Khim.* **13,** 617.

Kveseth, K. (1975): *Acta Chem. Scand.* **A29,** 307.
Lafferty, W. J., Lide, D. R. and Toth, R. A. (1965): *J. Chem. Phys.* **43,** 2063.
Lagowski, J. J. (1959): *Quart. Rev. (London)* **13,** 233.
Landolt–Börnstein (1976): New Series, Vol. II/7. *Structure Data of Free Polyatomic Molecules* (Ed. in Chief: Hellwege, K.-H.). Springer-Verlag, Berlin–Heidelberg–New York.
Landsberg, B. M. (1978): *Z. Naturforsch.* **33a,** 1333.
Landsberg, B. M. (1979): *Chem. Phys. Lett.* **60,** 265.
Langs, D. A., Silverton, J. V. and Bright, W. M. (1970): *J. C. S. Chem. Commun.* 1653.
La Placa, S. J. and Ibers, J. A. (1966): *Inorg. Chem.* **5,** 405.
Larsen, N. W. and Nicolaisen, F. M. (1974): *J. Mol. Struct.* **22,** 29.
Larsen, N. W., Pedersen, T. and Schulz, L. (1974): cited by: Larsen and Nicolaisen (1974).
Lazarev, A. N., Mirgorodskii, A. P., Ignat'ev, I. S. and Muldagaliev, Kh. Kh. (1975): *Zh. Strukt. Khim.* **16,** 578.
Lees, R. M. and Baker, J. G. (1968): *J. Chem. Phys.* **48,** 5299.
Legon, A. C. (1973): *J. C. S. Faraday Trans.* II. **69,** 29.
Lentz, D. and Seppelt, K. (1978): *Angew. Chem. Int. Ed. Engl.* **17,** 355.
Lett, R. G. and Flygare, W. H. (1967): *J. Chem. Phys.* **47,** 4730.
Levine, D. M., Krugh, W. D. and Gold, L. P. (1969): *J. Mol. Spectrosc.* **30,** 459.
Lide, D. R., Mann, D. E. and Fristrom, R. M. (1957): *J. Chem. Phys.* **26,** 734.
Lindley, P. F., Mahmoud, M. M., Dodd, C., Smith, C. H., Boyd, G. V. and Norris, T. (1977): *Acta Crystallogr.* **B33,** 2160.
Lindner, E. and Weber, H. (1967): *Z. Naturforsch.* **22b,** 1243.
Lindøy, S., Seip, H. M. and Seip, R. (1976): *Acta Chem. Scand.* **A30,** 54.
Lindøy, S., Seip, H. M. and Siebert, W. (1975): *Acta Chem. Scand.* **A29,** 265.
Loghry, R. A. and Simonsen, S. H. (1975): private communication.
Lowenstein, M. Z. (1965): *Diss. Abstr.* **26,** 2500.
Lu, C. S. and Donohue, J. (1944): *J. Amer. Chem. Soc.* **66,** 818.
Lucas, N. J. D. and Smith, J. G. (1972): *J. Mol. Spectrosc.* **43,** 327.
Luger, P., Bradaczek, H., Steudel, R. and Rebsch, M. (1976): *Chem. Ber.* **109,** 180.
Mamleev, A. H., Pozdeev, N. M. and Magdesieva, N. N. (1976): *J. Mol. Struct.* **33,** 211.
Manocha, A. S., Fateley, W. G. and Shimanouchi, T. (1973): *J. Phys. Chem.* **77,** 1977.
Markov, P. and Stølevik, R. (1970): *Acta Chem. Scand.* **24,** 2525.
Marsden, C. J. (1974): *J. Mol. Struct.* **21,** 168.
Marsden, C. J. and Bartell, L. S. (1976): *Inorg. Chem.* **15,** 3004.
Marsden, C. J. and Bartell, L. S. (1977): *J. C. S. Dalton Trans.* 1582.
Marsden, C. J., Bartell, L. S. and Diodati, F. P. (1977): *J. Mol. Struct.* **39,** 253.
Marsden, C. J., DesMarteau, D. D. and Bartell, L. S. (1977): *Inorg. Chem.* **16,** 2359.
Marsden, C. J. and Sheldrick, G. M. (1971a): *J. Mol. Struct.* **10,** 405.
Marsden, C. J. and Sheldrick, G. M. (1971b): *J. Mol. Struct.* **10,** 413.
Marsden, C. J. and Sheldrick, G. M. (1971c): *J. Mol. Struct.* **10,** 419.
Mastryukov, V. S., Osina, E. L. and Vilkov, L. V. (1975): *Zh. Strukt. Khim.* **16,** 850.
Mastryukov, V. S., Vilkov, L. V. and Dorofeeva, O. V. (1975): *Zh. Strukt. Khim.* **16,** 922.
Mathew, M. and Palenik, G. J. (1974): *J. C. S. Perkin Trans.* II. 532.
Matsushita, T., Osamura, Y., Misawa, N., Nishimoto, K. and Tsuno, Y. (1979): *Bull. Chem. Soc. Japan* **52,** 2521.
Maxwell, L. R. and Mosley, V. M. (1939): *Phys. Rev.* **57,** 21.
Maxwell, L. R., Mosley, V. M. and Hendricks, S. B. (1936): *Phys. Rev.* **50,** 41.
Mayer, I. and Hargittai, I. (1979): *Z. Naturforsch.* **34a,** 911.
McDonald, W. S. and Cruickshank, D. W. J. (1967): *Acta Crystallogr.* **22,** 48.
McKean, D. C. and Duncan, J. L. (1979): *J. Mol. Struct.* **51,** 149.
Meij, R., Oskam, A. and Stufkens, D. J. (1979): *J. Mol. Struct.* **51,** 37.

Meinwald, J., Knapp, S., Obendorf, S. K. and Hughes, R. E. (1976): *J. Amer. Chem. Soc.* **98,** 6643.
Meisingseth, E. and Cyvin, S. J. (1962): *Acta Chem. Scand.* **16,** 2452.
Mijlhoff, F. C. (1965): *Rec. Trav. Chim.* **84,** 74.
Mijlhoff, F. C. (1977): *J. Mol. Struct.* **36,** 334.
Milligan, D. E. and Jacox, M. E. (1968): *J. Chem. Phys.* **49,** 4269.
Mirri, A. M., Guarnieri, A., Favero, P. and Zuliani, G. (1962): *Nuovo Cimento* **25,** 265.
Momany, F. A. and Bonham, R. A. (1964): *J. Amer. Chem. Soc.* **86,** 162.
Moncur, N. K., Willson, P. D. and Edwards, T. H. (1974): *J. Mol. Spectrosc.* **52,** 380.
Morino, Y. and Iijima, T. (1962): *Bull. Chem. Soc. Japan.* **35,** 1661.
Morino, Y., Kikuchi, Y., Saito, S. and Hirota, E. (1964): *J. Mol. Spectrosc.* **13,** 95.
Morino, Y., Kuchitsu, K. and Moritani, T. (1969): *Inorg. Chem.* **8,** 867.
Morino, Y. and Matsumura, C. (1967): *Bull. Chem. Soc. Japan* **40,** 1101.
Morino, Y., Murata, Y., Ito, T. and Nakamura, J. (1962): *J. Phys. Soc. Japan* **17,** Suppl. BII, 37.
Morino, Y., Nakamura, Y. and Iijima, T. (1960): *J. Chem. Phys.* **32,** 643.
Morino, Y. and Saito, S. (1966): *J. Mol. Spectrosc.* **19,** 435.
Moritani, T., Kuchitsu, K. and Morino, Y. (1971): *Inorg. Chem.* **10,** 344.
Murdoch, J. D., Rankin, D. W. H. and Beagley, B. (1976): *J. Mol. Struct.* **31,** 291.
Murdoch, J. D., Rankin, D. W. H. and Glidewell, C. (1971): *J. Mol. Struct.* **9,** 17. Erratum: (1971): **10,** 496.
Murray, J. T., Little, W. A., Williams, Q. and Weatherly, T. L. (1976): *J. Chem. Phys.* **65,** 985.
Murty, K. S. R. (1965): *Bull. Nat. Inst. Sci. Ind.* **30,** 73.
Murty, K. S. R. and Mohanty, A. K. (1971): *Indian J. Phys.* **45,** 535.
Mustoe, F. J. and Hencher, J. L. (1972): *Can. J. Chem.* **50,** 3892.
Müller, A., Krebs, B., Fadini, A., Glemser, O., Cyvin, S. J., Brunvoll, J., Cyvin, B. N., Elvebredd, I., Hagen, G. and Vizi, B. (1968): *Z. Naturforsch.* **23a,** 1656.
Müller, A., Krebs, B. and Gattow, G. (1967): *Z. anorg. allg. Chem.* **349,** 74.
Müller, A., Mohan, N., Cyvin, S. J., Weinstock, N. and Glemser, O. (1976): *J. Mol. Spectrosc.* **59,** 161.
Müller, A. and Nagarajan, G. (1967a): *Z. phys. Chem. (Leipzig)* **235,** 57.
Müller, A. and Nagarajan, G. (1967b): *Z. phys. Chem. (Leipzig)* **235,** 113.
Müller, A., Peacock, C. J. and Heidborn, U. (1968): *Z. Naturforsch.* **23a,** 1687.
Nagarajan, G., Durig, J. R. and Perumal, A. (1969): *Monatsh. Chem.* **100,** 179.
Nagel, B. (1976): private communication.
Nagel, B., Stark, J., Fruwert, J. and Geiseler, G. (1976): *Spectrochim. Acta* **32A,** 1297.
Nagel, B., Steiger, Th., Fruwert, J. and Geiseler, G. (1975): *Spectrochim. Acta* **31A,** 255.
Nagel, B. and Stock, T. (1978): private communication.
Náhlovská, Z., Náhlovský, B. and Seip, H. M. (1969): *Acta Chem. Scand.* **23,** 3534.
Náhlovská, Z., Náhlovský, B. and Seip, H. M. (1970): *Acta Chem. Scand.* **24,** 1903.
Nair, K. P. R. (1978): *Indian J. Pure Appl. Phys.* **16,** 586.
Naito, T., Ohashi, O. and Yamaguchi, I. (1977): *J. Mol. Spectrosc.* **68,** 32.
Nakagawa, J., Kuwada, K. and Hayashi, M. (1976): *Bull. Chem. Soc. Japan* **49,** 3420.
Nakano, Y., Saito, S. and Morino, Y. (1970): *Bull. Chem. Soc. Japan* **43,** 368.
Nakata, M., Fukuyama, T., Kuchitsu, K., Takeo, H. and Matsumura, C. (1980): *J. Mol. Spectrosc.* **83,** 118.
Nakata, M., Fukuyama, T., Wilkins, C. J. and Kuchitsu, K. (1981): *J. Mol. Struct.* **71,** 195.
Nakata, M., Kohata, K., Fukuyama, T. and Kuchitsu, K. (1980): *J. Mol. Spectrosc.* **83,** 105.
Nakata, M., Kohata, K., Fukuyama, T., Kuchitsu, K. and Wilkins, C. J. (1980): *J. Mol. Struct.* **68,** 271.
Naumov, V. A. (1975): *Kém. Közlem.* **43,** 511.
Naumov, V. A. and Bezzubov, V. M. (1976): *Dokl. Akad. Nauk SSSR* **228,** 888.
Naumov, V. A., Garaeva, R. N., Geise, H. J. and Mijlhoff, F. C. (1979): *Dokl. Akad. Nauk SSSR* **245,** 650.
Naumov, V. A., Gulyaeva, N. A. and Pudovik, M. A. (1972): *Dokl. Akad. Nauk SSSR* **203,** 590.
Naumov, V. A. and Semashko, V. N. (1971): *Dokl. Akad. Nauk SSSR* **200,** 882.
Naumov, V. A., Semashko, V. N. and Shaidulin, S. A. (1973): *Zh. Strukt. Khim.* **14,** 595.

Naumov, V. A., Semashko, V. N. and Shatrukov, L. F. (1973): *Dokl. Akad. Nauk SSSR* **209**, 118.
Naumov, V. A., Semashko, V. N., Zav'yalov, A. P., Cherkasov, R. A. and Grishina, L. N. (1973): *Zh. Strukt. Khim.* **14**, 787.
Naumov, V. A. and Shaidulin, S. A. (1976): *Zh. Strukt. Khim.* **17**, 304.
Naumov, V. A., Zaripov, N. M. and Shatrukov, L. F. (1970): *Zh. Strukt. Khim.* **11**, 579.
Nave, C. R. and Sheridan, J. (1973): *J. Mol. Struct.* **15**, 391.
Nayar, V. U. and Aruldhas, G. (1968): *Indian J. Pure Appl. Phys.* **6**, 126.
Nogami, N., Sugeta, H. and Miyazawa, T. (1975): *Bull. Chem. Soc. Japan* **48**, 3573.
Noordik, J. H. and Vos, A. (1967): *Rec. Trav. Chim.* **86**, 156.
Nygaard, L., Asmussen, E., Høg, J. H., Maheshwari, R. C., Nielsen, C. H., Petersen, I. B., Rastrup-Andersen, J. and Sørensen, G. O. (1971): *J. Mol. Struct.* **8**, 225.
Nygaard, L., Hansen, L. and Sørensen, G. O. (1971): *J. Mol. Struct.* **9**, 163.
Nyquist, R. A. and Muelder, W. W. (1968): *Spectrochim. Acta* **24A**, 187.
Oberhammer, H. (1970): *Z. Naturforsch.* **25a**, 1497.
Oberhammer, H. (1975): *J. Mol. Struct.* **29**, 375.
Oberhammer, H. (1980): *J. Chem. Phys.* **73**, 4310.
Oberhammer, H. and Boggs, J. E. (1979a): *J. Mol. Struct.* **55**, 283.
Oberhammer, H. and Boggs, J. E. (1979b): *J. Mol. Struct.* **56**, 107.
Oberhammer, H., Glemser, O. and Klüver, H. (1974): *Z. Naturforsch.* **29a**, 901.
Oberhammer, H., Gombler, W. and Willner, H. (1981): *J. Mol. Struct.* **70**, 273.
Oberhammer, H., Haas, A. and Schlosser, K. (1979): *J. C. S. Dalton Trans.* 1075.
Oberhammer, H., Kumar, R. C., Knerr, G. D. and Shreeve, J. M. (1981): *Inorg. Chem.* **20**, 3871.
Oberhammer, H. and Seppelt, K. (1978): *Angew. Chem. Int. Ed. Engl.* **17**, 69; *Inorg. Chem.* **17**, 1435.
Oberhammer, H. and Zeil, W. (1969): *Z. Naturforsch.* **24a**, 1612.
Oberhammer, H. and Zeil, W. (1970a): *J. Mol. Struct.* **6**, 399.
Oberhammer, H. and Zeil, W. (1970b): *Z. Naturforsch.* **25a**, 845.
O'Connell, A. M. and Maslen, E. N. (1967): *Acta Crystallogr.* **22**, 134.
O'Connor, B. H. and Maslen, E. N. (1965): *Acta Crystallogr.* **18**, 363.
Ohashi, O., Ohnishi, M., Tagui, A., Sakaizumi, T. and Yamaguchi, I. (1977): *Bull. Chem. Soc. Japan* **50**, 1749.
Ohno, K., Mitsui, A. and Murata, H. (1979): *Bull. Chem. Soc. Japan* **52**, 2178.
Ohsaku, M., Bingo, N., Sugikawa, W. and Murata, H. (1979): *Bull. Chem. Soc. Japan* **52**, 355.
Ohsaku, M. and Murata, H. (1978): *Spectrochim. Acta* **34A**, 869.
Ohsaku, M., Shiro, Y. and Murata, H. (1973): *Bull. Chem. Soc. Japan* **46**, 1399.
Okiye, K., Hirose, C., Lister, D. G. and Sheridan, J. (1974): *Chem. Phys. Lett.* **24**, 111.
Olie, K. and Mijlhoff, F. C. (1969): *Acta Crystallogr.* **B25**, 974.
Onda, M., Kambayashi, S., Sakaizumi, T. and Yamaguchi, I. (1976): *J. Mol. Struct.* **34**, 299.
Orville-Thomas, W. J. (1965): *J. Chem. Phys.* **43**, 5244.
Otake, M., Matsumura, C. and Morino, Y. (1968): *J. Mol. Spectrosc.* **28**, 316.
Oyanagi, K., Fukuyama, T., Kuchitsu, K., Bohn, R. K. and Li, S. (1975): *Bull. Chem. Soc. Japan* **48**, 751.
Oyanagi, K. and Kuchitsu, K. (1978a): *Bull. Chem. Soc. Japan* **51**, 2237.
Oyanagi, K. and Kuchitsu, K. (1978b): *Bull. Chem. Soc. Japan* **51**, 2243.
Paetzold, R. (1968): *Spectrochim. Acta* **24A**, 717.
Paetzold, R. (1970): *Spectrochim. Acta* **26A**, 577.
Paetzold, R. and Amouloung, H. (1965): *Z. anorg. allg. Chem.* **337**, 225.
Paetzold, R. and Rönsch, E. (1970): *Spectrochim. Acta* **26A**, 569.
Palmer, K. J. (1938): *J. Amer. Chem. Soc.* **60**, 2360.
Pandey, G. K. and Dreizler, H. (1977): *Z. Naturforsch.* **32a**, 482.
Párkányi, C. and Herndon, W. C. (1978): *Phosphorus and Sulfur* **4**, 1.
Pascard-Billy, C. (1965): *Acta Cryst.* **18**, 827.
Pasinski, J. P. and Kuczkowski, R. L. (1971): *J. Chem. Phys.* **54**, 1903.

Paul, I. C. (1974): Structural Chemistry of the Thiol Group. *The Chemistry of the Thiol Group* (Ed.: Patai, S.). Wiley, London.

Pauling, L. (1949): *Proc. Natl. Acad. Sci. USA* **35,** 495.

Pauling, L. (1960): *The Nature of the Chemical Bond.* Cornell University Press, Ithaca, New York.

Pearson, E. F. and McCormick, R. V. (1973): *J. Chem. Phys.* **58,** 1619.

Pedersen, Th., Skaarup, S. V. and Pedersen, C. Th. (1977): *Acta Chem. Scand.* **B31,** 711.

Penn, R. E. and Curl, R. F. (1971): *J. Chem. Phys.* **55,** 651.

Penn, R. E. and Olsen, R. J. (1976): *J. Mol. Spectrosc.* **61,** 21.

Perahia, D. and Pullman, A. (1973): *Chem. Phys. Lett.* **19,** 73.

Peron, J. J., Saumagne, P. and Lebas, J. M. (1970): *Spectrochim. Acta* **26A,** 1651.

Perttilä, M. (1976): *Spectrochim. Acta* **32A,** 1011.

Petit, G., Lenstra, A. T. H., Van de Mieroop, W., Geise, H. J. and Hellier, D. G. (1978): *Recueil* **97,** 202.

Petrov, K. P., Ugarov, V. V. and Rambidi, N. G. (1980): *Zh. Strukt. Khim.* **21,** No. 1, 159.

Pichai, R., Krishna Pillai, M. G. and Ramaswamy, K. (1967): *Aust. J. Chem.* **20,** 1055.

Pickett, H. M. and Boyd, T. L. (1979): *J. Mol. Spectr.* **75,** 53.

Pierce, L. (1961): *J. Chem. Phys.* **34,** 498.

Pierce, L. and Hayashi, M. (1961): *J. Chem. Phys.* **35,** 479.

Pierce, L., Nelson, R. and Thomas, C. (1965): *J. Chem. Phys.* **43,** 3423.

Plato, V., Hartford, W. D. and Hedberg, K. (1970): *J. Chem. Phys.* **53,** 3488.

Pople, J. A. and Beveridge, D. L. (1970): *Approximate Molecular Orbital Theory,* McGraw-Hill, New York.

Pozdeev, N. M., Akulinin, O. B., Shapkin, A. A. and Magdesieva, N. N. (1970): *Zh. Strukt. Khim.* **11,** 869.

Price, C. C. and Gillis, R. G. (1953): *J. Amer. Chem. Soc.* **75,** 4747; 4750.

Radnai, T., Hargittai, I., Kolonits, M. and Gregory, D. C. (1975): unpublished results.

Raley, J. M., Wollrab, J. E. and Lovejoy, R. W. (1973): *J. Mol. Spectrosc.* **48,** 100.

Ralowski, W. and Ljunggren, S. (1977): *Acta Chem. Scand.* **A31,** 492.

Ralowski, W., Ljunggren, S. and Mjöberg, J. (1973): *Acta Chem. Scand.* **27,** 3128.

Ramaswamy, K. and Jayaraman, S. (1970): *J. Mol. Struct.* **5,** 325.

Ramaswamy, K. and Jayaraman, S. (1971a): *J. Mol. Struct.* **7,** 470.

Ramaswamy, K. and Jayaraman, S. (1971b): *J. Mol. Struct.* **10,** 183.

Ramaswamy, K. and Jayaraman, S. (1972): *Indian J. Pure Appl. Phys.* **10,** 72.

Rambidi, N. G. (1975a): *J. Mol. Struct.* **28,** 77.

Rambidi, N. G. (1975b): *J. Mol. Struct.* **28,** 89.

Randhawa, H. S. (1974): *Z. phys. Chem. (Frankfurt)* **88,** 11.

Rankin, D. W. H. and Cyvin, S. J. (1972): *J. C. S. Dalton Trans.* 1277.

Rao, V. M., Yardley, J. T. and Curl, R. F. (1965): *J. Chem. Phys.* **42,** 284.

Reed, P. R. and Lovejoy, R. W. (1968): *Spectrochim. Acta.* **24A,** 1795.

Reinhardt, R., Steudel, R. and Schuster, F. (1978): *Angew. Chem. Int. Ed. Engl.* **17,** 57.

Remizov, A. B. (1971): *Zh. Strukt. Khim.* **12,** 1101.

Remizov, A. B., Fishman, A. I. and Butenko, G. G. (1977): *Zh. Obshch. Khim.* **47,** 1204.

Remizov, A. B., Fishman, A. I. and Pominov, I. S. (1979): *Spectrochim. Acta* **35A,** 909.

Remizov, A. B., Kozlova, R. I., Vakhrushcheva, N. N. and Mannafov, T. G. (1973): *Zh. Prikl. Spektr.* **19,** 109.

Remizov, A. B., Kuramshin, I. Ya., Aganov, A. V. and Butenko, G. G. (1973): *Dokl. Akad. Nauk SSSR* **208,** 1118.

Remizov, A. B. and Mannafov, T. G. (1976a): *Zh. Prikl. Spektr.* **25,** 356.

Remizov, A. B. and Mannafov, T. G. (1976b): *Zh. Prikl. Spektr.* **25,** 467.

Remizov, A. B. and Mannafov, T. G. (1977): *Zh. Prikl. Spektr.* **26,** 765.

Remizov, A. B. and Samarina, O. A. (1975): *Zh. Prikl. Spektr.* **22,** 870.

Rérat, B. (1969): *Acta Crystallogr.* **B25,** 1392.

Richert, H. (1961): *Z. anorg. allg. Chem.* **309,** 171.

Rimmer, D. F., Smith, J. G. and Whiffen, D. H. (1973): *J. Mol. Spectrosc.* **45,** 114.

Robiette, A. G. (1973): private communication cited by Glidewell (1973).
Robinet, G., Crasnier, F., Labarre, J.-F. and Leibovici, C. (1972): *Theoret. Chim. Acta* **25,** 259.
Robinson, E. A. (1961): *Can. J. Chem.* **39,** 247 and references cited therein.
Robinson, E. A. and Aroca, R. (1978): *Spectrochim. Acta* **34A,** 1249.
Roesky, H. W., Diehl, M., Fuess, H. and Bats, J. W. (1978): *Angew. Chem. Int. Ed. Engl.* **17,** 58.
Romers, C., Altona, C., Buys, H. R. and Havinga, E. (1969): *Topics in Stereochemistry* (Eds.: Eliel, E. L. and Allinger, N. L.). Vol. 4, Interscience, New York.
Rosmus, P., Stafast, H. and Bock, H. (1975): *Chem. Phys. Lett.* **34,** 275.
Rozsondai, B. and Hargittai, I. (1980): *Kém. Közlem.* **54,** 268.
Rozsondai, B., Hargittai, I. and Pappalardo, G. C. (1979): *Z. Naturforsch.* **34a,** 752.
Rozsondai, B., Moore, J. H., Gregory, D. C. and Hargittai, I. (1977): *Acta Chim. (Budapest)* **94,** 321.
Rozsondai, B., Moore, J. H., Gregory, D. C. and Hargittai, I. (1979): *J. Mol. Struct.* **51,** 69.
Rozsondai, B., Schultz, Gy. and Hargittai, I. (1981): *J. Mol. Struct.* **70,** 309.
Rundle, R. E. (1950): *Acta Crystallogr.* **3,** 16.
Sadova, N. I., Vilkov, L. V., Hargittai, I. and Brunvoll, J. (1976): *J. Mol. Struct.* **31,** 131.
Saito, S. (1969a): *J. Mol. Spectrosc.* **30,** 1.
Saito, S. (1969b): *Bull. Chem. Soc. Japan* **42,** 663.
Saito, S. and Makino, F. (1972): *Bull. Chem. Soc. Japan* **45,** 92.
Sakaizumi, T., Kohri, Y., Ohashi, O. and Yamaguchi, I. (1978): *Bull. Chem. Soc. Japan* **51,** 3411.
Sakaizumi, T., Ohashi, O. and Yamaguchi, I. (1976): *Bull. Chem. Soc. Japan* **49,** 948.
Sakakibara, M., Matsuura, H., Harada, I. and Shimanouchi, T. (1977): *Bull. Chem. Soc. Japan* **50,** 111.
Salahub, D. R., Foti, A. E. and Smith, V. H. (1978): *J. Amer. Chem. Soc.* **100,** 7847.
Samdal, S. and Seip, H. M. (1971): *Acta Chem. Scand.* **25,** 1903.
Samdal, S. and Seip, H. M. (1975): *J. Mol. Struct.* **28,** 193.
Samdal, S., Seip, H. M. and Torgrimsen, T. (1979): *J. Mol. Struct.* **57,** 105.
Sands, D. E. (1963): *Z. Kristallogr.* **119,** 245.
Sands, D. E. (1972): *Acta Crystallogr.* **B28,** 2463.
Sands, D. E. and Day, V. W. (1967): *Z. Kristallogr.* **124,** 220.
Sastry, K. V. L. N., Dass, S. C., Brooks, W. V. F. and Bhaumik, A. (1969): *J. Mol. Spectrosc.* **31,** 54.
Sathianandan, K., Ramaswamy, K., Sundaram, S. and Cleveland, F. F. (1964): *J. Mol. Spectrosc.* **13,** 214.
Schmidt, R. E. and Quade, C. R. (1975): *J. Chem. Phys.* **62,** 2864.
Schmiedekamp, A. (1976): *Dissertation*, University of Texas, Austin.
Schmiedekamp, A., Cruickshank, D. W. J., Skaarup, S., Pulay, P., Hargittai, I. and Boggs, J. E. (1979): *J. Amer. Chem. Soc.* **101,** 2002.
Schomaker, V. and Stevenson, D. P. (1941): *J. Amer. Chem. Soc.* **63,** 37.
Schultz, Gy. (1976): unpublished calculations, private communication.
Schultz, Gy., Fellegvári, I., Kolonits, M., Kiss, Á. I., Pete, B. and Bánki, J. (1978): *J. Mol. Struct.* **50,** 325.
Schultz, Gy. and Hargittai, I. (1973): *Acta Chim. (Budapest)* **75,** 381.
Schultz, Gy. and Hargittai, I. (1974): *Acta Chim. (Budapest)* **83,** 331.
Schultz, Gy., Hargittai, I. and Hermann, L. (1972): *J. Mol. Struct.* **14,** 353.
Schultz, Gy., Hargittai, I., Martin, J. and Robert, J. B. (1974): *Tetrahedron* **30,** 2365.
Schultz, Gy., Serke, I. and Kapovits, I. (1979): *J. C. S. Faraday Trans.* II. **75,** 1612.
Scott, D. W. and El-Sabban, M. Z. (1969): *J. Mol. Spectrosc.* **31,** 362.
Seel, F. (1974): *Adv. Inorg. Chem. Radiochem.* **16,** 297.
Seel, F., Boudier, J. and Gombler, W. (1969): *Chem. Ber.* **102,** 443.
Seel, F., Heinrich, E., Gombler, W. and Budenz, R. (1969): *Chimica*, **23,** 73.
Seip, H. M. and Seip, R. (1973): *Acta Chem. Scand.* **27,** 4024.
Seip, H. M., Seip, R. and Siebert, W. (1973): *Acta Chem. Scand.* **27,** 15.
Seip, H. M. and Stølevik, R. (1966): *Acta Chem. Scand.* **20,** 1535.
Seip, R., Seip, H. M. and Smith, Z. (1976): *J. Mol. Struct.* **32,** 279.

Shen, Q. and Hedberg, K. (1974): *J. Amer. Chem. Soc.* **96,** 289.
Shibata, S. and Bartell, L. S. (1965): *J. Chem. Phys.* **42,** 1147.
Shriver, D. F. (1970): *Accounts Chem. Res.* **3,** 231.
Siebert, H. (1954): *Z. anorg. allg. Chem.* **275,** 210.
Silvia, C. J., True, N. S. and Bohn, R. K. (1978): *J. Chem. Phys.* **82,** 483.
Silvia, C. J., True, N. S. and Bohn, R. K. (1979): *J. Mol. Struct.* **51,** 163.
Sime, J. G. and Woodhouse, D. I. (1974): *J. Cryst. Mol. Struct.* **4,** 269.
Simon, A., Kriegsmann, H. and Dutz, H. (1956): *Chem. Ber.* **89,** 2390.
Simpson, K. O. and Beynon, E. T. (1971): *J. Phys. Chem.* **71,** 2796.
Sinnott, K. M. (1961): *J. Chem. Phys.* **34,** 851.
Smeyers, Y. G. and Hernandez-Laguna, A. (1978): *J. Chim. Phys.* **75,** 83.
Smith, Z. and Seip, R. (1976): *Acta Chem. Scand.* **A30,** 759.
So, S. P. and Richards, W. G. (1978): *J. C. S. Faraday Trans.* II. **74,** 1743.
Solouki, B. and Bock, H. (1977): *Inorg. Chem.* **16,** 665.
Spiridonov, V. P., Khodchenkov, A. N. and Akishin, P. A. (1965): *Zh. Strukt. Khim.* **6,** 633.
Spiridonov, V. P. and Lutoshkin, B. I. (1970): *Vestnik Mosk. Univ. Khim.* 509.
Spoliti, M., Chackalackal, S. M. and Stafford, F. E. (1967): *J. Amer. Chem. Soc.* **89,** 1092.
Steger, E., Řehák, J. and Faltus, H. F. (1965): *Z. phys. Chem. (Leipzig)* **229,** 110.
Steidel, J., Pickardt, J. and Steudel, R. (1978): *Z. Naturforsch.* **33b,** 1554.
Stephenson, C. V., Coburn, W. C. and Wilcox, W. S. (1961): *Spectrochim. Acta* **17,** 933.
Steudel, R. (1970): *Z. Naturforsch.* **25b,** 645.
Steudel, R. (1971): *Z. Naturforsch.* **26b,** 750.
Steudel, R. (1975): *Angew. Chem. Int. Ed. Engl.* **14,** 655.
Steudel, R. and Lautenbach, D. (1969): *Z. Naturforsch.* **24b,** 350.
Steudel, R., Reinhardt, R. and Sandow, I. (1977): *Angew. Chem. Int. Ed. Engl.* **16,** 716.
Steudel, R., Reinhardt, R. and Schuster, F. (1977): *Angew. Chem. Int. Ed. Engl.* **16,** 715.
Stevenson, D. P. and Schomaker, V. (1940): *J. Amer. Chem. Soc.* **62,** 1913.
Stone, R. G., Tigelaar, H. L. and Flygare, W. H. (1970): *J. Chem. Phys.* **53,** 3947.
Strand, T. G. (1967a): *Acta Chem. Scand.* **21,** 1033.
Strand, T. G. (1967b): *Acta Chem. Scand.* **21,** 2111.
Sugisaki, R., Tanaka, T. and Hirota, E. (1974): *J. Mol. Spectrosc.* **49,** 241.
Sugeta, H., Go, A. and Miyazawa, T. (1973): *Bull. Chem. Soc. Japan* **46,** 3407.
Sung, E.-M. and Harmony, M. D. (1977): *J. Amer. Chem. Soc.* **99,** 5603.
Sutter, D., Dreizler, H. and Rudolph, H. D. (1965): *Z. Naturforsch.* **20a,** 1676.
Sutton, L. E. (1958): *Tables of Interatomic Distances and Configuration in Molecules and Ions.* Spec. Publ. No. 11. The Chem. Soc., London.
Suzuki, S., Yamaguchi, M., Onda, M., Sakaizumi, T., Ohashi, O. and Yamaguchi, I. (1981): *J. Mol. Struct.* **73,** 41.
Szmant, H. H. (1971): *Sulphur in Organic and Inorganic Chemistry* (Ed.: Senning, A.). Vol. 1. Dekker, New York.
Szpiridonov, V. P. (1972): *Kém. Közlem.* **37,** 399.
Takagi, K. and Kojima, T. (1971): *J. Phys. Soc. Japan* **30,** 1145.
Takagi, K. and Oka, T. (1963): *J. Phys. Soc. Japan* **18,** 1174.
Takeo, H., Hirota, E. and Morino, Y. (1970): *J. Mol. Spectrosc.* **34,** 370.
Tamagawa, K., Iijima, T. and Kimura, M. (1976): *J. Mol. Struct.* **30,** 243.
Tanimoto, M., Almond, V., Charles, S. W., Macdonald, J. N. and Owen, N. L. (1979): *J. Mol. Spectrosc.* **78,** 95.
Tanimoto, M. and Macdonald, J. N. (1979): *J. Mol. Spectrosc.* **78,** 106.
Tchir, P. O. and Spratley, R. D. (1975): *Can. J. Chem.* **53,** 2311.
Thomas, C. H. (1973): *J. Chem. Phys.* **59,** 70.
Thomas, R., Shoemaker, C. B. and Eriks, K. (1966): *Acta Crystallogr.* **21,** 12.

Tiemann, E., Hoeft, J., Lovas, F. J. and Johnson, D. R. (1974): *J. Chem. Phys.* **60,** 5000.
Timofeeva, T. V., Kravtsov, D. N., Bokii, N. G. and Struchkov, Yu. T. (1977): *Zh. Strukt. Khim.* **18,** 882.
Tokue, I., Fukuyama, T. and Kuchitsu, K. (1974): *J. Mol. Struct.* **23,** 33.
Tolles, M. W. and Gwinn, W. D. (1962): *J. Chem. Phys.* **36,** 1119.
Towns, R. L. R. and Simonsen, S. H. (1974): *Cryst. Struct. Commun.* **3,** 373.
Towns, R. L. R. and Simonsen, S. H. (1975): *Cryst. Struct. Commun.* **4,** 469.
Török, F., Páldi, E., Dobos, S. and Fogarasi, G. (1970): *Acta Chim. Acad. Sci. Hung.* **63,** 417.
Tranquille, M., Labarbe, P., Fouassier, M. and Forel, M. T. (1971): *J. Mol. Struct.* **8,** 273.
Trinkaus, A., Dreizler, H. and Rudolph, H. D. (1973): *Z. Naturforsch.* **28a,** 750.
True, N. S. and Bohn, R. K. (1976): *J. Amer. Chem. Soc.* **96,** 1188.
True, N. S. and Bohn, R. K. (1979): *Chem. Phys. Lett.* **60,** 332.
Trueblood, K. N. and Mayer, S. W. (1956): *Acta Crystallogr.* **9,** 628.
Tsuchiya, S. and Iijima, T. (1972): *J. Mol. Struct.* **13,** 327.
Tuzova, L. L., Naumov, V. A., Galiakberov, R. M., Ofitserov, E. N. and Pudovik, A. N. (1981): *Dokl. Akad. Nauk SSSR* **256,** 891.
Tyler, J. K. and Sheridan, J. (1963): *Trans. Faraday Soc.* **59,** 2661.
Typke, V. (1978): *Z. Naturforsch.* **33a,** 842.
Uehara, H. and Morino, Y. (1969): *Mol. Phys.* **17,** 239.
Ugarov, V. V., Ezhov, Yu. S. and Rambidi, N. G. (1973): *Zh. Strukt. Khim.* **14,** 548.
Ugarov, V. V., Ezhov, Yu. S. and Rambidi, N. G. (1975): *J. Mol. Struct.* **25,** 357.
Ukshe, E. A. (Ed.) (1966): *Stroenie rasplavlennykh solei.* Mir, Moscow.
Uno, T., Machida, K. and Hanai, K. (1968): *Spectrochim. Acta,* **24A,** 1705.
Uno, T., Machida, K. and Hanai, K. (1971): *Spectrochim. Acta,* **27A,** 107.
Vahrenkamp, H. (1971): *J. Organometal. Chem.* **28,** 181.
Vajda, E. and Hargittai, I. (1975): unpublished calculations.
Vajda, E., Hargittai, I., Maltsev, A. K. and Nefedov, O. M. (1974): *J. Mol. Struct.* **23,** 417.
Vajda, E., Kolonits, M., Hargittai, I. and Szőke, S. (1976): *J. Mol. Struct.* **35,** 235.
Van der Helm, D., Childs, J. D. and Christian, S. D. (1969): *J. C. S. Chem. Commun.* 887.
Van Eijck, B. P. (1971): *Dissertation,* University of Utrecht.
Van Eijck, B. P., Hargittai, I. and Mayer, I. (1980): *J. Mol. Struct.* **69,** 301.
Van Eijck, B. P., Korthoff, A. J. and Mijlhoff, F. C. (1975): *J. Mol. Struct.* **24,** 222.
Van Schaick, E. J. M., Geise, H. J., Mijlhoff, F. C. and Renes, G. (1973): *J. Mol. Struct.* **16,** 23.
Van Wart, H. E., Cardinaux, F. and Scheraga, H. A. (1976): *J. Phys. Chem.* **80,** 625.
Van Wart, H. E. and Scheraga, H. A. (1977): *Proc. Natl. Acad. Sci. USA* **74,** 13.
Van Wazer, J. R. and Absar, I. (1972): *Advances in Chemistry* **110,** 20.
Van Woerden, H. F. and Bijl-Vlieger, S. H. (1973): *Rec. Trav. Chim.* **93,** 95.
Varsányi, Gy. (1973): *Assignments for Vibrational Spectra of 700 Benzene Derivatives.* Akadémiai Kiadó, Budapest.
Venkateswarlu, K. and Malathy Devi, V. (1965): *Indian J. Pure Appl. Phys.* **3,** 195.
Venkateswarlu, K., Thirugnanasambandam, P. and Balasubramanian, C. (1962): *Z. phys. Chem. (Leipzig)* **218,** 310.
Vilkov, L. V. and Hargittai, I. (1967): *Acta Chim. Acad. Sci. Hung.* **52,** 423.
Vilkov, L. V. and Khaikin, L. S. (1966): *Dokl. Akad. Nauk SSSR* **168,** 810.
Vilkov, L. V. and Khaikin, L. S. (1975): Stereochemistry of Compounds Containing Bonds between Si, P, S, Cl and N or O. *Topics in Current Chemistry* **53,** 25. Springer-Verlag, Berlin–Heidelberg–New York.
Vilkov, L. V., Khaikin, L. S. and Evdokimov, V. V. (1972): *Zh. Strukt. Khim.* **13,** 7.
Vilkov, L. V., Penionzhkevich, N. P., Brunvoll, J. and Hargittai, I. (1978): *J. Mol. Struct.* **43,** 109.
Vilkov, L. V. and Tarasenko, N. (1969): *J. C. S. Chem. Commun.* 1176.
Viswamitara, M. A. and Kannan, K. K. (1966): *Nature* **209,** 1016.
Vogt, L. H., Katz, J. L. and Wiblerley, S. E. (1965): *Inorg. Chem.* **4,** 231.

Wadhawan, V. K. (1976): *Acta Crystallogr.* **B32,** 397.

Wang, H. K. (1965): *Acta Chem. Scand.* **19,** 879.

Webb, N. S. and Gloss, R. A. (1967): *Tetrahedron Lett.* 1043.

Wehrung, T., Oberhammer, H., Haas, A., Koch, B. and Welcman, N. (1976): *J. Mol. Struct.* **35,** 253.

Weidner, J. U., Geisenfelder, H. and Zimmermann, H. (1971): *Ber. Bunsenges. phys. Chem.* **75,** 800.

White, W. F. and Wollrab, J. E. (1969): *Chem. Phys. Lett.* **3,** 25.

Wiberg, K. A. (1968): *Tetrahedron* **24,** 1083.

Wiese, J. and Sutter, D. H. (1977): *Z. Naturforsch.* **32a,** 890.

Wilkins, C. J., Hagen, K., Hedberg, L., Shen, Q. and Hedberg, K. (1975): *J. Amer. Chem. Soc.* **97,** 6352.

Williams, E. J. and Ladd, J. A. (1968): *J. Mol. Struct.* **2,** 57.

Williams, Q., Cox, T. and Gordy, W. (1952): *J. Chem. Phys.* **20,** 1524.

Winnewisser, G. (1972): *J. Mol. Spectrosc.* **41,** 534.

Winnewisser, G., Maki, A. G. and Johnson, D. R. (1971): *J. Mol. Spectrosc.* **39,** 149.

Winnewisser, G., Winnewisser, M. and Gordy, W. (1968): *J. Chem. Phys.* **49,** 3465.

Winnewisser, M. and Haase, J. (1968): *Z. Naturforsch.* **23a,** 56.

Wollrab, J. E. (1967): *Rotational Spectra and Molecular Structure.* Academic Press, New York.

Wollrab, J. E., Scharpen, L. H., Ames, D. P. and Merritt, J. A. (1968): *J. Chem. Phys.* **49,** 2405.

Yamaguchi, I., Ohashi, O., Takahashi, H., Sakaizumi, T. and Onda, M. (1973): *Bull. Chem. Soc. Japan* **46,** 1560.

Yamazaki, A., Mogi, K., Koyama, M. and Yamaguchi, I. (1979): *J. Mol. Struct.* **55,** 185.

Yokozeki, A. and Bauer, S. H. (1975): The Geometric and Dynamic Structures of Fluorocarbons and Related Compounds. *Topics in Current Chemistry* **53,** 71. Springer-Verlag, Berlin–Heidelberg–New York.

Yokozeki, A. and Bauer, S. H. (1976): *J. Phys. Chem.* **80,** 618.

Yokozeki, A. and Kuchitsu, K. (1971a): *Bull. Chem. Soc. Japan* **44,** 2356.

Yokozeki, A. and Kuchitsu, K. (1971b): *Bull. Chem. Soc. Japan* **44,** 2926.

Yow, H. Y., Rudolph, R. W. and Bartell, L. S. (1975): *J. Mol. Struct.* **28,** 205.

Zaripov, N. M. (1976): *Zh. Strukt. Khim.* **17,** 741.

Zaripov, N. M., Golubinskii, A. V., Chmutova, G. A. and Vilkov, L. V. (1978): *Zh. Strukt. Khim.* **19,** 894.

Zaripov, N. M. and Mannafov, T. G. (1975): *Zh. Strukt. Khim.* **16,** 832.

Zaripov, N. M., Naumov, V. A. and Tuzova, L. L. (1974a): *Phosphorus* **4,** 179.

Zaripov, N. M., Naumov, V. A. and Tuzova, L. L. (1974b): *Dokl. Akad. Nauk SSSR* **218,** 1132.

Zaripov, N. M., Popik, M. V. and Vilkov, L. V. (1980): *Zh. Strukt. Khim.* **21,** No. 6, 38.

Zaripov, N. M., Popik, M. V., Vilkov, L. V. and Mannafov, T. G. (1980): *Zh. Strukt. Khim.* **21,** No. 2, 37.

Zasorin, E. Z., Zharskii, I. M., Pinaev, G. F., Kupreev, V. N. and Spiridonov, V. P. (1974): *Zh. Strukt. Khim.* **15,** 691.

Zharskii, I. M., Zasorin, E. Z., Spiridonov, V. P. and Novikov, G. I. (1977): *Vestnik Mosk. Univ. Khim.* **18,** No. 2, 166.

AUTHOR INDEX*

*Page numbers in *italics* refer to the Bibliography.

FORMULA INDEX